SCIENCE IN THE AGE OF SENSIBILITY

JESSICA RISKIN

Science

IN THE AGE OF

Sensibility

⚜

THE SENTIMENTAL
EMPIRICISTS OF THE
FRENCH
ENLIGHTENMENT

THE UNIVERSITY OF CHICAGO PRESS
CHICAGO AND LONDON

JESSICA RISKIN is assistant professor in the Department of History at Stanford University.

The University of Chicago Press, Chicago 60637
The University of Chicago Press, Ltd., London
© 2002 by The University of Chicago
All rights reserved. Published 2002
Printed in the United States of America

11 10 09 08 07 06 05 04 03 02   1 2 3 4 5

ISBN: 0-226-72078-0 (cloth)
ISBN: 0-226-72079-9 (paper)

Library of Congress Cataloging-in-Publication Data

Riskin, Jessica.
   Science in the age of sensibility : the sentimental empiricists of the
French enlightenment / Jessica Riskin.
       p.   cm.
Includes bibliographical references and index.
   ISBN 0-226-72078-0 (cloth : alk. paper)—ISBN 0-226-72079-9 (paper :
alk. paper)
       1. Science—France—History—18th century. 2. Enlightenment—France.
3. Sensitivity (Personality trait). I. Title.
   Q127.F8 R57 2002
   509.44'09'033—dc21

2002005793

FOR MYRA JEHLEN AND CARL RISKIN,
PARENTS, TEACHERS, FRIENDS

CONTENTS

ILLUSTRATIONS

ACKNOWLEDGMENTS

This book's early stages were blessed by the mentoring of Roger Hahn, who taught a green graduate student the craft of history, and of Carla Hesse and Elisabeth Lloyd. John Heilbron's historical and stylistic voice, wisdom, and wit were my guides from the beginning. One of the great pleasures these paragraphs give me is an occasion for expressing my gratitude to him.

Many people offered vital help as the book unfolded: Alexi Assmus, Bernadette Bensaude-Vincent, Jed Buchwald, Carl Caldwell, Olivier Darrigol, Paula Findlen, Peter Galison, Charles Gillispie, Jan Golinski, Hazel Hahn, David Hollinger, Larry Holmes, Pierre Jacob, Evelyn Fox Keller, Daniel Kevles, Catherine Kudlick, the late Thomas Kuhn, Timothy Lenoir, Joseph Levine, Alan Marcus, Leo Marx, Sarah Maza, David Mindell, Timothy Moy, Katharine Norris, Alex Pang, Jean-Pierre Poirier, Jeffrey Ravel, François Recanati, Robert Richards, Carl Riskin, Margaret Schabas, Simon Schaffer, Sam Schweber, Stuart Strickland, Rebecca Ullrich, Lora Wildenthal, Pearce Williams, David Wilson, Elizabeth Wood, and Joseph Zizek. Emma Spary, as a reader for the University of Chicago Press, provided invaluable advice.

I owe a special debt of gratitude to Ken Alder, whom I thank with admiration for readings that have improved every part of this book, and have particularly helped to shape chapter 5. Myra Jehlen and Christopher Kutz too have read, and reread, the full text. The challenge of rethinking in light of these responses has been the hardest and best part of writing.

I have been continually grateful to Susan Abrams for the acumen and kindness with which she fostered my work and for her guidance into a field she has done so much to shape. I also thank the production team at the University of Chicago Press, particularly Erin DeWitt for her superb copyediting.

The editors of the Benjamin Franklin Papers at Yale University gave me full access to their materials, and an education in what was there and how to find it. My warm recognition goes especially to Ellen Cohn.

Richard Kim and Jehangir Malegam provided excellent research assistance.

Diana Wear performed an administrative coup to make such assistance available to me on the West Coast while I was still employed on the East. I thank Judy Spitzer, too, for her expert help.

I have been fortunate in receiving support for my research from the National Science Foundation, the Mellon Foundation, the Mabelle McLeod Lewis Foundation, the Doreen P. Townsend Center for the Humanities at the University of California at Berkeley, the History Department at Northwestern University, the Dibner Institute for the History of Science and Technology, the History of Science Department at Harvard University, and an Old Dominion Fellowship from the Massachusetts Institute of Technology.

Much earlier versions of some material in this book has appeared previously in *Historical Studies in the Physical and Biological Sciences* (parts of chapter 3), *Science in Context* (parts of chapter 5), and *Isis* (parts of chapter 7).

Finally, Sonia Jehlen has given unending support and encouragement, as have Naomi Allen, Steve Bloom, Alain Jehlen, Myra Jehlen, Pat Jehlen, Robert Kutz, Ruth Kutz, and Carl Riskin. In Christopher Kutz's intellectual companionship, I have had the benefit of his philosophical rigor and unerring eye for what is important. And Oliver Riskin Kutz arrived just in time to cheer the last pages.

ABBREVIATIONS

AGPC	Archives Générales de Pas-de-Calais, Arras
AN	Archives Nationales, Paris
AS	Académie des Sciences, Paris
BFP	Benjamin Franklin, *The Papers of Benjamin Franklin*
BIF	Bibliothèque de l'Institut de France, Paris
Coll. Bar.	Collection Barbier
FPA	Benjamin Franklin Papers Archives, Yale University
MAS	Académie des Sciences, *Mémoires de l'Académie des Sciences*

SCIENCE IN THE AGE OF SENSIBILITY

Figure 1.1. The frontispiece to Denis Diderot and Jean d'Alembert's *Encyclopédie, ou, Dictionnaire raisonné des sciences, des arts et des métiers* (1751–72). The engraving is by Benoît-Louis Prévost from a drawing by Nicolas Cochin-le-Fils. Diderot described the scene as follows: "We see on top Truth between Reason and Imagination; Reason is trying to snatch off her veil, Imagination is preparing to embellish her; below this group, a crowd of speculative philosophers; lower down, the troupe of artists. The philosophers have their eyes fixed on Truth; vain Metaphysics tries less to see her than to guess her. Theology turns her back to her, waiting for the light from on high." Diderot, *Salon* of 1765, in *Oeuvres complètes*, 10:448. Courtesy of the Bancroft Library, University of California, Berkeley.

Chapter One

INTRODUCTION: SENSIBILITY AND ENLIGHTENMENT SCIENCE

This book is about the sentimental youth of scientific empiricism, and about a time and place in which its dramas of self-definition were center stage: eighteenth-century France. Later, and for much of its four-century tenure at the heart of European culture, empiricism—the doctrine that natural knowledge originates in observation and experiment—came to have a hardnosed, unemotional reputation. Charles Dickens's Mr. Gradgrind—implacably demanding "nothing but Facts, sir; nothing but Facts!"—was recognizable to readers of the popular weekly in which he appeared in 1854 as a caricature of the dispassionate empiricist. Behold Gradgrind: a "square wall of a forehead"; a mouth that was "wide, thin, and hard"; a voice "inflexible, dry, and dictatorial"; and an "obstinate carriage, square coat, square legs, square shoulders—nay, his very neckcloth [was] trained to take him by the throat with an unaccommodating grasp, like a stubborn fact."[1] But scientific empiricism would not always have evoked Gradgrindian features. It appeared very different a century earlier to the philosophes who received the young doctrine from its seventeenth-century progenitors and were the guardians of its coming of age. To them, empirical knowledge was not a matter of impassive adherence to the hard facts of sensory experience, but rather one of sensibility.

"What is sensibility?" the playwright and philosophe Denis Diderot wrote to his friend Sophie Volland in 1760. "The vivid effect on our soul of an infinity of delicate observations." Sensibility is defined in Diderot's *Encyclopédie* as the capacity "to perceive impressions of external objects"; its first consequence, an answering "movement" of the sensible creature, is "sentiment." Throughout this book I will use "sentiment" in this sense, to describe an emotional "movement" in response to a physical sensation. Sensibility was the "first germ of thought" and "the most beautiful, the most singular phenomenon of nature." It attuned the animal to the world outside and governed its inner

1. Dickens, *Hard Times* (1854), chap. 1, 1–2.

processes—not only thought and emotion, but digestion, the secretions and excretions of the organs, the menstrual flux, and the functioning of the heart and lungs. In the more intimate setting of Diderot's letter to Mme. Volland, sensibility featured in some advice about the rearing of her young niece. He recommended that one be especially tolerant of small children. They were naturally "hard, even cruel" because experience had not yet developed their moral faculties. Moral sentiments arose from sensibility, the openness of the soul to vivid impressions of a delicate world.[2]

In contrast with the *Encyclopédie*, the *Dictionnaire de l'Académie française* (1694) had been matter-of-fact. It states plainly that sensibility was "that which has feeling. Stones are not sensible. The eye is a very sensible part." The *Dictionnaire* explains that "sensibility" can also be used "with regard to morals. This man is very delicate and sensible. He is sensible to the misfortunes of others."[3] "Sensibility," therefore, had already signified both physical sensation and moral sensitivity. But this dual signification did not become significant until the middle of the eighteenth century. Diderot was among the first to recast the old definition into a new meaning. He used an axiom that John Locke had proposed in his *Essay concerning Human Understanding* (1690). The wellspring of Locke's sensationist epistemology, this axiom posited that the mind at birth was a blank slate, and that all its thoughts were inscribed upon it by the outside world, with the instruments of inscription being the body's five senses. Diderot and his contemporaries joyously embraced this axiom, extended it in all directions, and erected their cities of philosophical light upon it. Indeed, they expanded Locke's notion of the sensory origins of ideas by applying it to the emotions and moral sentiments as well. Ideas, emotions, and moral sentiments alike were expressions of sensibility, movements of the body's parts in response to sensory impressions of the outside world.

By fusing sensation with sentiment, the inventors of the notion of sensibility transformed the meaning of scientific empiricism, for if knowledge arose from physical sensation, it must now originate equally in emotion. French natural philosophers began around 1750 to talk, write, and argue about the sentimental origins of knowledge. Georges Buffon's *Histoire naturelle* (1749), the first volumes of the *Encyclopédie* (1750–51), Diderot's *De l'interprétation de la nature* (1753), and Etienne Bonnot de Condillac's *Traité des sensations* (1754) are all filled with counsel that is sentimental, not methodological. Rather than techniques of observation or systems of experimentation, these epistemologi-

2. Diderot to Sophie Volland, 11 November 1760, in Diderot, *Lettres à Sophie Volland*, 1:195; Anon., "Locke, philosophie de," in Diderot and d'Alembert, eds., *Encyclopédie* (1765), vol. 9; and Fouquet, "Sensibilité, Sentiment," in ibid., vol. 15. Sophie Volland, whose given name was Louise-Henriette, was Diderot's lover during the 1760s and '70s. See Furbank, *Diderot* (1992), 186–90, 196–99, 202, 345.

3. *Dictionnaire de l'Académie française* (1694).

cal texts recommend emotional responses. One must not be "vain" or "frivolous" in building grand, rational systems and clinging to favorite hypotheses, but must follow one's "instincts" and "let the thoughts flow." (In the flowery, allegorical frontispiece to the *Encyclopédie* [fig. 1.1], arrogant Reason tries unsuccessfully to snatch away demure Truth's veil.) The texts depict the sensory and emotional discovery of a world outside the self: the "intimate blossoming" of hope and fear, horror and humility, love and hate, "vivid desire." Julien Offray de La Mettrie gave his *L'Homme-machine* (1748) a title that can mislead the modern ear: the human machinery he described was not cold but capable of all the passions. Sensibility marked the style as well as the content of these midcentury texts of natural philosophy. Diderot began his *Lettre sur les aveugles* (1749) with a poignant image of a blind man teaching his son to read.[4]

Over the following half-century, sensibility became crucial to ever more areas of scientific inquiry, beginning with physiology and psychology. During the 1760s the Swiss physiologist Charles Bonnet made sensibility the "great and the unique mobile" of animal life.[5] Others made it the central feature of human nature. The polemical materialist Claude-Adrien Helvétius announced in two notorious tracts, *De l'esprit* (1758) and *De l'homme* (1773), that all operations of mind, all ideas, interests, and passions reduced to the "only quality essential to the nature of man," that is, "physical sensibility." The baron d'Holbach, a frequent contributor to the *Encyclopédie*, similarly wrote that "mind is a product of . . . physical sensibility" and that from "sensibility flow all the faculties that we call *intellectual*."[6] The moral sciences too became projects in sensibility, beginning with the baron de Montesquieu's *De l'esprit des lois* (1748). Here he explained that laws must suit the "degrees of sensibility" of the people governed.[7] And Jean-Jacques Rousseau, the doyen of sentimental moralists, advised in *Emile* (1762) that as long as a child's "sensibility remains limited to his person there will be no morality in his actions." Only after his sensibility extended beyond himself would Emile acquire "senti-

---

4. Diderot, *De l'interprétation de la nature* (1753), 175–85; Buffon, *Histoire naturelle* (1750), 3:124–26, 133, 142–45; Condillac, *Traité des sensations* (1754), parts I, II; Diderot, *Lettre sur les aveugles* (1749), 82. On sensibility in Buffon's *Histoire naturelle*, La Mettrie's *L'Homme-machine*, Diderot's *Lettre sur les aveugles*, and Condillac's *Traité des sensations*, see chapter 2. On Buffon's sensibilities, see also chapter 3.

5. Bonnet, *Contemplation de la nature* (1764), part XI, 279. See also his *Essai de psychologie* (1754), *Considérations sur les corps organisés* (1762), and *La Palingénésie philosophique* (1769).

6. Helvétius, *De l'esprit* (1758), 2–7, 41, 324, 368; Helvétius, *De l'homme* (1773), 1:123–33, 2:60; Holbach, *Système de la nature* (1770), part I, 138; and Holbach, *Morale universelle* (1776), §1, 3. Diderot objected that sensibility was an essential property of matter, not a consequence of organization, and that it was therefore too primitive to explain animal and human capacities such as the capacity to feel pleasure and pain. He called sensibility a condition but not a cause of such capacities. See Diderot, *Réfutation d'Helvétius* (1774), 301, 310, 357.

7. Montesquieu, *Esprit des lois* (1748), bk. XIV, chap. 2, 192.

ments . . . of good and evil that constitute him a true man."[8] Moral learning, like natural philosophy, was a matter of fostering sensibilities.

Around the turn of the nineteenth century, Destutt de Tracy founded a new philosophical movement called Idéologie, the science of ideas. This new science studied the origins of ideas in sensibility and their combinations in the sentiments. Destutt de Tracy claimed that Idéologie underlay all the moral sciences: "grammar, logic, teaching, private morality, public morality (or the social art), education and legislation."[9] A fellow Idéologue, the philosopher and physiologist Pierre-Jean-Georges Cabanis, wrote that it was no longer possible to doubt that "physical sensibility is the source of all the ideas and all the habits that constitute the moral existence of man: Locke, Bonnet, Condillac, Helvétius, have carried this truth to the last degree of demonstration."[10] At the same time, the inventor of "biology," Jean-Baptiste Lamarck, devoted much of his *Philosophie zoologique* (1809) to "physical and moral sensibility," discussing their relations to "interior emotions" and the "interior sentiment" of one's own existence.[11] During the first years of the nineteenth century, a narrowing philosophical focus sharpened the contours of sensibility, reducing it to a phenomenon of the nervous system and the basis of psychology.

By the beginning of the nineteenth century, then, sensibility had itself become a fully established object of scientific inquiry. This book, however, concludes around the time that development took place. I am interested in the preceding five decades, during which sensibility was only gradually becoming a scientific subject but was already fully in place as a scientific tradition. By "scientific tradition," I mean a vocabulary and set of themes, and a repertoire of problems, methods, and principles, both tacit and explicit. As construed within the scientific tradition of sensibility, empiricism implied that knowledge grew not from sensory experience alone, but from a combination of sensation and sentiment. Therefore I propose to call this distinctive eighteenth-century mode of natural science "sentimental empiricism."

It is important to identify the sentimental component of Enlightenment empiricism for several reasons. First, doing so can enable us to understand the eighteenth-century interaction between the natural sciences, on the one hand, and moral and political thought and practice, on the other. A central effect of sentimental empiricism was a corresponding intimacy between the natural sciences and the emerging moral sciences. These precursors of the modern social sciences were defined by one of their leading inventors, the marquis

8. Rousseau, *Emile* (1762), 501.
9. Destutt de Tracy, *Eléments d'idéologie* (1796), 151, 227.
10. Cabanis, *Rapports du physique* (1815), 72.
11. Lamarck, *Philosophie zoologique* (1809), see especially bk. II, part III, chaps. III, IV. Lamarck argued that sensibility was a product of the nervous system, so that animals lacking a nervous system were without sensibility.

de Condorcet, as sciences that studied "either the human mind in itself, or the relations of men to one another," thereby joining epistemology and psychology with questions of proper behavior and good government. The moral sciences included political economy, civic education, and law.[12] A scientific, naturalizing approach to such moral subjects arose during the Age of Sensibility in unison with newly moralized natural sciences.

Both the naturalization of moral subjects and the moralization of natural subjects followed from a sentimental-empiricist understanding of thought and emotion. Sentimental empiricism, by tracing emotions to sensory experience, implied that moral sentiments might be subjected to empirical scrutiny and manipulation, which was the founding assumption of the moral sciences. However, by the same logic applied in reverse, sentimental empiricism also infused empirical experience, and therefore natural science, with sentiment and moral import. One of my principal interests in this study is to know how scientific and moral arguments interacted during a period in which the same people framed both sorts of argument and applied them to projects of political administration and activism. France in the half-century that led to the Revolution was a primary site for the genesis of moral sciences, whose descendants are our social sciences, and therefore also of the modern relations among scientific, moral, and social understanding and political application. I take the mutually transformative intimacy of natural and moral science to have been a defining feature of the Age of Sensibility, and this intimacy is a pivotal theme of the book.

A second reason for scrutinizing the sentimentalism of the empirical sciences during the Enlightenment is to show that these sciences were embedded within the contemporary culture, rather than acting upon it from outside. The "Age of Sensibility" has generally referred to developments in literature and the arts rather than in the sciences, and in particular to a moment in English cultural history defined by sentimental novelists such as Samuel Richardson and Laurence Sterne, who wrote during the period this book examines, from the middle of the eighteenth through the turn of the nineteenth century.[13] In the study of English letters, the "Age of Sensibility" refers to a literary tradition characterized by a set of features that appeared both together and separately, and whose combination can seem incongruous in retrospect. The first was emotionalism, ranging in tone from pathetic to extravagant. The second was a preoccupation with the bodily mechanisms of experience and emotion.

---

12. Condorcet, "Eloge de Bucquet" (1780), in *Oeuvres*, 2:410. See also Baker, *Condorcet* (1975), 197–98. On the eighteenth-century French origins of modern social science, see Baker, *Condorcet* (1975); Olson, *Emergence of the Social Sciences* (1993), chap. 9; and Manicas, *History and Philosophy of the Social Sciences* (1987), chap. 3.

13. On the Age of Sensibility in English literature, see Frye, "Towards Defining an Age of Sensibility" (1956); Bredvold, *Natural History of Sensibility* (1962); Hilles and Bloom, eds., *From Sen-*

And the third was a skeptical irreverence toward theories and institutions, including the same materialist, physiological theories of human nature that provided the backdrop for the first two features. Insofar as sciences such as physiology, medicine, and natural history, and scientific methods such as sensationist empiricism, have come into histories of sensibility, they have generally played the role of a source of technical information about sensibility—about the operation of the senses and their epistemological function—which artists and writers then transformed into their own artistic and literary notions of sensibility. Literary scholars and cultural historians have assumed that whereas sensibility was a florid style in literature and the arts, it was a plain object of research in the sciences.[14] They have also taken for granted that the influence flowed from sober science to fanciful culture.[15]

Historians of science have traditionally also studied sensibility as an object of scientific inquiry. Their work has contained only passing hints that sensibility might also have constituted a style of science.[16] Daniel Mornet ob-

sibility to Romanticism (1965); Birkhead, "Sentiment and Sensibility" (1925); Brissenden, Virtue in Distress (1974); Sambrook, Eighteenth Century (1986); Todd, Sensibility (1986); Mullan, Sentiment and Sociability (1988); Brown, Preromanticism (1991); Jones, Radical Sensibility (1993); Van Sant, Eighteenth-Century Sensibility and the Novel (1993); and Mullan, "Sentimental Novels," in Cambridge Companion, ed. Richetti (1996).

14. Walter Jackson Bate has traced "one of the major sources of the romantic stress on feeling" to "the mechanistic psychology of the seventeenth and eighteenth centuries." Bate, From Classic to Romantic (1961), 129–30. G. S. Rousseau has shown how theories of the functioning of the brain and nervous system informed the novels of sensibility and has argued, more generally, that one cannot understand the literary Age of Sensibility without reference to the history of physiology. Rousseau, "Nerves, Spirits, and Fibres" (1976) and "Discourses of the Nerve" (1989). On the origins of literary sensibility in natural philosophy, particularly physiology, see also Bredvold, Natural History of Sensibility (1962); Sambrook, Eighteenth Century (1986); Rodgers, "Sensibility, Sympathy, Benevolence" (1986); and Stephanson, "Richardson's 'Nerves'" (1988). Paul Ilie has made a parallel argument regarding the importance of physiology to the arts in the Age of Sensibility. Ilie, Cognitive Discontinuities (1995), 3. G. J. Barker-Benfield has identified a "gendered view of the nerves" as the "material basis" for the "culture of sensibility" of eighteenth-century Britain. Barker-Benfield, Culture of Sensibility (1992), vii. And John Brewer has similarly described English culture in the eighteenth century as having taken its shape by loosening the tight claims of scientific theories: "Sensibility and sentiment were technical terms employed in medicine, philosophy and psychology, but from the mid-eighteenth century they were widely and loosely used to describe the expression of heightened and intense human feelings, of a new sort of refinement." Brewer, Pleasures of the Imagination (1997), 113–14. On the scientific roots of literary sensibility, see also Todd, Sensibility (1986), 23–31.

15. Anne Vila, in her book on sensibility in medicine and literature, does define sensibility as "a joint literary and scientific tradition," in which literature and the sciences shared a reciprocal interaction. Vila, Enlightenment and Pathology (1998), 8. However, in her sections on sensibility in the sciences, she considers it rather as an object than as a style of investigation. Here I am suggesting that the sciences did not just provide technical accounts of sensibility that the arts drew upon to create a cultural style, but themselves participated in the style, and so were influenced by, as much as they influenced, the arts.

16. See, for example, Figlio, "Theories of Perception" (1975); Haigh, "Vitalism, the Soul, and Sensibility" (1976); Canguilhem, La Formation du concept de réflexe (1977).

served that "science and sentiment supported and explained one another," [17] but, as E. C. Spary has pointed out, Mornet separated his studies of science and sentiment into two separate works. Spary is one of a number of historians of eighteenth-century sciences who have begun to discern sentimentalism in their subjects. She argues that "'science' and 'sentiment' were not distinct" in French natural history and urges historians to recover the "links that existed ... between taste and reason, connoisseurship and utility, sensibility and scientificity." [18] Michael Hagner makes a parallel argument regarding German physiology. He describes the way in which physiologists who argued for the epigenetic theory of embryology represented that theory as "an aesthetic view of nature" and emphasized the importance of an "inner feeling" and a "sensibility" on the part of the anatomist and physiologist as "the key to [their] ... understanding of bodily processes." [19]

But the role of sentiment in Enlightenment empiricism seems to me deeper and more pervasive. Sentimentalism was not confined to physiology, which made animal sensibility one of its objects of inquiry, nor to natural history, whose panoramic view and descriptive methodology gave sentiment easy access. Sensibility operated even in fields that studied the inanimate rather than the animate, and that took an experimental rather than a narrative view of nature. [20] Sentimentalism characterized the methods of what are now considered the hardest sciences, physics and chemistry. This book therefore seeks, using these hard cases, to show that sentimentalism was integral to the method of Enlightenment science as a whole. Empirical science then takes its place within the culture of sensibility, not only acting, but acted upon. If scientific theories and results were a crucial ingredient of sensibility, it is reciprocally the case that the ideals of sensibility were constitutive of those very theories and results.

I hope the phrase "Age of Sensibility," which first calls to mind English sentimental novelists, though unexpected, will thus be persuasive as applied, in place of the more usual "Age of Reason," to French science. The engagement of Enlightenment French science with sentimental English literature should not be surprising. Natural science and literature were not far apart

17. Mornet, *Le Sentiment de la nature* (1907), 456.

18. Spary, "The 'Nature' of Enlightenment" (1999), 299. See also Spary, *Utopia's Garden* (2000), chap. 5, which discusses the role of sensibility in natural history. Spary writes, "The power to sense, *sensibilité*, was central to the making of natural historical knowledge at the Jardin in the second half of the century" (196). On "the rise of an intensified 'feeling for nature'" in "later-eighteenth-century France," see Charlton, *New Images of the Natural* (1984).

19. Hagner, "Enlightened Monsters" (1999), 198–99.

20. Cf. Spary, *Utopia's Garden* (2000), 6: "While experimentalists of the eighteenth century often portrayed nature as artful, coquettish, modestly covering her innermost secrets, naturalists proudly announced they perceived nature in herself, through a relationship of unmediated sensibility."

during the eighteenth century, nor were France and England. Diderot again personified the relevant intimacies. He was himself both natural philosopher and man of letters, and his favorite novelists were Richardson and Sterne—especially Richardson, whose novels Diderot called his "touchstone." Reading them he was utterly captivated, he said, like a child at the theater, his heart "in a state of permanent agitation," and afterward he felt "like a man who had spent the day doing good." He required the same susceptibility in all his friends, approving of one who "began to sob" over a passage from *Clarissa,* and firmly dismissing another—who laughed instead!—as one who could "never be my friend. I blush to think that she once was." Diderot himself was unable to finish reading Clarissa's deathbed scene aloud but burst into cries that became the seeds of *La Religieuse.* Meanwhile his *Jacques le Fataliste* was an homage to Sterne's *Tristram Shandy,* the "maddest, the wisest, the gayest of all books."[21]

Another reason for recognizing the sentimentalism of Enlightenment science, situated as it was within the culture of sensibility, is that it complicates our understanding of that culture. Sensibility is currently a principal theme of eighteenth-century studies. In particular, cultural historians and literary critics have been discovering sensibility at the heart of eighteenth-century notions of community and social union.[22] Their arguments about the centrality of sensibility to Enlightenment sociability have generally focused upon English polite society, and in that context they tend to associate sensibility essentially with femininity.[23] This book finds, on the contrary, that the principal elements of the culture of sensibility—sentimentalism, the notion that sentiments originated in physical sensations, and the conviction that sentiments were in turn the foundation of social life—were by no means restricted to feminine, English, polite society, but that these same beliefs and

21. Diderot, "Eloge de Richardson" (1761); Diderot to Sophie Volland, 26 September 1762, in Diderot, *Lettres à Sophie Volland,* 2:9; Furbank, *Diderot* (1992), 209–10, 267. Diderot was not the only French writer who could claim membership in the literary Age of Sensibility. Voltaire's *Candide* (1759), Rousseau's *Julie ou, La Nouvelle Héloïse* (1761), and Laclos's *Les Liaisons dangereuses* (1782) are novels of sentiment if ever there were such.

22. John Mullan, for example, has argued that authors such as Richardson, Sterne, and David Hume "discover in their writings a sociability which is dependent upon the communication of passions and sentiments." Mullan, *Sentiment and Sociability* (1988), 2. Adela Pinch, similarly, in *Strange Fits of Passion* (1996), proposes that the "circulation of feeling" in sentimental literature, from author to reader as well as among characters, provided a model for late Enlightenment understandings of the basis of social life.

23. Barker-Benfield defines sensibility as a specifically feminine "kind of consciousness." Barker-Benfield, *Culture of Sensibility* (1992), xvii, 36. His "culture of sensibility" has come to mean "a realm of polite action and discourse that partook of a general feminization of manners." Clark, Golinski, and Schaffer, eds., *Sciences in Enlightened Europe* (1999), 170. See Mary Terrall's contribution to that volume, in which she writes that "reason was frequently gendered as masculine in opposition to imagination, sensibility or feeling." Terrall, "Metaphysics, Mathematics and the Gendering of Science" (1999), 258–59. On the gendering of sensibility in literature, see Todd, *Sensibility* (1986), 17–21.

preoccupations were at work in the overwhelmingly male, French sciences.[24]

My final purpose for looking to recover the sentimental component of eighteenth-century science has been to contribute to a current historiographical venture. In this venture, historians have been examining the origins of a set of related notions: empiricism, materialism, fact, objectivity. Lorraine Daston has given a name to the endeavor, which she defines as "a history of the categories of facticity, evidence, objectivity and so forth": she has called it "historical epistemology."[25] Daston herself has chronicled the rise of probabilistic reckoning to its modern, central position in rational decision-making.[26] Theodore Porter and Mary Poovey have examined the emergence of related social uses of numbers (statistics, accounting, cost-benefit analysis, double-entry bookkeeping) to discover "how one kind of representation—numbers—came to seem immune from theory or interpretation,"[27] and so to epitomize the objective, scientific fact. By scrutinizing the genesis of such crucial elements of modern science and, more generally, of our modern epistemology, historians have discovered some surprising things. Robert Darnton, for example, in *The Forbidden Bestsellers of Pre-Revolutionary France* (1995), has disrobed erotic parts of eighteenth-century materialism and, correspondingly, a philosophical impulse in erotic literature. Sex and philosophy, he writes, "belonged together."[28] Daston has unveiled monsters and marvels, which she argues were the original, objective facts of early modern empiricism. And Steven Shapin, in *A Social History of Truth* (1994), has shown how gentlemanly etiquette informed the rules of empiricism in seventeenth-century Britain.[29]

24. Anne Vincent-Buffault has responded to historians' tendency to see sensibility as essentially feminine by noting that men wept too: "People enjoyed crying, the women in their boudoirs, the men in their studies." Vincent-Buffault, *History of Tears* (1991), 3 (see also pages vii–ix and 246). Vila suggests that sensibility was more gendered in England than in France: "French writers did not polarize sensibility in relation to sex and gender nearly as much as did their British counterparts . . . even the most hard-boiled philosophes prided themselves on their sensibility and saw nothing unmanly about cultivating this quality." Vila, *Enlightenment and Pathology* (1998), 3. On the androgynous nature of French sensibility, see also Ridgeway, *Voltaire and Sensibility* (1973). For a criticism of Barker-Benfield's gendering of sensibility, see Rousseau, "Sensibility Reconsidered" (1995).

25. Daston, "Moral Economy of Science" (1995), 24. Daston's other works of historical epistemology include "Marvelous Facts and Miraculous Evidence" (1991); "Baconian Facts, Academic Civility, and the Prehistory of Objectivity" (1991); "Objectivity and the Escape from Perspectivity" (1992); "Fear and Loathing of the Imagination in Science" (1998); and Daston and Galison, "Image of Objectivity" (1992). For others' contributions to this endeavor, see Porter, ed., "Social History of Objectivity" (1992), including Dear, "From Truth to Disinterestedness" (1992).

26. Daston, *Classical Probability in the Enlightenment* (1988).

27. Porter, *Rise of Statistical Thinking* (1986); Porter, *Trust in Numbers* (1995); Poovey, *History of the Modern Fact* (1998). The quotation is taken from Poovey, xii.

28. Darnton, *Forbidden Bestsellers* (1995), 95 (and all of chap. 3). On the various implications of erotic writing in eighteenth-century France, see also Hunt, ed., *Eroticism and the Body Politic* (1991).

29. Daston, "Marvelous Facts" (1991); "Moral Economy of Science" (1995), 12–18; "Baconian Facts, Academic Civility, and the Prehistory of Objectivity" (1991); and Shapin, *A Social History of Truth* (1994).

Manners, monsters, and sex, three unexpected ingredients for early modern science, all gained their parts in its genesis thanks to Lockean sensationism. By naming sensory experience as the only genuine source of knowledge, sensationism engendered a suspicious attitude toward abstract theory, which operated at a remove from immediate physical sensation. Suspicion toward theory expressed itself in a wry but substantive appreciation of sex, which banished the rational half of Descartes's mind/body dualism, allowing the material, sensory half to conquer the stage. Hostility toward abstract theory also legitimated a fascination with monsters, which, being resolutely anomalous, tamed theories by defying them. Finally, the early modern empiricists' conviction that a tendency to theorize must be well tamed went hand-in-hand with the emphasis they often placed upon civility, which, they maintained, demonstrated a gentleman-philosopher's humble willingness to relinquish his own theory when confronted by evidence against it.

In sum, historians have been learning that modern scientific empiricism, despite its claims to straightforwardness and neutrality, hides a baroque past. This book is a further exploration of that past, mapping a new area, that of sentiment. The sentimental element of the history of empiricism has in common with its erotic, monstrous, and polite elements a single ideal: receptiveness to a world outside the mind, a world that imposed its claims through the senses.

Each chapter examines an instance of the ideals of sensibility at work in eighteenth-century French science. These instances represent the ramifications of sentimental empiricism for the various branches of natural and moral science, their relations to one another, and their applications to political administration, reform, and revolution.

I begin with the study of sensation (theoretical and experimental, psychological, physiological, and moral) and its therapeutic and institutional applications. Vision, in particular, was the Enlightenment's supreme sense, and the philosophes' preoccupation with it activated a midcentury explosion of interest in blindness, specifically in the thoughts and sentiments of blind people. Chapter 2, entitled "The Blind and the Mathematically Inclined," traces the development of this interest, which began with a much-discussed thought experiment, published in Locke's *Essay concerning Human Understanding* (1690), featuring a hypothetical blind man.[30] Next came a tradition of actual experiment and observation with real blind people: cataract surgeries, interviews with the blind, and reports of their lives. One such examination, based on an interview with a blind man, was Diderot's *Lettre sur les aveugles* (1749). He argued that blind people thought like mathematicians and mathe-

30. Locke, *Essay* (1690), bk. II, chap. IX, §8.

maticians like blind people: both were unusually impervious to sensory experience, therefore lacking in sensibility. This insensibility stunted the moral as well as intellectual faculties, Diderot suggested, and so he "suspected blind people," and, by implication, mathematicians, of "inhumanity." [31] The chapter closes with institutional applications of these philosophical analyses of blindness. During the early 1780s Valentin Haüy founded the first school for the blind in Paris, its curriculum shaped by connections like the one Diderot drew between scientific epistemology and civic morality. Later this school and other schemes for blind education figured prominently in Revolutionary discussions of the sensory and sentimental bases of social harmony.

The morals of sentimental empiricism are also the subject of chapter 3, "Poor Richard's Leyden Jar." The chapter analyzes the conflict between proponents of two competing theories of electricity, those of Benjamin Franklin and of Jean-Antoine Nollet. Their conflict arose in 1752 when Buffon sponsored the translation into French of Franklin's letters on electricity. Buffon saw in Franklin's electrical theory a model of the sort of natural science he liked, which was the antithesis of Nollet's variety. Nollet followed Descartes in constraining his physics to a discussion of mechanical causation, matter in motion. But this mechanical model of natural philosophy had been acquiring a bad name as dogmatic and arrogant, disdainfully abstract from sensory experience. In contrast, Franklin's electrical science was teleological, in that it rested upon final causes rather than efficient causes. Franklinist physics portrayed a natural world guided by purposes rather than driven by mechanisms. Like his moralism, Franklin's natural science appealed to proportion and balance as ideals intrinsic to nature, and advertised sensibility, an openness of mind to experience, as the means to uncover these ideals. The chapter's thesis is that the opposition between Nollet's mechanical theory and Franklin's teleological one enabled Franklinism to represent sensibility in science, lending Franklinism both epistemological and moral authority. This double imprimatur, I believe, explains Franklin's extraordinary popularity and influence in France.

The French Economists—the Physiocrats—were among those who seized upon Franklinist philosophy and adapted it to their own purposes. They introduced Franklin to the emerging field of political economy and to a prominent position in French politics by enlisting him in their program of economic reform. Chapter 4, "From Electricity to Economy," begins with this program. The Economists cast wealth as a natural fluid akin to heat, light, or electricity, and subjected it to a conservation law like Franklin's conservation of electrical charge. The same natural motives that regulate the flow of electricity would, the Economists argued, also maintain a harmonious distribution

31. Diderot, *Lettre sur les aveugles* (1749), 92–93.

of wealth. They therefore recommended an empirical method of determining correct taxes from sensory "evidence," the annual agricultural surplus.

The Economists named their science "Physiocracy," meaning "rule of nature," in reference to the natural origins of order and value. They brought the notion of an opposition between empiricist sensibility and rationalist system-building from natural philosophy, where it had been much discussed, into political economy, where it became a powerful though fickle force. The "spirit of system," a favorite epithet among natural philosophers after midcentury and in political-economic debate during the last few decades of the century, represented the antithesis of sensibility, the dogmatic adherence to a system founded in abstract reasoning rather than sensory evidence. The accusation was so popular and effective that everyone, on all sides of any given dispute, used it against everyone else. Common sense might seem to dictate that any slogan launched in support of both sides of a controversy cannot have been a significant force in determining its outcome. On the contrary, however, by its very ubiquity, the phrase "spirit of system" indicates one of the most powerful, widespread, governing assumptions of the moment.[32] While administrators, members of the Parlement, and reformists differed in where they identified system-building, they were unanimous in condemning it.

Representing their own economic program as empiricist, the Physiocrats accused their mercantilist opponents of "system-building." Then, during the disastrous tenure of the physiocratic fellow traveler A.-R.-J. Turgot as finance minister from 1774 to 1776, the Physiocrats' opponents seized their rhetorical weapon and turned it back against them, making "physiocratic" synonymous with "systematic." The chapter follows charges and countercharges of system-building into the Revolution, when the Physiocrats were able to reclaim the accusation of system-building and use it to shape important decisions, for example, the decision in favor of a unicameral legislature. They repeatedly invoked Franklin's avoidance of dogmatic mechanism and system-building as their ideal of both scientific and political sensibility.

While chapter 4 examines the moral authority of sentimental empiricism in the context of economic policy-making, chapter 5, "The Lawyer and the Lightning Rod," looks at its influence in the context of legal decision-making. This chapter tells the story of M. de Vissery de Bois-Valé of Saint-Omer, who, in the spring of 1780, put a lightning rod on his roof. His neighbors feared

---

32. This seems to me worth emphasizing. The points of disagreement are the most apparent features of a dispute, but they are not necessarily the most telling. Historians looking back upon a controversy can learn as much from what the disputants agreed upon as from their divergences. A modern example: every Western politician would describe him- or herself as "pro-democratic." This does not mean that claims of democracy are empty, but, on the contrary, that they are the most powerful claims available. One cannot understand any political dispute of our age without understanding their power.

the rod, sued him to remove it, and won. The case was inherited on appeal by an unknown and junior member of the Arras bar, Maximilien Robespierre. To defend Vissery, Robespierre corresponded with jurisconsults and electricians. He then exploited the empiricist dogma shared by law and physics in 1780s France. According to this dogma, knowledge was founded in particular facts not because of their places in general theories, but, on the contrary, because of their irreducible particularity. A particular fact derived its importance precisely by resisting theory and tradition. Robespierre stated the facts of electrical behavior while denying that the theoretical explanations of those facts, demanded by Vissery's opponents, had any place in empirical science or in courts of law. If the courts would only set aside the theories of both physics and jurisprudence, Robespierre argued, the two sciences would meet in the truth.

Implicitly, Robespierre also proposed setting aside the experts whose job it was to spin the theories. He maintained that the law rested upon neither theory nor precedent nor expertise. It required only sensibility, receptiveness to the "inexplicable" dictates of experience. The triumph of this argument launched a career that would culminate ten years later in the abolition from French officialdom of experts and their theories. Robespierre would at the same time justify his establishment of a civic religion as the pragmatic acceptance and instrumental manipulation of an empirical fact of human nature, a passionate sensitivity to the "incomprehensible power" of nature.[33]

Manipulation of sensibility is the subject of chapter 6, "The Mesmerism Investigation and the Crisis of Sensibilist Science." The Viennese polymath Franz Anton Mesmer captured the imagination of Paris society in the early 1780s with his claim to have harnessed the imponderable fluid of human sensibility. He said he could channel this fluid using implements such as wands, tubs filled with water and metal, and his own fingers. His assertions were borne out by dramatic results: he could provoke powerful sensations, emotional agitation, and dramatic convulsions in his patients. Louis XV appointed two commissions of academicians and doctors to look into the matter, and the chapter tells the story of their investigation and its startling conclusion.

The commissioners designed a series of experiments to test Mesmer's claims, which they ultimately rejected. They attributed mesmeric effects to manipulations not of the fluid of sensibility, but instead of the power of "imagination." Though they admitted that they could not say precisely what imagination was or how it worked, the commissioners were nevertheless certain it was a powerful and manipulable force that could be used to override natural

33. Robespierre, "Second plaidoyer" (1783), in *Oeuvres*, 1:74–75; and Robespierre, "Contre le philosophisme et pour la liberté des cultes (Aux Jacobins, le 1er frimaire an II/21 November 1793)," in *Ecrits*, 284.

sensibility. Mesmer, by carrying sentimental empiricism to its logical extreme with his fluid of sensibility, had exposed its vulnerabilities, in particular, the problematic promiscuity of a subjectivist epistemology founded in feeling. If Mesmer's patients felt the fluid, and feeling was the ultimate test of truth, who could say that the fluid did not exist? Mesmerists thereby forced the commissioners to undermine the central tenet of sentimental empiricism, the principle that sensations were responses to a world outside the mind. Their new power of imagination meant that sensations did not always connect the sensible creature to the natural and social world outside, a possibility with frightening implications. Where sensibility was the basis of civic harmony, imagination, loosening the hold of the five senses, made people vulnerable to demagogues and was therefore responsible, the commissioners wrote on the eve of the Revolution, for revolts and seditions and all manner of fanaticism.

The relations among sensibility, system-building, and revolution are the concern of chapter 7, "Languages of Science and Revolution." The chapter studies two related disputes about language: the controversy surrounding Antoine Lavoisier and his collaborators' new system of chemical names, published in 1789; and the debate within the Revolutionary Committee of Public Instruction over the role of language in civic education. The scientific and political disputes about language overlapped, involving many of the same people and same arguments. I identify two conceptions of language at work in each dispute, a cultural conception, that of the sensibilists, and a competing social understanding of language in which the idealism and instrumental empiricism of social engineers replaced the romanticism and sentimental empiricism of the sensibilists. In keeping with the root meaning of the word "culture," advocates of the cultural conception of scientific language strove to keep their words natural, cultivating them from chemists' distinctive sensibilities of nature. In contrast, promoters of the social conception of language sought deliberate linguistic conventions that would provide the basis for a socially collaborative science. These instrumental empiricists had no interest in spontaneous expressions of sensibility. Formal rules, not feelings, they argued, were the origin of language, knowledge, and civic union.

Opponents of the new chemical nomenclature saw its authors' system-building as sinister and coercive. Many came to associate these reforms of chemical language with another contemporary set of linguistic changes, those introduced by the Jacobin party during their Reign of Terror. The analogy expressed a widespread conviction that the spirit of system was not only misguided but dangerous. At the same time, advocates of a sensibilist pedagogy in the Committee of Public Instruction opposed the use of technical vocabularies, and even of books, in teaching. Instead these sentimental-empiricist theorists of civic education called for a pedagogy that operated by the molding of pupils' sensibilities through the careful management of their sensory

experiences. Dominated by such sentiments, the Committee oversaw the closing of the learned academies in the summer of 1793 and the subsequent founding of a system of national education. As it went about establishing France's
modern system of public education and state-sponsored research, the Committee of Public Instruction and its correspondents maintained that the "whole
art of instruction" must be in the "linking of sensations." They proclaimed, "It
is by way of the senses that the virtues enter the heart ... [and] vices enter by
the same door." Because virtues originated, like natural philosophy, in sensibility, "physics [should] be always the guide of morality," and "a course of experimental physics ... [should] serve as an introduction to moral education."
In physics as in morals, education was less a matter of enlightening the mind
than of cultivating the "body and heart."[34]

Because chapter 7 is about language, it is the part of the book where I have
been most concerned to clarify the relations between actors' categories such
as "spirit of system" and my own analytical categories such as "sentimental
empiricism." Here, therefore, is the place for a brief word about the language
of this book. Mindful that one should be wary of neologisms, I hesitated before coining the term "sentimental empiricism" but decided it was the simplest way to say what I meant: a tradition of science founded on the assumption
that knowledge grew not from sensory experience alone, but from an inseparable combination of sensation and sentiment. This conviction informed
theories, infused vocabularies, shaped methods, and was voiced by many
people throughout the natural and moral sciences; so in the course of this
book, I apply the term "sentimental empiricism" to theoretical standpoints,
languages, methods, and people, as well as to institutions, political controversies, and legal arguments. Across all these applications, the meaning of the
phrase remains consistent. Diderot's understanding of blindness and Haüy's
school for the blind, Franklin's electrical physics, the Physiocrats' understanding of wealth, and the Parlement's opposition to the Physiocrats, Robespierre's
legal defense of a provincial lightning rod, the royal investigation of mesmerism, chemists' opposition to the new chemical language of 1787, and the discussions of the Revolutionary Committee of Public Instruction, all were shaped
by the sentimental empiricist conviction that sensation and emotion were inseparable, and that together they formed the basis of natural knowledge.

None of the people I talk about used the phrase "sentimental empiricism,"
but I do draw upon their language in coining it, for my protagonists spoke a
great deal about sensibility, experience, and sentiment. In using their understandings of these words to create my own interpretive category, I move back

---

34. Raffron, "Troisième discours" (1793), in Comité d'Instruction Publique, *Procès-verbaux*,
2:233; Manuel, *Etude de la Nature* (1793), 15–16, 36; Rabaut Saint-Etienne, "Projet d'Education Nationale" (21 December 1792), 2:232–33.

and forth between their terms and my own. Furthermore, eighteenth-century phrases such as "spirit of system" and "system-builder" play important roles in my story. These phrases are difficult to read because they were polemical, not applied with rigor and consistency, but rather launched by everyone against everyone else. Their ubiquity reflects their power, and the depth of the preoccupations to which they gave voice. Like most polemical phrases, these were neither straightforward descriptions of reality, nor utterly detached from actual practice. Such phrases are my raw materials, and they inform my own interpretive categories even as I use these to indicate where I think the actors' categories cannot be taken at face value. It was while trying to distinguish mine from the actors' categories that I realized that the two were necessarily inseparable. I discuss this methodological problem at the end of chapter 7.

The tradition of sensibility in French science shaped the way people understood physiological and psychological processes like sensation, emotion, and imagination; natural phenomena such as electrical action and chemical reactions; social and cultural institutions including taxation, legal decision-making, language, and moral behavior; and the forces and constraints of political life. This is no surprise; the principles and techniques of modern scientific empiricism continue to shape our understanding in all these areas. But in the chapters that follow, I trace the ramifications of an earlier, often unfamiliar scientific empiricism, construed and applied throughout a world in which sensation could not be detached from sentiment.

I hope it will be clear from these brief chapter sketches that although this is a book in which ideas play a central role, it is not intended as a traditional intellectual history. One of my premises is that ideas are not separable from scientific practice or social organization and cannot be treated separately. I mean this book to be an intellectual history, yes, but not an intellectual history instead of a history of practice or a social history. It is intellectual history driven by an interest in the engagement of ideas with cultural, social, and experimental practices. I take this engagement to be a reciprocal matter, in which neither side determines the other, and in which each side continually reshapes the other.

In the disputes between so-called internalists, who depict the history of science as propelled from within by intellectual forces, and externalists, who represent it as shaped from the outside by social and cultural factors, it is hard to find supporters of the notion that ideas are powerful agents in the social and cultural world, but, at the same time, they are powerful precisely because they are embedded in it. In contrast, as will be apparent throughout this book but with particular force in chapter 7, my eighteenth-century protagonists generally took for granted the mutual influence of scientific knowledge and social and cultural practice. Their having done so, I suggest in chapter 7 and in the conclusion, complicates the internalism/externalism debate, and its eruption in the science wars of the 1990s, by giving both sides a common ancestry in Enlightenment science. Participants in the internalism/externalism debate also generally project a sharp division between the history of intellectuals and powerful people, on the one hand, and the history of common folk, on the other. But in writing this book and examining the history of ideas in their interaction with cultural and social practices, I have always found intellectuals inextricably linked in the production of ideas with the people among whom they lived.

Interactions between philosophers and commoners, ideas and practices, philosophy and the facts of daily life are the heart of my story. In chapter 2, ideas about the senses and blindness directed, and were in turn reshaped by, cataract and other eye surgeries and figured centrally in the founding of a school for the blind. The blind subjects of philosophical experimentation defined the outcomes as actively as the philosophers and government officials who took a sudden interest in them. Chapter 3 treats ideas about electricity interacting with questions of moral behavior, etiquette, and the basis of community. Chapter 4 is about economic policy and, ultimately, the design of new governmental institutions, in relation to ideas about the naturalism of the social world, ideas that were given forceful expression in popular protests against economic reform. Chapter 5 looks at the theoretical uncertainties of Franklinist electrical physics in the context of the design and implementation of the lightning rod and uses these uncertainties to examine the relations between science and law and the problem of the relative authority of philosophers and magistrates to make public decisions. The townspeople of Saint-Omer were the motive forces in this episode, while Robespierre and his Parisian consultants were latecomers and peripheral players. Chapter 6 is about the popularization of scientific ideas and the state's authority to oversee scientific and medical practice. Once again philosophers and political actors were respondents: a popular craze provided the impetus for the investigation of mesmerism. Chapter 7 relates ideas about language and chemistry to programs of civic education.

In each case the ideas take shape precisely through their engagement with matters of scientific, social, and political practice—and vice versa. My argument throughout is that the theme of sensibility brings together ideas (about nature, scientific method, human psychology, and physiology) with practices (social, experimental) and with politics (the enactment of policies and founding of institutions). The book, then, is a study of the engagement of scientific ideas with scientific and social practices, as this engagement took an eighteenth-century shape in the philosophy and politics of sentimental empiricism.

NICHOLAS SAUNDERSON *LL.D.*
Lucasian Professor of Mathematicks in
the University of Cambridge
Died 19. Ap. 1739 Aged 56

I. Vanderbanck. pinx. 1718.   From the Original painted for Martin Folkes Esq.   G. Vander Gucht. Sculp.

Figure 2.1. Nicholas Saunderson (1682–1739), Lucasian Professor of Mathematics at the University of Cambridge and subject of Diderot's *Lettre sur les aveugles* (1749). The portrait is by I. Vanderbanck, and the engraving by C. F. Fritzsch, from the frontispiece to Saunderson's *Elements of Algebra* (1740). Courtesy of the Bancroft Library, University of California, Berkeley.

Chapter Two

THE BLIND AND THE
MATHEMATICALLY INCLINED

Mathematics . . . are the staff of the blind.
—Voltaire, *Le Siècle de Louis XIV* (1751)

On a spring day in 1728, Dr. William Cheselden—surgeon to the queen and to the just-deceased Sir Isaac Newton, renowned for his ability to remove a bladder stone in fifty-four seconds—operated on the cataracts of a "Young Gentleman, who was Born Blind, or Lost his Sight so Early, that he had No Remembrance of Ever Having Seen." Cheselden's report of the surgery, focusing on the child's immediate postoperative feelings and impressions, was published in the *Philosophical Transactions of the Royal Society of London.*[1] Another decade passed before news of the Cheselden experiment arrived in France. Voltaire, exiled in London, dispatched it across the Channel in 1738.[2] The philosophes responded vigorously, citing Cheselden's "boy born blind" in a small flood of treatises, developing a lively interest in the thoughts of the blind, and performing their own cataract surgeries.

Cheselden's procedure brought to life a thought experiment described in Locke's *Essay concerning Human Understanding* involving the sudden acquisition of sight by a man blind from birth. The thought experiment, known as "Molyneux's Problem" for the person who first proposed it to Locke, was meant to test the sensory origins of ideas. If, as Locke claimed, the mind at birth was a blank slate, then a man who had been blind from birth, lacking any visual experience, should have no visual ideas and therefore no ability to recognize objects visually. Cheselden confirmed Locke's prediction. Its place in history has therefore generally been that of a footnote to the story of the development and promulgation of Lockean epistemology.[3]

But the experiment of the newly sighted blind man was important in its own right. It transformed the terms of the epistemological discussion in which it arose. Molyneux's Problem implied such a transformation when it

1. Cheselden, "Account" (1728).
2. Voltaire, *Eléments* (1738), 319–20.
3. On Cheselden's life and career, see Condorcet, "Cheselden," in *Oeuvres*, 2:120–22; Morand, "Eloge de M. Cheselden" (1757); and Cope, *William Cheselden* (1953).

was still just a thought experiment, but Cheselden's procedure, making the thought experiment real, launched a new tradition that was primarily experimental rather than philosophical. Speculations about the sense of sight would henceforth involve cataract and other eye surgery and postoperative psychological examinations of newly sighted people.

This experimental study of blindness departed sharply from Locke's original philosophical purpose. It marked a turning point in the history of epistemology, from speculative to experimental, and from an older assumption that knowledge consisted of ideas composed from sensations, to a newer presumption that knowledge resided in a combination of sensation and emotion: that is, in feeling. Investigators of the thoughts of blind people abandoned the traditional epistemological question of the existence or nonexistence of innate ideas. Now they became primarily interested in the source of a receptive awareness of the world outside the mind, the non-self. To these experimenters, the salient feature of the mind at birth was not blankness but solipsism: obliviousness to the existence of an external world as such. Therefore, the questions that interested Cheselden and his followers concerned the origins of sensibility: How does one become sensitive to the existence of external things? How does one grasp that they are external to oneself? Do the senses themselves convey such understanding? Might one have sensation without sensitivity to the world outside? Are some senses more important than others in developing such sensitivity?

These questions about sensibility, tying together the moral and the physical, focused the new study of blindness, which consequently took the form of an indivisible mix of theory, therapy, and experiment. By examining this mix, I mean to lay bare the core logic of sensibilist science. Most historical writing on sensibility has tended to generalize and diversify its meaning, equating it variously with sentience, taste, emotionalism, politeness, imagination. G. J. Barker-Benfield writes that "sensibility" meant "the operation of the nervous system, the material basis for consciousness." But, he continues, the "flexibility of a word synonymous with consciousness, with feeling . . . permitted a continuous struggle over its meanings and values." His study of sensibility follows this struggle outward from the root meaning, taking as its own definition of the term a specifically feminine "kind of consciousness" whose central characteristics were "the powers of the intellect, imagination, the pursuit of pleasure, the exercise of moral superiority."[4]

My trajectory is the reverse: from the various uses of "sensibility" back toward their common source. I am receptive to Anne Vila's caution that "sensi-

---

4. Barker-Benfield, *Culture of Sensibility* (1992), xvii, 36. On the physiological basis of eighteenth-century sensibility, see Trahard, *Les Maîtres de la sensibilité française* (1931–33), vol. 2, chap. 4.

bility was a polysemous concept, a notion that not only cut across disciplinary boundaries, but represented several different things at once."[5] Indeed, sensibility sometimes had opposite implications from one context to the next, suggesting imaginativeness in one instance, for example, and its antithesis in another. (In chapter 6 I discuss an episode in which sensibility and imagination were opposed, the 1783–84 investigation of mesmerism.) My purpose is to recover, within the various meanings of "sensibility," their unifying logic, in order to show that this logic functioned as a medium of interaction among science, culture, and politics. The logic of sensibility operated at its clearest and most explicit in the discussion and treatment of blindness. To philosophers and experimenters in the Molyneux-Cheselden tradition, sensibility meant something that was at once specific and twofold: a physical, sensory receptiveness to the world outside oneself, whose consequence was emotional and moral openness. Its opposite was a physical insensitivity that brought solipsism.

The opposition between sensitivity and solipsism began to emerge in early discussion of Molyneux's Problem (§1). The subsequent experimental study of blindness developed this opposition and focused upon it (§2). Up to this point, the story takes place equally in France and England, and somewhat in Austria as well. However, following the emergence of solipsism as the central problem in the experimental study of blindness, solipsism and its moral and emotional implications came to dominate the philosophical discussion of blindness as well (§3). This final transformation of the philosophical study of blindness, in which blindness changed from a problem of ideas to one of feelings, was distinctively French. When one rereads Diderot's familiar *Lettre sur les aveugles* in the new light cast by these developments (§4), one sees that the emerging notion of the insensibility of the blind was crucial to it.

Drawing upon an interview with a blind man and a memoir on the life of the blind Nicholas Saunderson (fig. 2.1), late Lucasian Professor of Mathematics at Cambridge,[6] Diderot argued that the blind, because of their impoverished sensibilities, turned their minds inward and tended to think in abstractions. This made them natural mathematicians and rationalists: in a word, Cartesians. Conversely, Cartesians' abstract, inward focus made them insensible to the world outside their minds: philosophically blind. Diderot then suggested that blind men and Cartesians alike, because of their solipsistic cast of mind, were inhumane. This peculiar and pivotal point in Diderot's argument, his proposal of a moral equivalence between blindness and Cartesianism, which critics have surprisingly ignored, recasts our understanding of

5. Vila, *Enlightenment and Pathology* (1998), 1.
6. Anon., "Memoirs of the Life and Character of Dr. Nicholas Saunderson," in Saunderson, *Elements of Algebra* (1740), 1:i–xxvi.

Diderot's *Lettre*.[7] Recognizing it is essential to grasping one of Diderot's key purposes: to dramatize the opposition between sensibility and solipsistic rationalism, and the sensory basis of civic life.

To Diderot, sensibility grounded not only natural knowledge, but moral sentiment and civic engagement, leading preoccupations of the Enlightenment. The Enlightenment individual, beginning life as a blank slate, needed something to open him to the society of others. Political thinkers, challenging the legitimacy of tradition and inheritance as bases of civic life, sought a new basis in nature, and many, like Diderot, turned to the five senses and their action upon the mind within. Experimental and philosophical writers on blindness increasingly returned to the problem of sociability and its basis in sensibility, and their preoccupation had lasting institutional effects. It informed the programs for civic education that began to emerge at the end of the eighteenth century including, during the 1780s, the first school for the blind (§5). A decade after its founding, this school became a Revolutionary symbol of the brotherly love and community that could arise from a national program of moral education.

It should then be evident that the history of ideas about blindness, sensation, and thought, with its familiar sources, cannot be detached from the experimental tradition of Cheselden-type procedures. These little-remarked procedures transformed the epistemological discussion of the role of the senses in human intellectual and moral development, and the medical and institutional treatment of the blind, focusing both on the problem of insensibility and moral solipsism.[8] As a result of this interaction of philosophy, experimental and medical practice, and political application, the new social position of the blind finally became, at the hands of Revolution-era institution-builders, emblematic of the general task of civic education.[9]

7. See Furbank, *Diderot* (1992), 54–71; Ilie, *Cognitive Discontinuities* (1995), 346–47; Vartanian, *Science and Humanism* (1999), 97. According to Vartanian's reading of Diderot's *Lettre*, for example, its blind figures give direct voice to Diderot's own conception of nature. Nicholas Saunderson represents not the insensibility of the blind, but their "paranormal mentality," enabling them to see cosmic truths. This reading I believe overlooks critical points in the *Lettre*. Diderot's frequent suggestions that a blind man's view of the world was made of geometrical abstractions, and was therefore deeply problematic.

8. The literature on Molyneux's Problem gives Cheselden's surgery a passing mention at most; writing on Diderot's *Lettre sur les aveugles* deals similarly with Réaumur's cataract surgery; and other surgeries in the same tradition are not discussed at all. See, for example, Degenaar, *Molyneux's Problem* (1996), 58–60; Torlais, *Esprit* (1936), 250–52; Furbank, *Diderot* (1992), chap. 3; Roger, *Buffon* (1989), 216–17. An exception is von Senden, *Space and Sight* (1960).

9. For a cultural history of blindness in the eighteenth century, see Paulson, *Enlightenment, Romanticism and the Blind* (1987).

I. THE THOUGHT EXPERIMENT TRANSFORMED: FROM GEOMETRY TO SENSITIVITY

The thought experiment that Dr. Cheselden brought to life was a "jocose problem" submitted to Locke by an admirer, the "Learned and Worthy" William Molyneux: Irish politician, founder of the Dublin Philosophical Society, author of the first major treatise on optics in English, and the cousin and husband of blind women.[10] In 1688 Molyneux sent Locke the following query: If a man, blind from birth, suddenly gained vision, could he tell a sphere from a cube by sight alone on the basis of a lifetime of solely tactile experience?[11] Locke included this conundrum in the *Essay concerning Human Understanding*, along with Molyneux's own solution: "the acute and judicious Proposer answers: *Not.*" Locke added his warm endorsement: "I agree with this thinking Gent., whom I am proud to call my Friend."[12]

Molyneux's Problem inspired a philosophical debate in which the initial epistemological question about the origins of ideas was quickly abandoned in favor of a new question about sensitive openness to the outside world, even before actual eye surgeries made real the hypothetical blind man given sight. Molyneux had invoked in support of his hypothesis the Lockean principle that without visual experience there can be no visual knowledge. A man blind from birth would have no visual idea of a cube or a sphere. Also he could not know that an object affecting his sense of sight in a certain way would affect his sense of touch in a certain other way.[13] That was because the correlation of visual with tactile experiences, and the use of the five senses generally, relied upon a learned process of judgment "acquired by *Exercise.*"[14]

Locke, too, believed that ideas arose from sensations by means of a learned process of judgment. Accordingly, he offered Molyneux's Problem and its negative answer as an "occasion for [the reader] to consider, how much he may be beholding to experience."[15] The initial epistemological moral of Molyneux's Problem was therefore twofold: first, that ideas were not innate, but arose from physical sensation; and second, that ideas did not arise immediately from physical sensation, but only by means of an intellectual process that also developed through experience. Seeing was not merely a physical relation between the seen object, light, and the observer's retina, but an acquired

---

10. The treatise was Molyneux's *Dioptrica nova* (1692). See Degenaar, *Molyneux's Problem* (1996), chap. 2; and Simms, *William Molyneux* (1982), chaps. 1, 2.

11. Molyneux to Locke, 7 July 1688; cited in Degenaar, *Molyneux's Problem* (1996), 17.

12. Locke, *Essay* (1690), bk. II, chap. IX, §8.

13. Ibid.

14. Molyneux, *Dioptrica nova* (1692), 113.

15. Locke, *Essay* (1690), bk. II, chap. IX, §8.

skill. Molyneux's Problem originated as an empiricist fable about the acquisition of both ideas and mental processes from experience.

Not without some satisfaction, Molyneux reported to Locke that most of the people to whom he posed the problem got it wrong.[16] However, if he and Locke hoped for interlocutors who would deny their central claim and insist that the blind man had innate visual ideas, they were disappointed. None of their disputants took that route. Consider, for example, the first retort, which came from Edward Synge, later archbishop of Tuam. On a Thursday in the autumn of 1695, Synge met his friend Francis Quayle, prebendary of Brigown in Cloyne.[17] Quayle posed Molyneux's Problem to Synge, which greatly tickled Synge's curiosity. The next day he sent Quayle his reply, which Quayle dispatched to Molyneux, and Molyneux forwarded to Locke. Synge's positive answer turned on a distinction between *ideas* and *images*. An *idea* was "every notion of any thing which a man entertains," whereas an *image* was "that notion only, which a man entertains of a visible thing as it is visible." Molyneux's blind man, according to Synge, would have *ideas* of cubes and spheres gained through, but not specific to, his sense of touch. He could measure his new images against his old ideas to distinguish the cube from the sphere.[18]

Although Synge made no mention of innate ideas, Locke found him guilty of "anticipations of sense," contributing to "that infinite jargon and nonsense which so pesters the world."[19] Had he lived a bit longer, Locke would have been similarly pestered by the German rationalist G. W. Leibniz. At the time of Locke's death, Leibniz was about to publish a critical commentary on the *Essay concerning Human Understanding* in which he revisited Molyneux's Problem and assigned it a positive answer. But even Leibniz did not invoke innate ideas to justify his belief that a formerly blind man would immediately be able to identify shapes. Instead, like Synge, Leibniz distinguished "images," specific to the senses, from "exact ideas, which consist of definitions." Ideas, consisting of definitions, were built of images. Though each image belonged to an individual sense, the composite definition-idea belonged to "the common sense, that is to say, the mind itself." Geometry dealt in ideas not images. So a blind man using tactile images and a paralytic using only visual images would arrive at the same geometry, consisting of the same ideas. Both would know, for example, that a sphere has no distinct points, whereas a cube has eight. Molyneux's blind man could therefore count the distinct points in each of his new visual images to arrive at the correct identifications.[20]

---

16. Ibid.

17. The prebendary of a church is the canon who holds its prebend, that is, its clergy house and a portion of its revenue. Brigown remains a parish in the diocese of Cloyne.

18. Edward Synge to Francis Quayle, 6 September 1695, in Locke, *Correspondence*, no. 1984.

19. Locke to Molyneux, 5 April 1695, in ibid., no. 2059.

20. Leibniz, *Nouveaux essais* (1705), bk. II, chap. IX, §8. (When Locke died, Leibniz decided

Thus the first part of the shift in speculative writing on Molyneux's Problem occurred when Synge, Leibniz, and others, including the Irish moral theorist Francis Hutcheson and the Scottish commonsense philosopher Thomas Reid,[21] inaugurated a tradition of positive answers to Molyneux's Problem that invoked not innate ideas, but ideas common to the senses, a notion neither Locke nor Molyneux had rejected. Both Locke and Molyneux shared with their opponents the assumption that the senses worked together to transmit whole impressions of a multifaceted external world.[22]

Indeed, this belief was well entrenched not only among speculative philosophers, but in contemporary sensory physiology, where it informed the notion of a *sensorium commune*, a junction in the brain where the nerves from the five senses met and pooled their impressions. The sensory "integrity of the conscious self"[23] was axiomatic to physiologists including Hermann Boerhaave, Albrecht von Haller, Charles Bonnet, Georges Buffon, and ultimately P.-J.-G. Cabanis, J.-B. Lamarck, and Georges Cuvier.[24] Bonnet associated "this unity, this self of which I have so intimate and so clear a sentiment," with the soul, which he said was "*present* to" the body through its brain, at the point where the senses met one another: "It is always the same Me that sees, hears, tastes, smells, touches, acts."[25] The unity of the conscious self implied that beneath the particular sensations, each specific to one of the five senses, lay a common currency of sensibility.[26]

not to publish the *Nouveaux essais*, and they were ultimately published posthumously.) Reid, *Inquiry* (1764), chap. VI, §20; Jurin, "Dr. Jurin's Solution" (1738), 2:27–29; Boullier, *Essai philosophique* (1728), part II, chap. VI, §18.

21. Hutcheson, "Original Letter" (1727), 158–60; Reid, *Inquiry* (1764), chap. VI, §20.

22. "Like Molyneux, Locke probably assumed that visual and tactile ideas of shape have an essential relationship to one another which can be learnt by experience; nowhere did Locke defend the idea of complete heterogeneity of sight and touch, an idea which the Irish philosopher George Berkeley was most certainly to support." Degenaar, *Molyneux's Problem* (1996), 29.

23. Figlio, "Theories of Perception" (1975), 181.

24. Ibid., 179–86, 198. Figlio argues that the *sensorium commune* was a physiological application of the philosophical assumption of the "indivisibility of personality." On Bonnet's belief in the unity of the self, see Anderson, *Charles Bonnet* (1982), 107–10; and Rey, "Le Partie, le tout et l'individu" (1994), 69–71. On the *sensorium commune* in Hallerian physiology, see Duchesneau, *Physiologie* (1982), 201–15.

25. Bonnet, *Méditations sur l'origine des sensations*, cited in translation in Anderson, *Charles Bonnet* (1982), 109; *Essai analytique* (1760), v.27; *Contemplation de la nature* (1764), part IX, chap. I, 8. On the notion that the soul was the junction of the body's nerves in the brain, see La Mettrie, *Histoire naturelle* (1745), 182; *L'Homme-machine* (1748), 134; and *L'Anti-Sénèque* (1750), 111. On the *sensorium commune*, see also Jaucourt, "Sens interne," in Diderot and d'Alembert, eds., *Encyclopédie* (1765), 15:31–33; and Anon., "Sensorium," in ibid., 15:55.

26. The eighteenth-century notion of "common sense" grew out of the *sensorium commune*. Buffon referred to a "common sense" in its Aristotelian meaning as an "internal" faculty that integrated the sensations. Roger, *Buffon* (1989), 218; Aristotle, *De Anima*, III.2, 425b15–20; Buffon, "Discours sur la nature des animaux" (1753), in *Oeuvres philosophiques*, 323. Voltaire tells us, in the *Dictionnaire philosophique*, that when men invented the expression "common sense," they "ex-

Physiologists took this oneness of sensibility literally. Boerhaave, for example, located the sensory union in the cerebral cortex. Bonnet considered that it might be the corpus callosum. Haller and his followers placed the *sensorium commune* "in the medulla ... where the origin of every nerve [lay]" and demonstrated the medulla's extreme sensitivity by means of vivisection experiments.[27] Lower organisms provided a test of the essential unity of sensation: they had sensibility itself, although they lacked particular senses. The crayfish and the polyp, for example, had no eyes but were sensitive to light. In them, sensibility was reduced to its most basic form, the form it took in the *sensoria* of higher animals. Some physiologists therefore considered the polyp to be all *sensorium*.[28]

Thus when Locke's and Molyneux's opponents argued that a newly sighted man could readily make visual identifications of cubes and spheres, they did not appeal to innate ideas, but to the generally accepted notion, at once philosophical and physiological, of a *sensorium commune*. They thereby displaced the problem from innate ideas to common-currency ideas. It was this new assumption, of ideas common to the senses, that George Berkeley addressed when he took up Molyneux's Problem in his *Essay towards a New Theory of Vision* (1709). Berkeley gave the problem a negative answer, reiterating Molyneux's and Locke's thesis: the newly sighted man would not be able to make visual identifications. But, because Berkeley's argument was directed against ideas common to the senses rather than innate ideas, he meant something quite different by his negative answer.[29]

Ideas common to the senses, Berkeley assumed, were necessarily abstract. For the brain to convert particular sensations into a common denominator of sensibility, as philosophers and physiologists assumed it did, it must strip these sensations of their individuating characteristics. It must, for example, filter from visual impressions all that made them visual rather than tactile,

---

pressed the opinion that nothing enters into the soul except by the senses." Voltaire, "Sens commun," in *Dictionnaire philosophique* (1764), 351–53. Diderot's *Encyclopédie* article "Bon-sens" explains that "good sense supposes experience. . . . [I]t is the faculty of deducing from experiences." This faculty was so homely that one could boast of it without vanity. Diderot, "Bon-sens" (1751), in Diderot and d'Alembert, eds., *Encyclopédie*, 2:329. Common sense, then, was not initially a manner of reasoning, but a stance of sensitive openness to the world.

27. Boerhaave, *Praelectiones de morbis nervorum* (1730–35), ii.256–57; Bonnet, *Essai analytique* (1760), 19; Haller, *First Lines of Physiology* (1786), ix.372; and Haller, "On the Sensible and Irritable Parts of Animals" (1755), 674. See also Godart, *La Physique de l'âme* (1755), 1:3–4; and Hatfield, "Remaking the Science of Mind" (1995), 205.

28. Diderot, *Eléments de physiologie* (1784), 96–99; Baker, *An Attempt* (1743), 81; Hartley, *Observations* (1749), 1:32. See Ilie, *Cognitive Discontinuities* (1995), 251–63.

29. On Berkeley and the history of Molyneux's Problem in the Enlightenment, see Roger, *Buffon* (1989), 215–18; Furbank, *Diderot* (1992), 54–71; Hankins, *Jean D'Alembert* (1970), 74–75.

and vice versa. It was this notion, that the mind acquired its ideas by a process of abstraction, that Berkeley rejected.

His disbelief in a mental process of abstraction, like his opponents' belief in ideas common to the senses, found some support in contemporary physiology of the senses. Along with the *sensorium commune*, a central feature of eighteenth-century sensory physiology was what Karl Figlio has called the "impression theory of sensation." In this model, an "external object impressed an effect upon a sense organ by impact" through the action of an ethereal medium. The model originated in Newton's explanation of vision as the vibrations provoked by light in the retina's optic nerve, which were then propagated to the brain by ethereal matter in the filaments of the nerve. More generally, according to the impression theory, objects "impress[ed] themselves upon the nervous system, which in turn impress[ed] ideas . . . by some action upon the sensorium." Hallerian physiologists adopted this Newtonian model as an experimental guide, testing the propagation of sensory impulses through the nerves to the medulla.[30]

The impression theory of the senses depicted the mind as passive, the "recipient of a series of impressments."[31] It performed no action, no abstractive process upon the senses' inputs, but was merely imprinted by them. Physiologists who developed and subscribed to this view, like Berkeley, were reacting against Descartes's explanation of sensation, in which the mind was an active agent, forming ideas rather than simply receiving them. Descartes had emphasized that there need be "no resemblance at all between the ideas [our soul] conceives, and the movements that cause these ideas." Body and soul, nerves and ideas, were adjacent rather than united in Descartes's physiology. A crucial gap remained between "the movements that take place in the areas of the brain from which the little filaments of the optic nerve come" and the soul's "sentiment of light."[32]

Descartes filled this gap with the soul's capacity to reason and with God's guarantee that the images it formed of the external world were accurate. But, as Figlio observes, "the sensualist philosophers of the Enlightenment shunned mediation of any sort." No divinely assured rational process "could stand between the mind and the world." The impression theory of sensation was the physiological expression of this insistence upon an "unmediated" relation

30. Figlio, "Theories of Perception" (1975), 178, 196; Newton, *Opticks* (1730), query 12, 23, 24; Haller, *First Lines of Physiology* (1786), xviii.56, xviii.57. See also Wolff, *Psychologia rationalis* (1740), §§136–41; Krüger, *Naturlehre* (1740–49), 2:§§314–22; and *Versuch einer Experimental-Seelenlehre* (1756), iii–iv. On Wolff's and Krüger's theories of sensation, see Hatfield, "Remaking the Science of Mind" (1995), 201–7.

31. Figlio, "Theories of Perception" (1975), 197.

32. Descartes, *Dioptrique* (1637), 100–1, 117–18.

between mind and world: the world impressed itself directly upon the brain, which responded by receiving the impression.[33] A philosophical expression of the same preoccupation—albeit a troubled and complex one—was Berkeley's rejection of the notion of ideas common to the senses and the process of mental abstraction he believed would have to generate them. Already in the *New Theory of Vision*, as we will see, the notion that all sensations were direct imprints of an external world was problematic to Berkeley, and he would soon afterward embrace an immaterialist philosophy that called into question the very existence of a world independent of perception. But in the *New Theory of Vision*, he tried to preserve an immediate connection between a real, outside world and one of the five senses, the sense of touch.[34] He did so by denying that the mind formed ideas by abstracting them from sensations.

Berkeley's targets were those who invoked a geometry common to both sight and touch, a geometry whose object was "abstract extension." This chiefly meant Descartes and his followers, but included, as we have seen, Synge, Leibniz, and others who answered Molyneux's Problem positively.[35] Berkeley disbelieved the Cartesian supposition that one could separate geometrical figures from their tangible and visible qualities "and form thereof an abstract idea." He reasoned by introspection: "After repeated endeavors to apprehend the general idea of a triangle, I have found it altogether incomprehensible." A similar logic ruled out any idea of extension common to the two senses of sight and touch. Similar quantities must be combinable into a single, continuous whole. One could add two visible lines to get a longer one, and one could similarly add two tangible lines. But the project of adding a visible to a tangible line was as mysterious, Berkeley thought, as the apprehension of an abstract triangle. He concluded that figures *"perceived by sight are specifically different from the ideas of touch, called by the same names."*[36]

Berkeley's dislike of the notion of a mentally derived, abstract geometry, common to the five senses, drove him to depart not only from the Cartesians, but from every previous writer on Molyneux's Problem—Molyneux, Locke, and their opponents alike—as well as from the consensus among philosophers and physiologists regarding the unity of sensory experience. Berkeley

33. Figlio, "Theories of Perception" (1975), 193, 195.

34. Shortly after the *New Theory of Vision*, Berkeley published his first idealist work, *A Treatise concerning the Principles of Human Knowledge* (1710), in which he argued that *all* sensible qualities, tactile included, existed in the mind alone. But the distinction he drew in the *New Theory of Vision* between real, tactile qualities and conventional, visual qualities remained influential in the experimental and philosophical literature on blindness.

35. Leibniz's response to Locke had not yet been published; but Synge's letter had (see above, n. 18), as had a similar argument, relying upon ideas common to sight and touch, by Lee, *Anti-Scepticism* (1702).

36. Berkeley, *New Theory of Vision* (1709), §122, 124, 125, 127, 131.

now claimed that the tangible and the visible were two utterly distinct worlds.[37] And one was real, the other not. The tangible world was genuine: it existed outside the mind of the feeling subject and impressed itself unproblematically upon the mind through the tactile sense. In contrast, visible objects shifted and transformed in a way that any object "which exists outside the mind" could never do. An object looked larger when closer, smaller when farther, distorted when seen from an angle. Because visible dimensions had "no fixed and determinate measures," when people spoke of magnitude or distance, they actually meant tactile dimensions only. One considered that an object remained constant in size although the visible object changed depending on one's position with respect to the tangible object. Thus a man at a distance of ten feet was thought to be the same size as he would be at five feet, although his visible magnitude was much smaller. Tactile magnitudes were manifestly the only real ones.[38]

The changefulness of the visible world implied it was an illusion, existing only in the eyes and mind of the observer. Berkeley was not the only one to make such a case against the authenticity of vision. Philosophers worried about the trustworthiness of the senses had often noted that optical illusions were more common than, for example, tactile illusions, and that vision was especially prone to subjective impressions. Newton had found that by pressing his fingers against his closed eyelids in a darkened room, he could cause himself to see colors. David Hartley and Erasmus Darwin would later remark upon the same thing, as, in the nineteenth century, would Goethe and the German physiologist Franz Joseph Gall. Darwin and Gall would both conclude that the eye was capable of seeing "in consequence of inward irritations only, and without the concurrence of the external world." The prominence of optical illusions in the performances of public science lectures, especially "magic lanterns" that projected images of demons and devils, dramatized and made popular the notion that vision was unreliable.[39]

37. See Descartes, *Regulae* (1628), bk. XII: "shape which is both seen and felt"; and Locke, *Essay* (1690), bk. II, chap. XIII, §2: "We get the idea of space both by our sight and touch."

38. Berkeley, *New Theory of Vision* (1709), §55, 60. For a physiological application of this principle, see Le Cat, *Traité des sensations* (1767), 2:441–84. Le Cat, following Berkeley, argued that geometry could not explain the visual perception of dimensions and distances, which must therefore rely upon intuitive judgment.

39. See Vartanian, *Science and Humanism* (1999), 101, concerning the greater incidence of optical over tactile illusion. For the history of subjective visual impressions, see Wade, *Natural History of Vision* (1998), chap. 4 (discussion of Newton's, Hartley's, Darwin's, Goethe's, and Gall's views of the phenomenon can be found on 177–78); Newton, *Opticks* (1730), 347 (query 16); Hartley, *Observations* (1749), 198; Darwin, *Zoonomia* (1794), 21; Goethe, *Theory of Colors* (1810), 42–43; Gall, *On the Functions of the Brain* (1825), 200. On visual illusions in eighteenth-century popular-science lectures, see Stafford, *Artful Science* (1994), 45–46, 58, 73. On the magic lantern in partic-

But the conclusion that Berkeley drew from the variability of visual impressions was the most radical possible: he argued that one did not *see* an outside world at all. One saw but "lights and colors . . . degrees of faintness and clearness, confusion and distinctness. All of which visible objects are only in the mind, nor do they suggest aught external." Certainly, things in the fictional, seen world communicated information about corresponding objects in the real, felt world. But Berkeley proposed that these visible manifestations were related to their tangible corollaries as word to object: that is to say, purely by convention. The visible world was a world of signs, the "universal language of the Author of nature," meant to instruct people how to navigate and exploit the tangible world. In order to make use of this language, one must first learn it. It made as little sense to speak of a common nature between a cube seen and a cube touched, as it made to speak of a common nature between the word "cube" and the object it denoted. By Berkeley's logic, expecting Molyneux's blind man to recognize a cube by sight alone was like expecting an English speaker to recognize the Chinese word for "cube."[40]

Thus Berkeley's theory of vision encompassed two extremes: an utterly unmediated, material relation between the tactile sense and the thing touched; and an entirely conventional, divinely given relation between sight and the thing seen. Whereas touching the world required only that one come up against it, seeing required that one first learn the language of sight. But even this divinely mediated, visual connection to the world did not operate by a rational process. Sighted people understood the language of sight thanks to a "habitual connection" between sight and touch, and not, as Leibnizians claimed, by geometry. An understanding of the visual world resided in experience, habit, and a sort of sensitive intuition: "We see both [distance and magnitude] in the same way that we see shame or anger in the looks of a man. Those passions are themselves invisible; they are nevertheless let in by the eye along with colors and alterations of countenance which are the immediate object of vision."[41]

---

ular, see Hankins and Silverman, *Instruments and the Imagination* (1995), chap. 3. On the cultural prominence of optical illusions in the Romantic period, see Burwick, "Romantic Drama" (1990).

40. Berkeley, *New Theory of Vision* (1709), §77, 141, 145, 151. Locke's distinction between primary and secondary qualities gave some precedent to Berkeley's claim about vision: "The ideas of primary qualities of bodies are resemblances of them, and their patterns do really exist in the bodies themselves; but the ideas produced in us by . . . secondary qualities have no resemblance of them at all. There is nothing like our ideas existing in the bodies themselves. They are, in the bodies we denominate from them, only a power to produce those sensations in us: and what is sweet, blue or warm in ideas, is but the certain bulk, figure, and motion of the insensible parts in the bodies themselves, which we call so." Locke, *Essay* (1690), bk. II, chap. VIII, §15. But Berkeley's claim was more radical: he maintained that visual qualities existed purely in the mind, with no physical correlation whatsoever to the outside world. On the relation between Berkeley's and Locke's understandings of vision and the problem of solipsism, see Frye, "Varieties of Eighteenth-Century Sensibility" (1990–91), 162–63.

41. Berkeley, *New Theory of Vision* (1709), §77, 65.

In his treatment of Molyneux's Problem, then, Berkeley first of all made problematic what had previously been taken for granted by all parties: the ability of the senses to convey an understanding of external objects as such. Next, the problems he identified in the process of sensing external objects—the discrepancy between sight and touch, and its implication that some, but not all, of the senses conveyed direct imprints of an external world—meant that the world outside could not be captured in the mind by means of ideas abstracted from the operations of the five senses. On the contrary, the world must be apprehended by means of an ever-active, intuitive sensitivity. Berkeley transformed the problem from an empiricist fable about the origins of knowledge in experience to a cautionary tale against abstracting from experience. Only through a continual sensitive engagement with the world could one understand it. The moral of Berkeley's tale was that no abstract geometry could combine sight with touch. Instead, people saw the outside world in the same way that they saw inside one another's minds, by a combination of experience, intuition, and sympathy. Sight required sensibility, not geometry.

II. THE THOUGHT EXPERIMENT MADE REAL: DO THE NEWLY SIGHTED THINK THAT ALL THEY SEE IS TOUCHING THEIR EYES?

Shortly after Berkeley published his essay, Molyneux's Problem began its transformation from speculative to experimental. In August 1709 the *Tatler* reported that an English surgeon named Roger Grant had restored the sight of one William Jones of Newington Butts, a young man of twenty who had been blinded in infancy by cataracts. The psychological rather than the surgical procedure constituted the innovation. "Couching" cataracts—that is, using a needle to depress the occluded lens below the line of sight—is an ancient technique (figs. 2.2 and 2.3). The superior method of actually extracting the cataract was developed in France in the early eighteenth century by Jacques Daviel, *oculiste ordinaire* to Louis XV, but Grant's surgery involved couching, as did the main experimental cases cited in discussions of Molyneux's Problem.[42]

So the surgical restoration of vision was not new; but recording the mental and emotional experience of the patient was, and was likely inspired by philosophical interest in Molyneux's Problem. The *Tatler* reported that the minister of Newington, who had attended Grant's operation upon the young William Jones, was "a Gentleman particularly curious," who had requested "the whole company to keep secret, and let the Patient make his own observations, without the direction of any thing he had received by his other senses." The journalist began his report promisingly by extolling the interest in seeing

42. See Degenaar, *Molyneux's Problem* (1996), 58–60.

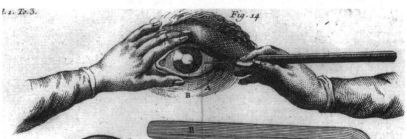

Figures 2.2 and 2.3. A cataract operation, from Robert James, *Dictionnaire universel de la médecine* (1746–48), vol. 3, plate 1, fig. 1. Courtesy of the Bancroft Library, University of California, Berkeley.

what would happen "when one of full age and understanding received a new sense." The story mounted from there toward a sentimental moral. It came finally when the hero declared to his love, who feared he would find her plain: "Dear Lidia . . . I wished for [these eyes] . . . but to see you; pull them out if they are to make me forget you."[43] That was moving but not much use to Berkeley. Almost twenty years later, however, he did get what many of his contemporaries regarded as empirical confirmation for his theory of vision: the Cheselden experiment.

Cheselden had recently attended Newton through his final illness and, as his doctor, famously vouched to Voltaire, Newton's French image-maker, for his patient's lifelong chastity. It was Voltaire who later informed the French public of Cheselden's experiment (and Newton's chastity) in his *Eléments de la philosophie de Newton* in 1738.[44] Cheselden's operation seemed to have brought Molyneux's hypothetical blind man dramatically from the page to the flesh. Just as Molyneux had predicted, Cheselden found his postoperative patient utterly bewildered by the appearance of shapes and distances. The boy could not initially identify objects by sight but had to learn their identities with great difficulty. He "learn'd to know, and again forgot a thousand Things in a Day." Animating Molyneux's fable, Cheselden left its pristine, geometrical landscape behind; the hypothetical cubes and spheres gave way to warmer, fuzzier objects: "Having often forgot which was the Cat, and which the Dog, he was asham'd to ask; but catching the Cat (which he knew by feeling) he was observ'd to look at her stedfastly, and then setting her down, said, So Puss! I shall know you another Time."

As the story in Cheselden's report unfolds, however, it begins to fit Berkeley better than Molyneux. According to Cheselden, his boy born blind was bewildered not for lack of the learned process of visual "judgment" to which Molyneux had alluded, but because of the utter capriciousness of the visual. world, which precluded judgment. Instead of reasoning, Cheselden recounted, his patient seemed to need simply to memorize correlations that felt bafflingly arbitrary. He was continually amazed by the visual world's apparent detachment from the world made up of his other senses and feelings: "He was very much surpriz'd, that those Things which he had lik'd best, did not appear most agreeable to his Eyes, expecting those Persons would appear most beautiful that he lov'd most, and such Things to be most agreeable to his Sight that were so to his Taste."

Of all the senses from which it strayed, vision was most disloyal to touch. The patient's mystification at pictures epitomized this faithlessness. His

43. *Tatler*, no. 55 (16 August 1709): 124–26. Grant's credentials and operation were discredited in the philosophical community. See Degenaar, *Molyneux's Problem* (1996), 52.

44. Voltaire, *Eléments* (1738), 319–20.

doctor and family soon thought he understood pictures, "but we found afterwards that we were mistaken; for about two Months after he was couch'd, he
discovered at once, they represented solid Bodies; when to that Time he consider'd them only as Party-colour'd Planes." Now able to see the three dimensions represented in pictures, the boy was nevertheless "no less surpriz'd, expecting the Pictures would feel like the Things they represented, and was
amaz'd when he found those Parts, which by their Light and Shadow appear'd
now round and uneven, felt only flat like the rest." In a perfectly Berkeleyan
quandary, the boy "ask'd which was the lying sense, Feeling or Seeing."

But it was immediately after his surgery that Cheselden's patient provided
the most dramatic confirmation of Berkeley's theory. Upon first opening his
eyes to the world, the boy—according to his doctor—was not merely unable
to judge distances: he simply could not see them. All that was visible seemed
splayed across the surfaces of his eyes. The doctor wrote: "He thought all Objects whatever touch'd his Eyes, (as he express'd it) as what he felt, did his
Skin."[45] Berkeley would have said that what the blind boy first saw was closer
to the truth than what most people see. The visible world was indeed no farther away than the surfaces of his eyes; it existed there and in his mind alone.

Whether the tactile and the visual shared a common, abstract geometry,
and especially whether visual impressions conveyed any sense of dimension
or distance—whether one could see external objects as external without the
aid of touch—these became the crucial questions in the experimental literature on blindness. They were questions not about the origins of ideas, but
about the sensory sources of sensibility.

Besides Grant and Cheselden, during the eighteenth century three other
doctors published reports in which they described having given sight to congenitally blind patients by means of surgery and tested their first visual experiences. Daviel, the French royal oculist, claimed twenty-two such procedures, all taking place in Paris around midcentury. In 1764 a doctor in Lyon
named Jean Janin bribed and cajoled a reluctant twenty-two-year-old peasant woman into letting him experiment on her cataract-clouded eyes. And
the Viennese oculist Georg Josef Beer operated upon fourteen patients between 1783 and 1813.[46] Three more doctors working in the first third of the
nineteenth century, all in London, continued the tradition. James Ware, a surgeon, restored the sight of a seven-year-old boy in 1800 and an eighteen-year-
old woman in 1801. Everard Home, a surgeon and fellow of the Royal Society,
operated upon two young boys—William Stiff, aged twelve, and John Salter,
aged seven—at St. George's Hospital in 1806. Finally James Wardrop, an eye
doctor, operated on a fourteen-year-old boy in 1810 to remove cataracts and

45. Cheselden, "Account" (1728).
46. Daviel, "Réponse" (1762), 249; Janin, *Mémoires* (1772), 214; Beer, *Das Auge* (1813).

on a forty-six-year-old woman in 1826 to create an artificial pupil. The last was the only case that did not involve cataract surgery; and before her operation, Wardrop's lady was the blindest of all the patients, unable to see colors nor anything but the difference between very bright and very dim light.[47]

A satisfactory scientific interpretation of the results of these examinations would be complicated by two factors: many of the patients had surgery only on one eye, at least initially, which would have hindered their depth perception; and the removal of the lens in cataract surgery means that the eye cannot vary its refraction of light and renders the patient farsighted.[48] However, the doctors, though surely aware of these factors,[49] paid little or no attention to them in their reports. They set aside questions of physiology and physics, to focus instead upon the psychology and philosophy of vision. Wardrop wrote of one of his cases that although it had established a "curious physiological fact" (the optic nerve could remain fit even through years of disuse), it claimed "a much higher interest in a philosophical point of view," as a test of Berkeley's theory of vision and to "throw light on the operations and development of the human mind."[50] This philosophical bias was typical of the late-eighteenth- and early-nineteenth-century case literature on cataract surgery.

All six doctors confirmed Molyneux's original claim: their patients could not immediately identify forms by sight alone. Daviel wrote that "not a single one" of his twenty-two patients had been able to recognize objects. And he added that any doctor who claimed the contrary must be overlooking the fact that his patients "were not really blind from birth; for the latter have no real idea at all of even the meanest objects."[51] But if a negative answer to Molyneux's Problem had become uncontroversial, other questions remained salient, in particular the two indicated by Berkeley and Cheselden: Could the patients correlate the visual with the tactile; and could they perceive distances? Here, the results were sufficiently equivocal to propel further debate. Wardrop's 1826 account of a lady born blind, the latest of the reports considered here, was written about a century after Cheselden's experiment and over a century after Berkeley's *New Theory of Vision*. That Wardrop located his story in the Berkeley-Cheselden tradition reflects the longevity of interest in that

47. Ware, "Case of a Young Gentleman" (1801); Home, "Account of Two Children" (1807); Wardrop, "Case of a Lady Born Blind" (1826). For an exhaustive list of reports of cataract operations from 1020 through 1958, see von Senden, *Space and Sight* (1960), 326–27.

48. See Degenaar, *Molyneux's Problem* (1996), 58–60, 83–86.

49. The surgeon Antoine Louis, for example, in his *Encyclopédie* article "Cataracte," raised the problem of the removal of the lens in cataract surgery. See Louis, "Cataracte" (1752), in Diderot and d'Alembert, eds., *Encyclopédie*, vol. 2.

50. Wardrop, "Case of a Lady Born Blind" (1826), 529.

51. Daviel, "Réponse" (1762), 249.

tradition's central claim: that the eye does not see either dimension or distance, except by a learned convention.[52]

Wardrop's patient had a sealed pupil in her left eye and a collapsed eyeball in her right, probable results of a botched cataract operation in her infancy. In three surgeries in January and February 1826, Wardrop created an artificial pupil in her left eye. In the first, he "introduced a very small needle through the cornea, passing it through the center of the iris," and in two subsequent operations, he widened the opening. On the way home from the doctor's house on the day of the last surgery, the patient saw a hackney coach and exclaimed, "What is that large thing that has passed by us?" But six days later, she insisted unhappily that although she could see, "I cannot tell what I do see. I am quite stupid."

According to Wardrop's report, his patient could distinguish and associate visual shapes, but she could not correlate them to tactile images, and therefore could not identify them. Thus she could see that a pencil case and a key were different shapes but could not tell which was which. Wardrop judged her to be "bewildered from not being able to combine the knowledge acquired by the senses of touch and sight." Colors, in contrast, which had no correlation with touch, were easier to learn. Upon asking a lady the color of her dress and being told it was blue, Wardrop's patient replied, "'So is that thing on your head' . . . which was the case; 'and your handkerchief, that is a different colour,' which was also correct."[53]

Similarly, a patient of Beer's, a sixteen-year-old boy, could recognize his father only by his red coat, while another, a fourteen-year-old girl, proved unable to distinguish a hat from a shoe of the same color. Beer followed Berkeley in concluding that, at least for three-dimensional objects, his patients could learn to correlate sight with touch only by touching the objects they saw. Judgment of solid forms "presumes a comparison of ideas of bodily form derived from feeling, with those acquired of sight." Ware put the point in more particularly Berkeleyan terms: "Ideas derived from feeling can have no power to direct the judgment, either to the distance or form of visual objects." Janin concurred, though he misattributed the divorce of vision and touch to Molyneux, who, Janin wrote, had "understood perfectly that each sense has its particular faculty, and the soul cannot transmit the ideas among them."[54]

Pictures epitomized the caprice of the visual world, for Wardrop's and Beer's patients as for Cheselden's. They repeated the stages of Cheselden's boy: incomprehension followed by disbelief. Two weeks after her surgery,

52. On the persistence of Berkeley's framing of the problem, and of the relevance of Cheselden's experiment, in the study of visual perception through the late nineteenth and even the early twentieth century, see Pastore, *Selective History* (1971), especially 178–91.

53. Wardrop, "Case of a Lady Born Blind" (1826), 530–34.

54. Beer, *Das Auge* (1813); Ware, "Case of a Young Gentleman" (1801); Janin, *Mémoires* (1772), 218.

Wardrop's lady born blind saw pictures only as patches of color. Meanwhile, a twenty-two-year-old patient of Beer's, having learned to see objects in pictures as three-dimensional, could not fathom that the painted images were in fact two-dimensional. Confronted by a still life with food, he "leapt up hastily towards the picture and would just as speedily have pierced the canvas with his hand." The doctor explained that the things in the picture were "merely painted on flat canvas." But the young man was "quite unable to grasp this and had to be suffered to convince himself of the fact by repeatedly feeling with the greatest care the length and breadth of the painting."[55]

Whereas Wardrop's, Beer's, and Ware's accounts squared with Berkeley's theory of the detachment of sight from touch and the essential two-dimensionality of vision, Home described his patients seeking and finding correspondences between the visual and the tactile. He thought they could reason their way to correct visual identifications of objects by a process like the one that Synge, Leibniz, Reid, and others had indicated, and he directed them accordingly in his questions. Seven-year-old John Salter, ten minutes after his left eye was couched, judged a circular card to be "round." He said the same of a square and a triangle. A couple hours later, the child was shown a square again and asked "if he could find any corners to it." After examining the square, he "said at last that he had found a corner, and then readily counted the four corners of the square; and afterward, when a triangle was shown him, he counted the corners in the same way ... his eye went along the edge from corner to corner, naming them as he went along." Thirteen days later, though the boy still could not identify shapes without counting their corners one by one, he had developed a "great facility" at this, "running his eye quickly along the outline."[56]

Reports diverged, too, regarding whether and how patients saw distances. Daviel said his patients were so little able to judge distances that "when they want to grasp an object, they always put their hands a foot above or beside it." Similarly, Wardrop's lady born blind had "the greatest difficulty in finding out the distances of any object; for, when an object was held close to her eye, she would search for it by stretching her hand far beyond its position, while on other occasions she groped close to her own face for a thing far removed from her." However, Janin reported that his patient had immediately been able to judge distances. When shown an object, she reached directly for it.[57]

Within the discussion about how well the patients judged distances, the more particular question of whether they initially thought the objects they saw were touching their eyes, as Cheselden reported of his patient, inspired a lengthy controversy. This indeed became the pivotal question in the case

55. Wardrop, "Case of a Lady Born Blind" (1826), 538; Beer, *Das Auge* (1813).

56. Home, "Account of Two Children" (1807), 87–89.

57. Daviel, "Réponse" (1762), 249–50; Wardrop, "Case of a Lady Born Blind" (1826), 538; Janin, *Mémoires* (1772), 220.

literature on the newly sighted. The experimental question neatly captured the central philosophical problem: Did vision in itself convey a sense of external things as external, or could one see things but fail to see them as existing outside oneself?

The idea that vision might feel like touch to a newly sighted person—like the notion of ideas common to the senses, and the understanding of sensations as the direct imprints of an active world on a passive mind—had an analogue in contemporary physiology. A corollary of the impression theory of sensation—and also of the associated physiological principle, discussed above, of the essential unity of all sensory experience—was the notion that the basic mechanism of sensory perception was touch: the impact of a material, external agent against a sense organ. Accordingly, the other four senses differed from the tactile only by being more or less refined: "That the eye could not hear, nor the skin taste, resulted from . . . [the] varying degrees of shielding or blunting of sensitivity. . . . The very delicate impulses of light could not affect the tongue, but could impinge on the eye, because the nerves of the tongue were wrapped in a sheath, while those of the eye were naked."[58] The primacy of touch was a commonplace among eighteenth-century physiologists, reflected in Louis de Jaucourt's *Encyclopédie* article on "Touch," in which he called that sense "the most general sensation" and the "foundation" of all the others.[59]

So when Cheselden and other doctors considered that their newly sighted patients might believe the objects they saw were touching their eyes, they made implicit reference to the physiological consensus that vision, like all the senses, was a form of touch. However, as we have seen, the doctors were less interested in the physiological significance of their experiments than in their philosophical and psychological implications. That is, they were not concerned with whether vision was a form of touch, but with what it would mean for vision to feel like touch to the newly sighted. To perceive the whole visual world as though it were crowded onto the surfaces of one's eyes: what would that imply regarding the "development of the human mind"?[60] It

58. Figlio, "Theories of Perception" (1975), 199, 186. On the primacy of touch in eighteenth-century sensory physiology, see also Smith, "Background of Physiological Psychology" (1973), 88–101; Ilie, *Cognitive Discontinuities* (1995), 336–52.

59. Jaucourt, "Tact" (1765), 819b. See also Buffon, *De l'homme* (1749), 213; Diderot, *Le Rêve de d'Alembert* (1767), 320–21; Delaroche, *Analyse* (1778), 1:93–100; Dumas, *Principes de la physiologie* (1800), chaps. VI, VII; Richerand, *Nouveaux éléments* (1804), 2:63; Cabanis, *Rapports du physique et du moral* (1805), 1:211; Cuvier, *Leçons sur l'anatomie comparée* (1805), 2:426; Bichat, *Anatomie générale* (1812), 1:117–18. Haller's "erectile hypothesis" concerning the expansion of the iris is an example of how touch served as a model for all sensation. Haller proposed that the iris's expansion, like the penis's, was vascular and caused by mechanical stimulus, in the case of the iris, the irritation of the retina by light. Haller, *Elementa physiologiae* (1763), 5:379. See Mazzolini, *The Iris in Eighteenth-Century Physiology* (1980), chaps. 9, 11.

60. Wardrop, "Case of a Lady Born Blind" (1826), 529.

would mean that newly sighted patients could not see distant objects as distant. It would prove that Berkeley was right: Vision takes place in the eyes and by itself can give no sense of an extended, outside world.

It was with this possibility in mind that doctors in the Cheselden tradition set out to test Cheselden's claim regarding his patient's conflation of vision with touch. Ware disbelieved Cheselden based on experiments with two blind children who had not been operated on but were able to see colors. Asking the children whether colors touched their eyes, he got negative responses. Home set out to settle the conflict between Ware and Cheselden. Even before operating on twelve-year-old William Stiff, who could see some light through his cataracts, Home questioned the boy about whether the sun seemed to touch his eye (yes); and whether a candle seemed to touch his eye (only at a distance of under twelve inches). Having thus prepped the patient for his postoperative interview, Home extracted the lens from William Stiff's left eye. "The capsule . . . was so very strong as to require some force to penetrate it. When wounded, the contents, which were fluid, rushed out with great violence. Light became very distressing to his eye, and gave him pain." The child closed his eye but remembered his lines. "On my asking him what he had seen, he said, 'your head, which seemed to touch my eye.'"

John Salter, though, was less obliging. When Home asked him after his surgery if objects seemed to touch his eye, he answered, "No." The younger boy apparently suffered less and saw far better after the operation and was perhaps distracted by the excitement of seeing. He was "highly delighted . . . and said it was 'so pretty.'" Also John Salter, unlike William Stiff, had been able to distinguish colors through his cataracts. Home observed that three months after his original surgery, William Stiff could see white, red, and yellow, and no longer thought objects were touching his eyes. Home concluded that colors were enough to give "an imperfect knowledge of distances."[61]

Janin, too, had difficulty getting the right responses from his patient concerning whether objects seemed to touch her eye. He wrote that the "sensations felt by this girl on the presentation of objects, did not in the least incline her to suppose that they touched her organ of vision, but rather that they were situated at a certain distance." Nevertheless, Janin maintained that one must "absolutely conclude" from his experiment that "distances, etc., are not, properly speaking, visible things." He had learned what he could, he said, from the "very limited and very capricious" young *paysanne* whose "indifference . . . to receiving a new sense augmented with the approach of the instruments." She got her revenge by thwarting Janin's hypothesis: "The sensation inside her organ could not have alerted her to what was going on outside:

---

61. Ware, "Case of a Young Gentleman" (1801); Home, "Account of Two Children" (1807), 84–85, 88–89, 86, 90. William Stiff's right eye had been couched in the interim.

nevertheless, as soon as she perceived an object, she brought one of her hands forward, and always in the direct line of sight." [62]

The most prominent of the patients in the Molyneux tradition was not a surgical patient. She was Maria-Theresia von Paradis, the blind daughter of a personal secretary to Emperor Francis I and goddaughter of the empress. Paradis was a gifted musician, and her performance of Pergolesi's *Stabat Mater* at the age of eleven moved the empress to become the patron of her goddaughter and namesake. Paradis became a concert harpsichordist and pianist, and later a composer, despite her blindness. On a tour of Europe at the age of twenty-four, she would perform in Paris to great acclaim and become involved in the founding of the Institution royale des jeunes aveugles. [63] Her doctor was Franz Anton Mesmer, ambitious polymath and inventor of the theory and therapy of animal magnetism, the subject of chapter 6. Mesmer's treatment of Fräulein Paradis set off the scandal that drove him from his native Vienna, unleashing him on Paris and later Europe.

One winter morning when she was almost three, the little Fräulein Paradis woke up unable to see and soon afterward suffered from "spasms" that caused her eyes to bulge, leaving only the whites visible. At the hands of the most renowned oculists in Vienna, the child suffered every cure in the catalog: suppuration, blistering, leeches, cauterization, and electricity, "administered to the eyes in over three thousand shocks." To Mesmer's credit, he judged this use of electricity "most unsatisfactory." Diagnosing a blockage in the flow of the animal magnetic fluid through the girl's head, he brought her to his house for treatment, where his procedures must surely have been the most pleasant she had endured. In one, for example, Mesmer pointed a stick at her reflection in a mirror. After several weeks he reported that she had recovered some vision. But a controversy began when members of the Faculty of Medicine who came to witness the cure contested its validity. In the drama that followed, Maria-Theresia's parents demanded the return of their daughter, who refused to come until her father arrived sword in hand to reclaim her. [64]

The account of Mesmer's treatment of Fräulein Paradis, included in his first memoir on animal magnetism, devotes little attention to the new therapy and focuses instead on the process of teaching the patient to see. In particular, Mesmer said he had schooled his patient to distinguish and correlate sight and touch, to "touch what she saw and to combine the two faculties." Placing light and dark objects alternately before her, he had "besought her to

62. Janin, *Mémoires* (1772), 217, 221, 214, 220.

63. See infra, §5.

64. The sketch of Maria-Theresia von Paradis's life and involvement with Mesmer is taken from Mesmer, "Dissertation" (1779), 71–76; Vinchon, *Mesmer* (1971), 42–45; and Didier-Weygand, *Cécité* (1997–98), 140–43, 145–51.

pay attention to the sensation in her eyes." In response, she had described the feeling of dark objects as being like "fine points . . . inserted in the eyeball." She had thought solid things touched her eyes, and Mesmer had had to make her understand that "the cause of these sensations was external."[65]

Mesmer and the other experimenters in the Cheselden tradition may not have agreed with one another about results, but the terms in which they conceived and performed their experiments reflected a general consensus about what the problem was. This consensus surrounded Berkeley's, not Molyneux's, version of Molyneux's Problem. Molyneux's claim—that the patients would not immediately be able to distinguish shapes—was so thoroughly accepted as to be uninteresting. Instead, the interesting problems were two related ones. The first was whether patients could pass from sight to touch by means of a process of reasoning, or whether the relations between the visual and the tactile were beyond the reach of reason, in the realm of intuitions born of experience. The second was whether patients could see the external world by means of sensation alone or whether, again, they needed experienced intuition as well as sensation to extrapolate beyond the surfaces of their own eyes.

When cataract surgeons of the eighteenth and early nineteenth centuries asked their patients to identify hats and boots, followed carefully their responses to trompe l'oeil paintings, and asked them again and again if the objects before them seemed to touch their eyes, these surgeons had no interest in demonstrating the existence or nonexistence of innate ideas. They were testing how one saw a world outside oneself. Was it by sensation alone? By sensation combined with geometry? Or was sensation inadequate and geometry irrelevant—as Berkeley had claimed—in the same way they were inadequate and irrelevant, respectively, to the process of empathy by which we discern our fellows' passions?

III. PHILOSOPHICAL RESPONSES TO CHESELDEN: HOW ONE FEELS THE OUTSIDE WORLD

Voltaire confidently deemed his horse the same size at any distance. When it was far away, he did not believe it the size of a sheep, although its image was smaller. And he was convinced that "geometry [would] never solve" this problem, and "physics [was] equally powerless": "the way we see things, does not at all follow immediately from the angles formed in our eyes; since these mathematical angles were in the eyes" of Cheselden's patient too. Bringing Cheselden to a French audience, Voltaire was faithful to Berkeley, repeating almost verbatim his assertion that people see distance and dimension not by

65. Mesmer, "Dissertation" (1779), 71–76.

geometry, but by sensitive intuition, "in the same way that we imagine the passions of men, by the colors these paint in their faces." [66]

In one respect, the other French philosophers did not queue neatly behind Voltaire in his reading of Cheselden and Berkeley. The three primary French commentators on Cheselden's experiment—Julien Offray de La Mettrie, Etienne Bonnot de Condillac, and Denis Diderot—were all believers in the sensory origins of thought. Nevertheless, they answered Molyneux's question messily with a "yes"; a "yes" followed by a "no"; and an "it depends," respectively. All three commentaries did, however, share one element. They made explicit what had previously been implicit in philosophical and experimental treatments of Molyneux's Problem: the emotional and moral dimension of the problem of solipsism and sensitivity at its heart. Here was the distinctively French contribution to the Molyneux-Cheselden tradition: emotional and moral solipsism and sensitivity grew out of, and replaced, the correlation of sight with touch and the ability to see distances as the crux of interest in blindness.

The story of Cheselden's experiment was the third in a chapter of La Mettrie's *Histoire naturelle de l'âme* (1745) entitled "Stories that Confirm that All Ideas Come from the Senses." We have "not one single innate idea," La Mettrie stated in the chapter's conclusion, in case the reader had missed the point; "they are all the product of bodily sensations." He spelled it out with three easy-to-grasp propositions: "1. no education, no ideas; 2. no senses, no ideas; 3. the fewer senses one has, the fewer ideas."

However, La Mettrie's version of "The Blind Man of Cheselden" did nothing to prove the point upon which he so insisted. For La Mettrie believed, despite his fervent commitment to Locke's formula, that Molyneux's blind man *would* be able to tell a sphere from a cube by sight alone. He reasoned like the other proponents of a positive answer, by appeal not to innate ideas, but to ideas common to the senses: "Ideas received by the eyes find themselves in touching, and those of touch, in seeing." Because the senses worked collaboratively, "one sense always profits from the lack of another," so the nerve endings in a blind man's fingers had "an exquisite sensitivity." Through his fingers the blind man would acquire "ideas of figures, distances, &c. . . . engraved in the brain." He could easily compare these ideas of globes and cubes to the new images received through his eyes and identify objects successfully without touching them. To explain Cheselden's negative results, La Mettrie decided that either the boy's optical organs were "disturbed" and needed time to settle into their proper positions, or else his questioners had

<hr />

66. Voltaire, *Eléments* (1738), part II, chap. VII, ll. 65–135. Cf. Montesquieu, *Mes pensées* (1720–55), 1071. Montesquieu drew the decidedly non-Berkeleyan moral that the senses worked collaboratively to convey whole ideas of the outside world.

"tormented" him into yielding the answers they expected, "since we have a greater skill at pressing errors than discovering the truth."

La Mettrie's proposition that "the fewer senses one has, the fewer ideas" thus did not apply to geometrical ideas. But the two stories preceding "The Blind Man of Cheselden" in La Mettrie's *Histoire* did in fact associate sensory handicap with a particular sort of mental inadequacy: the lack of a moral sense. The first was excerpted from Fontenelle's *Histoire de l'Académie Royale des Sciences*. It told the tale of a deaf man of Chartres who one day heard the bells of the cathedral, received a gush of liquid from his left ear, and soon afterward learned to speak. Interrogation by theologians proved the ex-deaf man to have no idea of "goodness or moral malice. . . . He lived a purely animal life." Fontenelle's moral, which La Mettrie adopted, was that ideas arose from the senses and through "the commerce of others," which was permitted by sight and hearing. Had the deaf man been blind as well, "it follows that he would have been [entirely] without ideas." The next story, "Of a Man without Moral Ideas," La Mettrie presented as the "*duplicata*" of "A Deaf Man of Chartres." It told the tale of a man about whom "nothing is deaf . . . except for his mind," which was impervious to the sufferings of others. This deafness of the mind had the same effect on the moral faculty as the other man's blockage of the ears.[67] Geometry might enter the soul equally through the eyes or the tips of the fingers. Moral ideas, to be fully developed, required sensitivity in all the organs and the mind itself.

Like La Mettrie, Condillac, who was Locke's leading advocate and interpreter in France, also initially answered Molyneux's question positively in his criticism of the Cheselden experiment in the *Essai sur l'origine des connaissances humaines* (1746). Just as Leibniz had distinguished ideas from images,

67. La Mettrie, *Histoire naturelle* (1745), 344–64, 389–92. Cf. Buffon, *Histoire naturelle* (1750), 3:137. Buffon wrote that hearing was the sense by which "we live in society and we receive the thoughts of others." Consequently a deaf-mute "must have no knowledge of abstract things and generalities." There is a parallel story to be told about philosophical and experimental interest in deafness during the same period, including Diderot's *Lettre sur les sourds et muets* (1751) and culminating in the founding of the first educational institution for the deaf, the Institut royal des sourds et muets, in 1755. The abbé Charles-Michel de l'Epée was its founder and the inventor of French sign language; the abbé Roche-Ambroise Cucurron Sicard took over the Institut after de l'Epée's death in 1789 together with Jules-Michel Duhamel. Just as Valentin Haüy, the founder of the Institut national des jeunes aveugles, responded to the philosophical literature on blindness, Epée, Sicard, and Duhamel responded to the literature on deafness, which was at once epistemological and moral. See infra §5 on both Epée's and Haüy's institutions. I mention the story about deafness only in passing in this chapter because although its issues overlapped with those involved in the story about blindness, they were also importantly divergent; the problem of language was paramount in discussions of deafness. The cultural and philosophical history of deafness in eighteenth-century France has been well studied. See Rosenfeld, "Language and Deviancy" (1997); Rosenfeld, *A Revolution in Language* (2001); and Lane, *The Wild Boy of Aveyron* (1976).

Condillac differentiated the impression of an object in the soul from its image in the eye. Locke had written that a globe would appear as a flat circle if it were not for the act of judgment involved in seeing. People recalled from experience the impressions of light rays from convex bodies and corrected their visual idea from a flat circle to a solid globe. Condillac did not believe in this process of judging, which he thought belied the immediate and involuntary nature of perception. Instead, he argued that when the eye saw a circle, the soul understood it to be a globe.

Impressions that objects made upon the soul of attributes such as distance, size, surface, or line were the sense organs' common currency. Touching a straight stick would impress upon a blind man's soul the idea of a straight line; a curved stick would give the idea of a curved line; and similarly for an angle, a cube, or a globe. The ideas of these figures were the same "whether I see or I touch" them. When the blind man first began to see, therefore, the world would not appear to him "as a point." He would perceive "an expanse in length, width and depth." He would know these dimensions, and the geometrical figures, by recognizing "the same ideas he had of them from touch." Like La Mettrie, Condillac assumed that Cheselden's boy had merely needed to exercise his eyes, whose muscles and fibers were unused to moving in response to light. Similarly, his interrogators had needed to exercise their minds, rather than inflexibly clinging to the hypothesis supplied by Berkeley.

On one point, however, Condillac agreed with Berkeley, and that was the "insufficiency of optics. We can measure as much as we like the angles that the rays of light form at the back of the eye, we will not at all find them in proportion with the manner in which we see objects."[68] Vision was a matter of intuition, not geometry (see fig. 2.4).

In the years between the *Essai* (1746) and the *Traité des sensations* (1754), however, Condillac changed his mind about the capacity of the soul to see dimensions and distances.[69] The central conceit of the *Traité* is an exhaustive generalization of Molyneux's Problem, a thought experiment more systematic than any actual experiment could be. In the *Traité*, Condillac imagined endowing a statue with each of the five senses separately and in various combinations, in order to settle the question of which senses conveyed the impression of an external world. He wanted to know how it was possible to "see objects outside ourselves" given that "our sensations are but manners of being."[70]

By asking this question, Condillac turned on its head the problem that had preoccupied sensationist philosophers and physiologists. They had focused

68. Condillac, *Essai* (1746), part I, 265.
69. Condillac was persuaded to change his views by Elisabeth Ferrand. See Degenaar, *Molyneux's Problem* (1996), 70; Bongie, *Diderot's femme savante* (1977), 150; and Bongie, "A New Condillac Letter" (1978), 92 n. 34.
70. Condillac, *Traité des sensations* (1754), parts II, III; "Extrait raisonné du traité," 298.

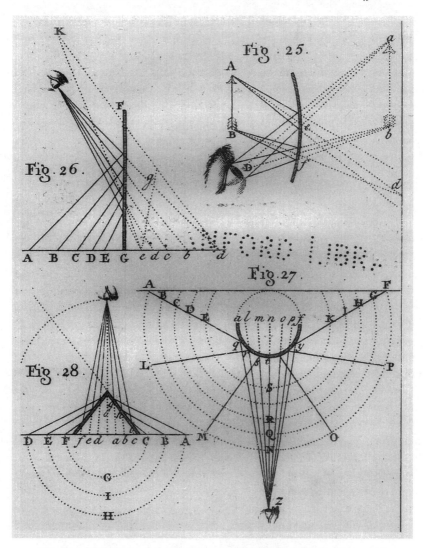

Figure 2.4. The plate from Nollet's *Leçons de physique expérimentale*, 5th ed. (1759–66) represents the geometrical approach to vision that Berkeley, Voltaire, Condillac, and others found inadequate.

on the inward journey of externally produced sensations into the sensible creature's brain, asking how the activities of the external senses comprised an internal, integral self. They had wanted a conscious mind that was at once coherent and utterly open to, indeed formed by, outside influence. For Boerhaave, Haller, and other proponents of a *sensorium commune*, its "internal," underlying senses had provided the answer: external senses merged to make internal ones, which constituted a single, sensible consciousness.[71] But Condillac reversed the problem. He pointed out that (according to the prevailing model) all sensations were in an important sense internal sensations. The impression theory of sensation held, as we have seen, that sensations were "but manners of being," physical states of the sensible creature caused by external forces. The passivity of the mind on this model, while guaranteeing its openness to outside influences, also meant that it did not actively perceive external things, but simply existed in whatever state they put it in. But how could such a mind know anything outside of its own states? The problem was not, therefore, how to build inward from external sensations to an internal self, but how to build outward from a sensible but purely passive self to an external world.

Imagine a statue that can only smell. Waft a rose under its nose. To an observer, it will be a statue that smells a rose. But "to itself, it will be simply the odor of this flower itself. It will be thus the odor of rose, of carnation, of jasmine, of violet, according to the objects that stimulate its organ." The odors the statue smells will seem to it not as properties of an external object, but rather as its own "manners of being." Think now that the statue can only hear. Again, it cannot suspect that there exists anything outside itself. It can only feel itself to change, to vary in tone and intensity. Thus "we transform it, as we like, into a noise, a sound, a symphony." A seeing statue similarly "cannot judge that there is something outside itself." It feels itself to be all light and color.[72] Here was a powerful image for what was in fact a common notion in post-Lockean psychological and physiological writing, the full transformation of the passive, impressionable mind into each successive impression. Diderot, for example, represented all sensation as akin to the experiences of the statue: "We cannot think, see, hear, taste, smell, touch at the same time. We can be at but one of these things at a time." He maintained that each separate sensation took over the entire being of the sensitive creature: "[Man] is always entirely at the place of sensation: he is but an eye when he sees . . . but a small part of the finger when he touches."[73]

71. On internal versus external senses, see Descartes, *Dioptrique* (1637), 98; Boerhaave, *Praelectiones de morbis nervorum* (1730–35), xi.372; Haller, "Of the Internal Senses," in *First Lines of Physiology* (1786), xviii.56. Emanuel Swedenborg, who was briefly but very productively interested in neurophysiology before he took up theology, wrote in one of his several treatises on the brain: "The external sensations do not travel to any point beyond the cortical cerebellula. This is clear since these are the origin of the nerve fibers." Swedenborg, *The Economy of the Animal Kingdom* (1740–41); cited in Gross, *Brain, Vision, Memory* (1999), 125. Swedenborg called the cerebral cortex "the general maintainer and regulator of the internal sensories" and therefore "the common or general sensory." Swedenborg, *The Brain Considered* (1882–87) 1:56–57.

72. Condillac, *Traité des sensations* (1754), part I, chap. 1, §2; chap. 7, §1; chap. 8, §1; chap. 11, §1, §2.

73. Diderot, *Eléments de physiologie* (1784), 239, 53.

Condillac's statue, then, could have sensation without sensibility. It would simply feel itself to become whatever it sensed. Even a statue with four of the five senses—smell, hearing, taste, and sight—although it would be capable of much thought, would never suppose that its manners of being have external causes. Condillac granted that the statue with four senses could reflect, remember, compare, judge, discern, imagine; it would have abstract ideas of number and duration; it would know general and particular truths; it would have desires, passions, loves, and hates; it would be capable of hope, fear, and surprise. But all of these mental operations and feelings would regard the statue itself and itself alone.[74]

When it acquired touch, however, Condillac's statue would soon discover its own boundaries and the existence of a world beyond. In touching its own body, it would recognize itself in all its parts. But when it touches "a foreign body, the *me*, that it feels modified in its hands, does not feel modified in this body. If the hand says *me*, it does not receive the same response." The feeling of solidity in an external object represents to the statue "two things that exclude each other. . . . *Voilà* therefore a sensation by which the soul passes from itself outside itself." This journey of the soul would be attended by fearsome emotions. Condillac imagined his statue "quite astonished not to find itself in all that it touches," and from this astonishment "is born the anxiety of knowing where it is, and if I dare express it thus, to what extent it is." The statue's disturbing discovery of its own borders would produce a new, outward focus for its thoughts and feelings. Its "love, hate, desire, hope, fear no longer have only its own manners of being for their sole object: there are palpable things that it loves, hates, hopes, fears and wants. It is thus not limited to loving only itself."[75] The sense of touch, and touch alone, was the source of sensibility.

Just as he had inverted the problem of external sensation and internal self, asking how to extrapolate a world from the self's states of being, rather than a self from the world's imprints, Condillac reversed the significance of the sense of touch. On the impression theory of sensation, touch was the model for all sensation because it represented the continuity between the world and the sensible creature: it was the world connecting with the creature's senses. But in Condillac's hands, touch became the crucial *dis*continuity between the world and the sensible creature, enabling it to realize where it ended and the outside world began. In describing this process of realization, Condillac also

74. Condillac, *Traité des sensations* (1754), part I, chap. 12.

75. Condillac, *Traité des sensations* (1754), part II, chap. 5, §2, §3, §5, §7, §8; chap. 7, §7. Cf. Bonnet, *Essai analytique* (1760), 81, 535. Bonnet uses the same device, the sensible statue, to trace the sense of self to the use of language to form abstract thoughts.

ultimately mitigated his sensible statue's passivity by allowing it some rudimentary actions: moving through the world and touching things. What Condillac saw was the paradox at the heart of sensibilist psychology and physiology: a creature comprised entirely of openness to the world's influences would be utterly solipsistic. It would have no way of distinguishing between itself and the whole world. It was thus not enough to *be* open to the world's influences; one must also *recognize* one's openness, which meant first recognizing the opposite: one's boundaries.

Georges Buffon had proffered a similar fantasy to Condillac's statue in the third volume of his *Histoire naturelle de l'homme*; indeed, the idea was timely and popular enough to provoke some priority disputes.[76] Buffon had cited Cheselden as proof that "we can have by the sense of vision no idea of distances." Like Condillac, Buffon claimed that only touch conveyed a "complete and real knowledge" of the non-self. Thus "animals who have hands seem the most spiritual." Buffon imagined his own sensory awakening: "I believed at first that all objects were in me. . . . I heard sounds, the song of the birds, the murmur of the breeze. . . . I was persuaded that this harmony was me. . . . I fixed my regard on a thousand divers objects. . . . I thought I recognized that all was contained in a part of my being." When the breeze brought him "perfumes that caused in me an intimate blossoming," it gave rise to "a sentiment of love for myself."

Agitated by these feelings, the imaginary Buffon began to move and to touch. At first he touched only himself, and "all that I touched . . . seemed to return sentiment for sentiment," but then he touched something that did not respond. "I turned with a sort of horror, and knew for the first time that there was something outside of me." A stormy period followed in which he was "profoundly occupied with myself." At last he found humility: "My knees bent and I found myself in a situation of rest." The fantasy came to its logical conclusion with the reversal of solipsism in Buffon's final discovery "of a form similar to my own. . . . [I]t was not me, but it was more than me, better than me, and I thought my existence would change places and pass entirely into this second half of myself." Buffon described the loving knowledge of another's existence in sensationist terms: "I wanted to give it all my being; this vivid desire took over my existence, and I felt the birth of a sixth sense."[77]

For Buffon, as for Condillac, the sensible creature needed one crucial piece of knowledge: the difference between itself and the world. This knowledge was as much a matter of emotion as of sensation. The epistemological problem with which the story began was now transformed. Lockeans, reacting against the doctrine of divinely given innate ideas, had been at first con-

---

76. See Condillac, "Réponse à un reproche" (1754), in *Oeuvres philosophiques*, 1:318–19.
77. Buffon, *Histoire naturelle* (1750), 3:124–26, 133, 142–45.

cerned to describe a being shaped entirely by its experiences, an utterly open, permeable being. They had carried this idea of openness to its logical extreme with their notion of the mind as an initially blank slate and their impression model of sensation, in which the mind received the world's imprints through the senses. But once their description of a supremely open creature had been accomplished, philosophers and physiologists began to worry about civilizing this blank, passive entity. It would be open to the world's influences; but would it recognize its own openness? Would it know its place in a larger world?

Experimenters and philosophers in the Molyneux-Cheselden tradition began to ask whether and how one could trace one's sensations to their origins in a world outside oneself. They raised the possibility that sensationists, by describing an absolutely open creature, and by erasing the boundaries between that creature's mind and its surroundings, had produced just the opposite of what they had intended: a subjectively boundless being, a being that knew only its own feelings, an absolutely solipsistic entity. Clearly, then, sensation was not sufficient to guarantee an understanding that one was part of a larger world. Along with sensation, one needed an awareness that one's sensations had external sources. French commentators on the Cheselden experiment understood this awareness to be an essentially emotional matter.

Or to be precise, they understood its cognitive and emotional aspects to be inseparable. Here they contributed to a more general change that has recently been receiving attention from historians, the shifting status of emotion during the late seventeenth and eighteenth centuries.[78] Whereas an earlier generation of philosophers and political writers, including Descartes, had treated the passions as innate tendencies of the human constitution, the antithesis of the rational faculty, and inherently destructive, Locke and his successors traced emotions to the operations of the five senses, assumed an allegiance between the emotional and the rational faculties, and put the passions to work alongside reason in both philosophy and social life.[79] Primary examples of emotion put to work are to be found in Adam Smith's moral and economic writing, in which sentiments such as sympathy and self-love provide the motive forces.[80] "The passions are to the moral," Claude-Adrien Helvétius observed, "as movement is to the physical; it creates, annihilates, conserves,

78. Some examples of recent works that treat this topic are Mullan, *Sentiment and Sociability* (1988); Williams, *The Physical and the Moral* (1994); Pinch, *Strange Fits of Passion* (1996); and Vila, *Enlightenment and Pathology* (1998).

79. Pinch, *Strange Fits of Passion* (1996), 18–19.

80. On the pragmatic harnessing of the passions in classical liberal economics, and on the alliance between reason and passion, see Hirschman, *The Passions and the Interests* (1977), especially 16–17, 43–44. On sentiment in Smith's moral and economic philosophy, see ibid., 107–11; and Rothschild, *Economic Sentiments* (2001), esp. chap. 8.

animates everything, and without it all is dead: it is they likewise that vivify the moral world."[81] In epistemology, too, reason began to require the corrective aid of passion.

This new allegiance had a physiological basis in the nerves, the source "for all human passion and reason."[82] One's capacity for both depended upon one's nervous sensitivity; so "in general," La Mettrie observed, "the more mind one has, the more one has a penchant for pleasure and voluptuousness.... [F]ools, limited minds, are commonly the most indifferent and the most restrained."[83] Without this new partnership of emotion, sense, and reason, Julie, the heroine of Rousseau's *La Nouvelle Héloïse*, could never have characterized her attenuated state by saying it "renders my soul almost insensible and leaves me the use neither of my passions nor my reason."[84]

Emotion thus gained a transformed reputation by the middle decades of the eighteenth century. "We finally agree today," Helvétius rejoiced, "about the necessity of the passions." Similarly for La Mettrie, the passions were "as necessary to man as the air he breathes."[85] No longer antithetical to the rational faculty, they became, on the contrary, essential to its proper functioning. A new generation of empiricists viewed the passions as "states of understanding" and emotion "as a way of knowing."[86] Specifically, emotion was the form of knowledge that most intimately and actively connected the knower to the known. The century's leading analyst of the passions, David Hume, found that they were the unique force that bound a person to external objects and other people, thereby also giving him a sense of self and serving as the inducement behind all his activity in the world. Book 1 of Hume's *Treatise of Human Nature* (1739–40), "On the Understanding," leaves its author in a pitiful state, "affrighted and confounded with that forelorn solitude, in which I am plac'd in my philosophy," having demonstrated the insufficiency of reason to answer "Where am I, or what? From what causes do I derive my existence,

81. Helvétius, *De l'esprit* (1758), 297. Helvétius also called the passions "the celestial fire that vivifies the moral world" (319) and estimated that "if men were without passions, there would be no way to make them good" (376).

82. Haller, *First Lines of Physiology* (1786), iv. Helvétius claimed similarly that "physical sensibility is the unique cause of our actions, our thoughts, our passions, and our sociability." Helvétius, *De l'esprit* (1758), 368. On the widespread eighteenth-century assumption of a physiological basis for the passions, see Williams, *The Physical and the Moral* (1994), 126–30; and Gardiner, *Feeling and Emotion* (1937), 248–55.

83. La Mettrie, *L'Ecole de la volupté* (1746), 120. La Mettrie also wrote: "Only man, that reasonable being, can elevate himself to voluptuousness" (139).

84. Rousseau, *Julie* (1758), 1:228.

85. Helvétius, *De l'homme* (1773), 82, 40; La Mettrie, *L'Ecole de la volupté* (1746), 146.

86. Pinch, *Strange Fits of Passion* (1996), 18–19. On the relations between empiricism, the shifting status of the emotions, and the origins of Romanticism, see Tuveson, *The Imagination as a Means of Grace* (1960); and Bate, *From Classic to Romantic* (1961), 94. On the utility of emotion, see Hirschman, *The Passions and the Interests* (1977).

and to what condition shall I return? Whose favour shall I court, and whose anger must I dread? What beings surround me? and on whom have I any influence, or who have any influence on me?"[87]

Book 2, "Of the Passions," provides the solution: here are the forces that attach one's internal ideas to external sources and propel one from ideas to actions. Without feeling, reason was inert; therefore, Hume declared, "Reason is, and ought only to be the slave of the passions."[88] Helvétius too made the passions the mind's motive force, judging that "the mind stays inactive if it is not set in motion by the passions."[89] This view, that the passions alone brought the mind into active engagement with the world, found frequent physiological expression from the mid-eighteenth century. An example is the work of Antoine Le Camus, a doctor and physiologist on the medical faculty of the University of Paris, who assigned the passions the role of stimulating the mind into action by setting the animal spirits in motion through the fibers of the nerves.[90] Bonnet similarly called the passions "the unique Motor of Sensible Beings, and of intelligent Beings,"[91] while La Mettrie wrote that pleasure "speaks to me through my organs, and attaches me to life."[92]

87. Hume, *Treatise of Human Nature* (1739–40), 1:544, 548. On the relations between the emotions and social life in Hume's philosophy, see Mullan, *Sentiment and Sociability* (1988), chap. 1. On his view of the passions as constitutive of identity, see McIntyre, "Personal Identity and the Passions" (1989).

88. Hume, *Treatise of Human Nature* (1739–40), 1:544, 548; 2:193–98. On the role of the passions in Hume's epistemology, see Pinch, *Strange Fits of Passion* (1996), 21–44; Stroud, *Hume* (1977), chap. 8; and Cléro, *La Philosophie des passions* (1985), 157–76. On the notion that the passions provided the mind's impetus, see Gardiner, *Feeling and Emotion* (1937), chaps. 8, 9; and Rorty, "From Passions to Emotions and Sentiments" (1982). Hutcheson wrote that "*Reason*, or knowledge of the relations of external things to our bodies, is so inconsiderable" that it must be that "beings of such degrees of understanding, and such avenues to knowledge as we have . . . need these additional forces, which we call *Passions*." Hutcheson, *On the Nature and Conduct of the Passions* (1728), 34.

89. Helvétius, *De l'esprit* (1758), 471. Helvétius also called the passions "the seed of the mind" (303) and wrote that the passions' "force alone can counterbalance in us the force of laziness and inertia, tearing us from the rest and stupidity toward which we ceaselessly gravitate" (320).

90. Le Camus, *La Médecine de l'esprit* (1753), 314. See Vila, *Enlightenment and Pathology* (1998), 87.

91. Bonnet, *Essai analytique* (1760), 253. More precisely, Bonnet wrote that self-love was the unique motor of sensible and intelligent beings because it was the source of all the passions, including compassion and charity. Like Condillac and Buffon, Bonnet thus made love of others contingent upon a recognition of self.

92. La Mettrie, *L'Ecole de la volupté* (1746), 140. Vartanian views the same attachment from the other side: life to La Mettrie rather than La Mettrie to life: "For its author, the man-machine was not only a factual truth in support of materialism, but an occasion to extract from science a full-blown, if controversial, humanism. . . . La Mettrie's science is double-edged, for, in mechanizing the human being, it conversely 'humanizes' mechanism . . . thereby restoring humanity, by a new dispensation, to its coveted place as the epitome and measure of all things." Vartanian, *Science and Humanism* (1999), 47–48. Once again, openness to the outside world and radical egoism are flip sides of the same coin.

Condillac's statue was "attached to life" ultimately by a similar fusion of the cognitive and the passionate. The epiphany brought by touch, a discovery at once of the self and the non-self, was equal parts idea and emotion, each necessary to the other. Condillac and the other French commentators on the Cheselden experiment advanced the process that Berkeley had begun. Berkeley had taken Molyneux's Problem and made it into a question of how we use vision to understand the external world of touch. He had claimed this visual understanding was a matter of sensitive intuition, not geometry. La Mettrie, Condillac, and Buffon focused on this part of Berkeley's story and largely dismissed the question of the identification of shapes, moving on to what really interested them. To them, the lesson of the Cheselden experiment was that one could have sensation without sensibility; one could perceive the external world but fail to perceive it *as* external. In that case, one would be a being for whom nothing but oneself existed. Insensibility meant isolation from "the commerce of others"; it meant solipsism. Sensibility was therefore not reducible to sensation, nor was it about the identification or nonidentification of shapes. It was about sentiment and morality, and its end was love.

IV. THE SOLIPSISM OF THE BLIND AND THE MATHEMATICALLY MINDED

In the spring of 1749, a Prussian surgeon named Hilmer, operating under the auspices of the Cartesian naturalist René-Antoine Ferchault de Réaumur, couched the cataracts of one Mademoiselle Simoneau, a young girl who had been blind from birth.[93] Diderot fervently sought to be admitted to the removal of the girl's bandages and the initial testing of her innate visual capacities, but Réaumur rebuffed him. "Persons of the highest distinction have had the honour of sharing this refusal," Diderot commented, for Réaumur "wanted to part the veil before a few unimportant eyes only."[94]

No matter; Diderot set himself to "philosophize with my friends." He also reported having found, on the same day as the Simoneau operation, his own subject for empirical study, the "man-born-blind of Puiseaux."[95] The results of Diderot's interview with this man, together with the memoir on the life of Saunderson, provided material for Diderot's *Lettre sur les aveugles*, published

93. Réaumur published no report of the surgery. The primary accounts are Diderot, *Lettre sur les aveugles* (1749), 81–82; Clément, *Les Cinq années littéraires* (1755), lettre XXXIII (20 juin 1749), 1:184–86; Trublet, *Mémoire* (1759); Vandeul, *Mémoires* (1830–31), 1:27. For secondary accounts, see Torlais, *Esprit* (1936), 250–52; Furbank, *Diderot* (1992), chap. 3; Roger, *Buffon* (1989), 216–17.

94. Diderot, *Lettre sur les aveugles* (1749), 81–82. According to Diderot's daughter, Mme. de Vandeul, the unimportant eyes to which he referred belonged to Mme. Dupré de Saint-Maur, whose outrage at the insult led to Diderot's imprisonment at Vincennes in the summer of 1749. See Torlais, *Esprit* (1936), 251.

95. Diderot, *Lettre sur les aveugles* (1749), 82.

the following June. Over thirty years later, as one of the last things he wrote, Diderot would add a postscript to the *Lettre*, mostly about Mélanie de Salignac, the niece of Diderot's lover Sophie Volland, who had been blind from the age of two.[96]

The *Lettre sur les aveugles* completed the redirection of Molyneux's Problem, begun by Berkeley and continued by La Mettrie, Buffon, and Condillac, from ideas to sensibilities. Diderot argued first that the blind, because of their sensory deprivation, were necessarily abstract, mathematical thinkers; and second that abstract, mathematical thinking amounted to emotional and moral solipsism. Neither piece of this argument, taken alone, was original. Indeed, the notion that the blind were mathematically minded was very nearly a cliché in the eighteenth century. Leibniz and Reid, as discussed above, rested their solutions to Molyneux's Problem upon the heightened capacity of the blind for mathematical reason. Others did the same, including the Dutch philosopher David-Renaud Boullier and the English doctor James Jurin. Boullier alluded to what he said was a generally acknowledged talent of the blind for geometry; and Jurin cited the example of Saunderson.[97] The stereotype of the blind rationalist was so pervasive that even Buffon, neither blind nor conspicuously orderly in his methodology—and indeed, as we have seen and as will emerge with even greater clarity in chapter 3, explicitly hostile toward rationalist approaches to nature—was nonetheless vulnerable to it. A critic, charging Buffon with exhibiting a misguided spirit of system, attributed it to his notoriously poor eyesight.[98]

Meanwhile in literature and the arts, the self-involvement and social isolation of the blind was as common a theme as their penchant for abstraction was ubiquitous in philosophy (see fig. 2.5). Pierre-Simon Ballanche, whose *Du sentiment considéré dans ses rapports avec la littérature et les arts* (1801) calls for "a new poetics" devoted to fostering sensibility, described blindness as "an enforced and continual solitude" and attributed to blind poets, notably Homer and Milton, a poetry whose dominant emotion was melancholy: the poetry of exile and isolation.[99] In a similar vein, the blind hero of the Chevalier de Cerfvol's *L'Aveugle qui refuse de voir* (1771) turns down the chance of being cured because he prefers a self-regarding to a social life. He explains to the eager

96. Mélanie de Salignac died in 1766 at the age of twenty-two. See Diderot, "Additions à la lettre" (1782), in Vernière, ed., *Oeuvres* (1998), 154–55 n. 2.

97. Leibniz, *Nouveaux essais* (1705), bk. II, chap. IX, §8; Reid, *Inquiry* (1764), chap. VI, §20; Jurin, "Dr. Jurin's Solution" (1738), 2:27–29; Boullier, *Essai philosophique* (1728), part II, chap. VI, §18.

98. The critic was the comte d'Angivillier, and he made the remark in his *Mémoires* for the year XIII (1805), in Bobé, ed., *Mémoires de Charles Claude Flahaut* (1933), 56. See Spary, *Utopia's Garden* (2000), 202.

99. Ballanche, *Du sentiment* (1801), 84, 122. On Ballanche's theory of the poetry of the blind, see Paulson, *Enlightenment, Romanticism and the Blind* (1987), 131–34.

oculist that, "concentrated upon ourselves," the blind devote "almost all our attention . . . to developing ourselves." Later he adds, "As a Blind Man, I judge things only by their relations to myself. I call them beautiful or ugly, good or bad, according to the manner in which they affect me."[100]

Diderot's *Lettre* thus combined a philosophical commonplace, that the blind were abstract thinkers, with a literary one, that they were self-involved and socially isolated, to argue that the intellectual disposition entailed the social and emotional one. As well as drawing upon both a philosophical and a literary theme, the *Lettre* also formally straddled the blurry line between philosophy and literature, as the French commentaries on the Cheselden experiment had done.[101] Their authors, arguing that ideas and emotions were inseparable, applied the principle by dramatizing the knowledge they meant to impart. Hume wrote admiringly of "easy philosophy" that borrowed from "poetry and eloquence" in order to "make us *feel* the difference between vice and virtue,"[102] while Diderot similarly described reading Richardson's novels as an experience of being made to feel moral truths: "I had heard the true tones of passion; I had seen the springs of self-interest and self-love acting in a hundred different ways."[103] While sentimental empiricists joined knowledge with feeling in their new sensibilist epistemology, they carried out the same union of ideas with emotions on a formal level by marrying philosophy with literature.

The result, in Diderot's case, was a parable showing the emotional solipsism of an abstract approach to nature. Diderot's primary interest being the moral implications of a manner of thinking, many elements of the discussion of blindness since Berkeley were no longer relevant. Briefly dismissing the

100. Cerfvol, *L'Aveugle* (1771), 42, 57. Paulson writes: "The blind man and the oculist, in their discussions, consider sight to be not a sensory or psychological matter but a means of participating in a certain kind of social and economic system. . . . Ostensibly a dialogue on the merits of seeing, *L'Aveugle qui refuse de voir* is in fact almost totally devoted to the advantages and disadvantages of being in society, as opposed to remaining in the solitary but uncomplicated place of the blind beggar." Paulson, *Enlightenment, Romanticism and the Blind* (1987), 88–90.

101. Vartanian characterizes Saunderson's cosmological views, as expounded in the *Lettre*, as an "unclassifiable text" made of "science, philosophy, fiction," and the *Lettre* as a whole as a "particular synthesis of science and humanism." To explain this synthesis, Vartanian draws upon Diderot's own definition of poetry, from the *Lettre sur les sourds et muets*, as writing in which "things are at once said and represented." Diderot, *Lettre sur les sourds et muets* (1751), 169; Vartanian, *Science and Humanism* (1999), 95–96. If things must be felt to be known, then science cannot be communicated without poetry.

102. Hume, *Enquiry* (1748), 1–2. On Hume's approval of "easy philosophy," see Pinch, *Strange Fits of Passion* (1996), 19–21.

103. Diderot, "Eloge de Richardson" (1761), 30. See Vila, *Enlightenment and Pathology* (1998), 155–60. In *Pleasures of the Imagination* (1997), John Brewer defines "high culture" as the "special collective identity" that "theatre, music, literature and painting" acquired during the eighteenth century as a result of artists' new, overarching purpose of instructing their audiences by instilling certain emotions in them (xvi).

Figure 2.5. Jean-Baptiste Greuze's *L'Aveugle trompé* (1755) poignantly depicts the social isolation and moral vulnerability of the blind man, whose faithless wife holds his hand in one of her own, while her other arm is wrapped around the shoulders of her lover. Pushkin State Museum of Fine Arts, Moscow.

Berkeleyan divorce of the visual world from the tactile, the related dependence of sight upon touch, and the impossibility of seeing distance and dimension, Diderot made Molyneux's Problem a question of character, not to say class. The patient's ability to identify geometrical shapes would depend entirely upon who the patient was. "Rough people, without education, without knowledge, unprepared" would see the shapes distinctly but would not be able to name them. The metaphysician would immediately make correct identifications but would surround these with caveats about the impossibility of knowing for certain that the visual world existed, or that it corresponded to the tactile world. Finally, the geometer would recall that when he had been blind, he and sighted people had agreed in their identifications of shapes. Therefore, not only would he name the shapes correctly, but he would be entirely confident of his answers.[104] Whether the newly sighted patient could recognize what he saw depended ultimately not on his eyes or his rational faculty, but on his sensibilities.

As for the blind man of Puiseaux, Diderot was once again interested not in what he knew, but in who he was: in his character, his sensibilities. The blind man, Diderot wrote, was "not lacking in good sense, a man who is known to many people, who knows a little chemistry and who has attended . . . the lecture-courses on botany at the *Jardin du roi*. His father taught philosophy with acclaim at the University of Paris." He was, in other words, a member of the Enlightenment, the "blind man of good sense" that Diderot recommended above the ignorant, uneducated, and suffering cataract patient as the better source for understanding the relations between vision and knowledge.[105] The cultural status of both the blind man of Puiseaux and Saunderson helped Diderot to use the ideas of the blind as an allegory for the philosophes' own manners of thought.

Particularly useful was the fact that Saunderson had been a mathematician since, as mentioned above, the *Lettre sur les aveugles* proposed an equivalence between sensory deprivation and abstract thinking, between blindness and geometry. Diderot wrote of the blind man of Puiseaux that he was a lover of order and "a very good judge of symmetries." A quaint logic, which Diderot termed Cartesian, characterized the blind man's definitions of objects in the sighted world. An eye, for example, was "an organ, on which the air has the same effect as my stick has on my hand." Diderot advised, "Open Descartes's *Dioptrique*, and you will see the phenomena of sight related to those of touch, and optical figures full of men engaged in seeing with sticks" (see figs. 2.6 and 2.7). A mirror was a machine for reversing things. Diderot exclaimed, "Descartes, blind from birth, would have, it seems to me, applauded such a definition."[106] Cartesian rationalism was equivalent to the naturally abstract way of thinking of the blind.

When Jean d'Alembert, Diderot's co-editor of the *Encyclopédie* and a mathematician, wrote the article "Aveugle" for its first volume, he turned Diderot's comparison of the blind man to Descartes into a compliment. He pronounced the blind man's definition of a mirror "very subtle" and credited the blind with

104. Diderot, *Lettre sur les aveugles* (1749), 136, 138–39, 142.
105. Ibid., 82, 126–27.
106. Ibid., 83–87, 113–14.

Figure tirée de la Dioptrique de Descartes.

Figures 2.6 and 2.7. Figure of a blind man seeing with sticks, from Diderot's *Lettre sur les aveugles* (1749) and described by Diderot as "taken from the *Optics* of Descartes"; and Descartes's figure illustrating his analogy between a blind man's use of two sticks to locate a single object in space and a sighted person's use of the light coming from the object into each eye, from Descartes, *La Dioptrique* (1637). Courtesy of Department of Special Collections, Stanford University Libraries.

a heightened capacity to concentrate their minds due to their relative isolation from distractions. D'Alembert often defended mathematicians against the popular idea that mathematics and sensibility were mutually exclusive. He hoped, for example, that his own collaboration on the *Encyclopédie* would demonstrate that one could "be a geometer and have common sense."[107] But to Diderot, the inward focus of an overactive rational faculty constituted a form of sensory impairment. "Reason tends to live within itself," Diderot wrote, since "unfortunately it is easier and quicker to consult oneself than to consult nature."[108] He had not intended the comparison as a compliment, either to Descartes or to the blind. Diderot dismissed Berkeleyan idealism along with Cartesianism as an "absurd" and "extravagant philosophy." With its "very abstract metaphysics," he decided it could "owe its birth only to blind people."[109]

Diderot reasoned that since "abstraction consists only of separating by thought the sensible qualities of bodies," the blind had a head start on the process, and so "the blind man perceives things in a much more abstract manner than we do." Saunderson was a perfect case in point, having been "drawn by taste to the study of mathematics." Indeed, the memoir on Saunderson's life and character that prefaced his *Elements of Algebra* and upon which Diderot partly based his *Lettre*, contains the same proposal: "The blind are by necessity more abstracted than others," thus perhaps "there is no Branch of Science more adapted to their Circumstances" than mathematics. It made sense that a blind person, who would "more frequently and more closely retire into himself," would "excel in these abstract sciences," and even that sighted people, in learning mathematics, might find it useful to "think and reason in the Circumstances of a blind Man." Mélanie de Salignac, too, had been talented in geometry and had herself believed "that geometry was the true science of the blind." She added, *"The geometer spends almost all his life with his eyes closed."*[110]

The fact that Saunderson had taught optics and lectured on light, color, and the theory of vision without ever having seen epitomized to Diderot the blind person's abstract approach to physical subjects. The blind were geometers in their very beings, collapsing "the passage from physics to geometry" so that all physical questions became, to them, "purely mathematical." Saunderson's mathematical manner of thought, Diderot wrote, led him to suppose "a geometrical relationship between things and their uses." Had he been given sight, like Molyneux's hypothetical blind man, he would have been baffled at the vagaries of the visual world. He would, for example, have been able to as-

---

107. D'Alembert, "Aveugle" (1751), 1:870–71; D'Alembert to Mme. Du Deffand; cited in Pappas, "L'Esprit de finesse" (1972), 1229.

108. Diderot, *De l'interprétation* (1753), 185.

109. Diderot, *Lettre sur les aveugles* (1749), 83–87, 113–14.

110. Ibid., 98, 110–12; Anon., "Memoirs," in Saunderson, *Elements of Algebra* (1760), x–xi; Diderot, "Additions à la letter" (1782), 161.

sociate his skullcap with his head but would have failed to recognize his square or tasseled caps. "To what end this tuft? Why four angles rather than six?" Diderot imagined the geometer exclaiming in perplexity.[111]

Diderot was, as I have said, by no means alone in associating mathematical thinking with blindness. But beyond developing the association of blindness with mathematics, his innovation was to add another element into the equation: a moral cast of mind specific to the blind and, by implication, the philosophically blind, partisans of theoretical abstractions. Oddly, the corpus of critical and historical writing on Diderot's *Lettre* has paid no attention to this argument regarding the moral equivalence of blindness and abstract, geometrical thinking. Historians have taken the central characters in the *Lettre* as blind seers, giving straightforward expression to Diderot's own views. Aram Vartanian, for example, describes the figure of Saunderson in the *Lettre* as a spokesman for Diderot's atheistic, mechanistic materialism, and blindness as "a global metaphor for the powers of intuition and conjecture" of the "scientist of genius."[112]

Of course, Diderot did use his blind characters to voice certain of his cosmological views. In particular, Saunderson describes a Diderotian cosmos—a dynamic, unstable, indeterministic swirl of matter—familiar also from *Le Rêve de d'Alembert* (1767). As Vartanian points out, the darkness and tactility of the blind man's world are useful for this purpose. In this use of blindness, Diderot reversed the significances of sight and touch, making sight, rather than touch, the sense of geometry and rational order. Here, sight distanced the viewer from what he saw and allowed him to order it into coherent (but false) patterns, while touch did the opposite, connecting the toucher immediately and intimately to the world's disorderly truth. Thus in this instance sight was the misguided, rational sense and touch the true, experimental one.[113] Clearly, then, Diderot was inconsistent in his construal of the nature of sight and touch, and in his use of the blind. Elsewhere in the *Lettre*, as we have seen, the blind man's groping knowledge of the world made him overly rational; his relevant feature was then sensory impoverishment rather than an experimenter's sensitive

111. Diderot, *Lettre sur les aveugles* (1749), 110–12, 143–44. Ware would later discern a similarly schematic approach in his seven-year-old patient, whom he said described a letter "as square, because it had corners; and an oval silver box . . . as round, because it had not corners; he likewise knew, and called by its name, a white stone mug, on the first day he obtained his sight, distinguishing it from a basin, because it had a handle." Ware, "Case of a Young Gentleman" (1801).

112. Vartanian, *Science and Humanism* (1999), 97–98.

113. Diderot, *Lettre sur les aveugles* (1749), 123–24. See Vartanian, *Science and Humanism* (1999), 93–95; Ilie, *Cognitive Discontinuities* (1995), 115–16, 233–34, 322–24. On the contradictory representations of vision and blindness in eighteenth-century French literature, in which knowledge is associated sometimes with the clear light of vision and sometimes with the groping touch of the blind, see also Swain, "Lumières et Vision" (1988); Trottein, "Diderot et la philosophie du clair-obscur" (1988).

tactility; and Diderot used him to personify not the sort of philosophy he liked, but the sort he rejected—abstract, geometrical, deterministic—and to make a moral indictment of such philosophy.

As blind people were intellectually different from sighted people, Diderot argued, they were also morally different. Having "never doubted that the state of our organs and senses had much influence on our metaphysics and morals," he hastened to interrogate his blind philosopher on the question of vices and virtues. He found that the blind man of Puiseaux had "a prodigious aversion to theft," being such an easy victim of it. For the same reason, Diderot reckoned that "women would be common in a land of the blind, or else the laws against adultery would be very strict." On the other hand, the blind man could not fathom physical modesty. Blind people were also immune to the "external signs of power that affect us so vividly." The blind man of Puiseaux, threatened with being put in prison after some youthful scrapes, replied to the police lieutenant, "Eh! monsieur . . . I have been there for the past 25 years." Diderot exclaimed, "What a text for a man who loves to moralize as much as I do." By the same logic, the blind must be less susceptible to religious arguments that appealed to the grandeur of nature. Diderot had his dying Saunderson tell a priest, "If you want me to believe in God, you will have to make me touch him." [114]

Virtues, Diderot concluded, "depend on our manner of sensing" and upon "the degree to which external things affect us." The more affected one was by external things, the better developed one's moral faculty. Because the vividness and intensity of one's sensory experience of an event shaped one's moral response to it, Diderot estimated that most people would have a much easier time "killing a man from a distance at which he appears the size of a swallow." [115] His application of this principle to a moral appraisal of blindness was quite new. The only prior allusion to the patients' moral ideas in the case literature on blindness was a brief moment in the 1709 Tatler article, when the philosophical minister asked the young patient, regarding the servant who guided him, "what sort of thing he took Tom to be before he had seen him." The young man had reportedly answered that he believed "there was not so much" of the servant as of himself, but "he fancied him the same sort of creature." [116]

Diderot suggested that the blind would be less inclined to view others as the same sort of creature, having less evidence to go by. He wrote that the

114. Diderot, *Lettre sur les aveugles* (1749), 92, 89–90, 119. In his later postscript to the letter, Diderot remarked that Mlle. de Salignac was as delicate and modest as could be. See Diderot, "Additions à la lettre" (1782), 158.

115. Diderot, *Lettre sur les aveugles* (1749), 92–93.

116. *Tatler*, no. 55 (16 August 1709): 125. Cf. Wardrop, "Case of a Lady Born Blind" (1826), 535. Here the blind patient was disturbed to see herself from the outside for the first time. When her brother asked her to look into the mirror and "tell him if she saw his face in it," she replied, "evidently disconcerted," "I see my own; let me go away."

morals of the blind and the deaf were inferior to those of people with all five senses, which were in turn topped by the sensibilities of a hypothetical being enjoying six. "Ah, madame! How the morality of blind people is different from ours! How that of a deaf person would differ from that of a blind person, and how a being who had one more sense than we do would find our morals imperfect, to put it mildly!"[117] This proposal has been characterized by P. N. Furbank as relativist,[118] but it seems to me quite the opposite: Diderot posits a moral spectrum with a perfect society at its vanishing point. His placement of the society of five-sensed humans somewhere in the middle of this spectrum is not relativism but radicalism. That most notorious of moral radicals, the Marquis de Sade, a convinced sensationist ("Without senses, no ideas"), made the same argument. Just as a "society of the blind" would seem imperfect to us, he wrote "our laws, our virtues, our vices, our gods, would be contemptible in the eyes of a society having two or three senses more than us, and a sensibility double what ours is." Sade was a revolutionary but no relativist: "the most perfect being we can conceive, would be the one that would remove itself farthest from our conventions."[119] Even in its naughtiest guise, sensibility was as absolutist as the moral traditions it replaced.

Blind people were the central figures in the working out of this new sensibilist moral scheme. Above all, according to Diderot, as with blind philosophy, the central characteristic of blind morality was abstraction. The blind were impervious, he suggested, to most signs of suffering in others. Their abstract manner of experiencing others' pain must weaken their sense of sympathy, and so Diderot "suspected" blind people of "inhumanity." He asked rhetorically, "What difference is there, for a blind man, between a man who urinates and a man from whom, without his complaining, blood pours forth?"[120] Sensory impairment equaled abstract thinking equaled solipsism. Diderot gave his moral repudiation of Cartesianism a literal, physiological meaning in the *Lettre sur les aveugles*. Cartesians thought like blind men, and the blind were inhumane.[121] The conclusion was plain to see.

117. Diderot, *Lettre sur les aveugles* (1749), 92–93.

118. "It is important to Diderot's strategy that he should begin with an ethical question. For if it can be shown that the blind possess a different code of ethics from the sighted, this is already quite a victory for relativism." Furbank, *Diderot* (1992), 58.

119. Sade, "Pensée inédite" (1782), 2–3.

120. Diderot, *Lettre sur les aveugles* (1749), 92–93. Diderot reported in the 1782 "Additions à la lettre" that Mélanie de Salignac had "not pardoned me for writing that the blind were deprived of the symptoms of suffering" and had insisted that she could perceive others' pain as well as he. See Diderot, "Additions à la lettre" (1782), 156.

121. The inward focus of Cartesian rationalism, whose moral consequences so worried Diderot, was intended, by its author, as a safeguard against social activism. Descartes believed his philosophy would shield the state against radical impulses, by redirecting such impulses inward toward the ordering of the individual's own life. Rather than demolishing "all the houses of a city for the sole purpose of rebuilding them," Descartes recommended another sort of "archi-

Diderot's *Lettre* constituted an allegorical, moral condemnation of Carte-sianism. The sensibilist science that he and his fellow sentimental empiricists recommended put a sympathetic identification with the things they studied in place of rational systems. The result, Diderot promised, was something much better than Cartesians' cold-blooded schemes: a science founded in compassion, a humane science.

V. THE INSTITUTION
ROYALE DES JEUNES AVEUGLES

Civilizing the impressionable mind described by Lockean sensationists, we have seen, became the central concern of the Molyneux-Cheselden tradition at the hands of French commentators. These commentators were no longer primarily interested in proving that the world inscribed itself upon an initially blank mind through the five senses, or in explaining how this inscription took place. Instead they wanted to know how an entity, so inscribed, would know the world outside itself and its proper place in that world. This concern with integrating sensationism's creature into a social and moral universe ultimately had institutional results. How the minds of the sensory-impaired worked had begun the century as a problem in epistemology and a subject for thought ex-periments. At the end of the century, the problem had become one of sensibil-ity and social engagement, and a matter for civic education. Philosophical and experimental interest in blindness brought about and shaped the first educa-tional institution for the blind. Its mission was to civilize its beneficiaries, and it became emblematic of the civilizing mission of civic education in general.[122]

The school's founder was Valentin Haüy, a translator of Italian, Spanish, and Portuguese for Louis XVI and the brother of René-Just Haüy, inventor of crystallography and member of the Académie des sciences. Valentin Haüy re-

---

tecture": to "change my desires rather than the order of the world." Descartes, *Discourse on Method* (1637), 117, 122–23. See Jacobs, *Cultural Meaning* (1980), 58.

122. The pedagogical implications of the philosophical and experimental literature on blind-ness extended beyond the blind. In a series of essays chronicling the history of Molyneux's Prob-lem written during the 1770s, Jean-Bernard Mérian, a philosopher and member of the Academy of Berlin, designed an educational program inspired by Condillac's fantasy and Cheselden's ex-periment. The ideal schooling, he wrote, would develop the five senses independently and in succession, so that none would detract from the full flowering of another. If this were possible, "what a force of mind and character would result? What men we would be!" But Mérian con-fessed that the idea was impractical. He proposed a less ambitious alternative: "to take children from the cradle, and raise them in profound darkness until the age of reason." They would learn to read raised print and would be taught "the Sciences, Physics, Philosophy, Geometry and Op-tics above all." Finally, the children would be brought, as fully developed rational beings, into the light of day. "What a torrent of delights will inundate them, what will be their transports . . . is there anything comparable to such an instant?" Mérian, "Huitième mémoire" (1778), in *Sur le prob-lème de Molyneux*, 184, 180, 187.

ported that he had been inspired to create the Institut national des jeunes aveugles, founded in 1784, primarily by Diderot's *Lettre sur les aveugles*, as well as by a glowing review in the *Journal de Paris* of the concerts of Maria-Theresia von Paradis. The review called Paradis a "phenomenon," citing her ability to read raised letters, her tactile knowledge of geography and arithmetic, her musical memory, and her qualities of "heart and mind." Haüy rushed to interview Paradis and based many of his tactile teaching techniques on her information.[123]

Traces of the philosophical discussion of blindness are everywhere apparent in the Institut's founding documents. Haüy assumed that the senses were so many conduits through which knowledge entered the soul, and that sensory impairment was the lack of one conduit among several mutually alternative ones. Like the abbé de l'Epée—whose school for the deaf and mute, founded in 1755, served as Haüy's model—Haüy adopted the non-Berkeleyan but instrumental view that certain senses could assume one another's functions. Epée had written that teaching the deaf was simply a matter of "sending through their eyes into their mind what enters into ours by our ears." Haüy admired this project of "impressing the soul with new ideas, in speaking to the eyes alone." In a letter to the *Journal de Paris*, the paper that chronicled his Institut's founding and early days, Haüy wrote that his own students would "replace sight by touch."[124]

This idea of replacing one sense by another was less obvious than it might seem in retrospect—as revealed by the prior philosophical controversy. To think of putting one sense in the place of another, one had to assume that the relevant senses were not merely mechanisms for experiencing and interacting with the world, but were channels of an information separable from the sensations conveying it. One had, in other words, to believe that knowledge could be abstracted from the sensory experiences that were its source, an idea that, as we have seen, caused much conflict in the early history of Molyneux's Problem. Hutcheson, for example, had argued that it would make far less sense to propose translating between sight and smell than between sight and touch. So the visual and the tactile must share ideas whose commonality to the two made them independent of either.[125] Berkeley and his followers had rejected the notion of information separable from sensation, but others, however they answered Molyneux's Problem, accepted it. The resulting controversy had forcefully raised the possibility that blindness was merely a blockage in one among several sources of knowledge. Here was the central axiom of Haüy's pedagogy.

123. Haüy, letter dated 18 septembre 1784, *Journal de Paris*, no. 274 (30 septembre 1784): 1158–59; La Blancherie, "Variété," *Journal de Paris* (24 avril 1784): 504; Haüy, *Essay* (1786), 229, 240, 244–45, 255–56.

124. L'Epée, *La Véritable manière d'instruire* (1784), 19; Haüy, letter dated 18 septembre 1784, *Journal de Paris*, no. 274 (30 septembre 1784): 1159; Haüy, *Essay* (1786), 251. See Didier-Weygand, *Cécité* (1997–98), 171–72. On l'Epée, see supra, n. 67.

125. Hutcheson, "Original Letter" (1727), 159.

The blind lacked a primary conduit for the acquisition of knowledge; a replacement could be found in their sense of touch and in new kinds of literacy. Reading was a "channel through which every different kind of knowledge is communicated to us." A new technology of raised printing would accordingly give the blind "the power of receiving by touch ... [an] education." One could then use blind people's sense of touch to open their minds to the world outside. Haüy intended this opening to be cultural and emotional as much as epistemological. "Let them only, by means of reading," he wrote, find a remedy "against that intolerable melancholy, which corporeal darkness, and mental inactivity" previously produced in them.[126] "Thus," commented an admirer, "the solitude in which they lived will be brightened."[127]

The committee from the Académie des sciences that granted formal approbation to Haüy's project drew the same connection between the senses and social membership, reporting that Haüy's pedagogical methods would not only give the blind "knowledge that the deprivation of the most necessary sense has refused them," but in so doing, would "open them ... to the society of other men."[128] Later the same committee praised Haüy for "returning to society beings who seemed destined to live separated from it for ever."[129] This emphasis upon recovering the blind from a state of moral isolation would remain central to blind education. In 1817 Sébastien Guillié, who directed the Institut national des jeunes aveugles during the Restoration, would write that for a blind child, "the moral world does not exist ... he acts as if he were alone; he relates everything to himself. It is from this deplorable state that we must try to draw him."[130] The project of teaching the deaf was also meant to recover its subjects from social and moral vacuity. "What is a deaf-mute by birth," asked the abbé Sicard, who succeeded l'Epée in 1789 as director of the Institut royal des sourds et muets, "before some education has begun to link him by several relations to the great family of which ... he is a part?" The answer: "a being perfectly useless in society.... [A]s for morality, this results and comprises itself from elements placed so far from him that we may well doubt if he even suspects their very existence."[131]

126. Haüy, *Essay* (1786), 258, 229, 238. "He devoted himself above all to releasing [the blind] from the moral and intellectual isolation in which they were plunged." Léon Le Grand, "Les Quinze-Vingts depuis leur fondation," in *Mémoires de la Société de l'Histoire de Paris et de l'Ile de France XIV*, 151.

127. M. Gleize, *Règlement de Vie* (1787), 93.

128. Procès-verbaux, 16 février 1785, AS.

129. La Rochefoucauld, Desmarest, Vicq-d'Azyr, Demours, "Rapport des Commissaires, nommés pour examiner un ouvrage de M. Haüy, qui a pour titre Essai sur l'enseignement des Enfans aveugles," Pochette de séance, 13 janvier 1787, AS.

130. Guillié, *Essai* (1817), 13. Armand Dufau, Guillié's successor as director of the Institut, would write: "With respect to the formation of reason ... nothing replaces language; but for social relations ... nothing could replace sight." Dufau, *Des aveugles* (1850), 85–86.

131. Roch-Ambroise Sicard, *Cours d'instruction d'un sourd-muet de naissance. Et qui peut-être utile à l'Education de ceux qui entendent et qui parlent.* An VIII, vi–ix, xi–xiv.

Blindness was to Haüy and to his supporters, in keeping with the experimental and philosophical construal of sensory impairment, a problem of sensitive isolation from the outside world, which meant cultural, emotional, and moral isolation. Indeed, the equivalence worked both ways: sensory isolation amounted to social isolation, and social isolation impaired the senses. The Wild Boy of Aveyron, captured and brought to Paris in 1800 at the age of about twelve, was reported to have poor use of his sight and tactile sense. The physiologist Philippe Pinel wrote, "The sense of touch is the sense of intellect, and it is easy to see how imperfect this sense is in the so-called wild boy of Aveyron." Jean Itard, resident physician at the Institut royal des sourds et muets and the doctor who cared for the wild child, reported that the boy could not recognize shapes by touch, nor could he distinguish between actual objects and painted representations. His "entire system had only the smallest amount of sensibility," such that "one could indeed say that he differed from a plant only in that he could move about and shout." [132] By 1800 sensibility was not only the medium of social interaction, but depended upon that interaction for its own development.

Thus the reduced sensibility of the blind meant social isolation, and it also, Haüy believed, brought with it a certain epistemological style, again in keeping with the research tradition on blindness: a tendency toward abstract thinking. In his *Essay on the Education of the Blind*, Haüy described blind people's "astonishing" talents for mental gymnastics. It was "well known," he claimed, "what a clear conception the greatest number of them discover in the difficult operations of the mind." These abilities included "so great a propensity for calculation that we have often seen them following an arithmetical process, and correcting its errors, by memory alone." [133] Haüy followed d'Alembert in proposing that the blind were better able than the sighted to concentrate their minds, not being distracted by "that crowd of images the impressions of which cross ceaselessly through our brain despite us." A "gentle calm so favorable to study," available to sighted people only in the dead of night, graced the blind at all hours. [134]

That the blind were naturally abstract thinkers was a truism in philanthropic and institutional writing on blindness, as it had been in philosophical discussions. A correspondent of the *Journal de Paris*, writing on the decision of the Société philanthropique (which supported Haüy's Institut) to focus its attention on the blind from birth, listed occupations he believed would be

132. Philippe Pinel, "Report to the Société des Observateurs de l'Homme concerning the child known by the name of 'Sauvage de l'Aveyron'" (1800?), in Lane, *The Wild Boy of Aveyron* (1976), 57–69 (the passage cited is on 59–60); Itard, "The Wild Boy of Aveyron" (1799–1806), 149, 97, 143, 142.

133. Haüy, *Essay* (1786), 247 n. 243.

134. Haüy, "Mémoire" (1784), 42.

well suited to the blind. The list, which began with tobacco grating and cement grinding, concluded with "geometry, and even optics, witness Saunderson . . . finally all the abstract sciences." [135] This axiom, that the blind had an abstract cast of mind, like the notion of their moral isolation, would have a long life among pedagogues. Guillié would write that the blind "see things in a more abstract manner than we do," and Pierre-Armand Dufau, who succeeded Guillié as director of the Institut, characterized them as analytical, rational, methodical, and dispassionate, "strangers to the passions that swirl about them." [136]

The blind personified the collusion of abstract thinking and social isolation. By using their sense of touch to teach them, Haüy dramatized the union of sensibility and social harmony. His school was a living symbol, to which the enlightened and fashionable came in throngs every Wednesday and Friday at noon to witness the blind students' "public exercises." [137] Alexandre Rodenbach, a Belgian student of Haüy's who became a writer and public official, later wrote that during the 1780s "there was a throng of subscriptions for the new school, and by the magic effects of fashion, the blind became the objects of every conversation." [138]

Ten years after the founding of the Institut, and seven years into the Revolution, Haüy made the school's larger moral mission explicit when he used it to house a religious cult. This sect, infelicitously named "Theophilanthropy" for the love of God and one's fellow man, was the successor to Robespierre's cult of the Supreme Being: it was the only successful natural religion of the Directory. Its original "manual" was the work of a Freemason, bookseller, and moderate republican named Chemin-Dupontès. But Haüy, upon reading it, seized hold of the project, formed a committee to study and rewrite the text, and developed from it a practicing order.

Haüy had recently moved his school into the convent of the Catherinettes in the rue Saint-Denis, made vacant by a decree of the Committee of Public Safety, and had been holding the school's public exercises in its chapel. In January 1797, when the Theophilanthropists held their first meeting, their temple was the same chapel, its walls now painted with moral inscriptions. A choir of Haüy's blind students sang hymns to a congregation comprised mostly of their fellows. Within two weeks, however, the flock had grown such that the Theophilanthropists were forced to hold two meetings each Sunday morn-

---

135. Régnier, "Aux auteurs du Journal" (16 janvier 1784), 234.

136. Guillié, *Essai* (1817), 47; Dufau, *Des aveugles* (1850), 43, 32. See Paulson, *Enlightenment, Romanticism and the Blind* (1987), 112–16.

137. *Journal de Paris*, no. 8 (31 janvier 1787), 31.

138. Rodenbach, *Lettre* (1828), 4; translation taken from Paulson, *Enlightenment, Romanticism and the Blind* (1987), 230.

ing, as well as a class of "moral instruction" and a Wednesday evening committee meeting to plan the events.[139]

In the Institut des jeunes aveugles, the moral scheme implicit in sentimental empiricism had found institutional expression in a kind of social reform, a program of public instruction intended to socialize blind children. With Theophilanthropy, the sentimental-empiricist moral vision achieved another kind of institutional apotheosis: it became the basis of an organized religion. The Theophilanthropists announced that their mission was to replace dogma with "beliefs of sentiment." In a phrase perfectly attuned to the song sentimental empiricists had been singing since midcentury, they asked, "Are there no other truths than those subject to rigorous demonstrations and exact sciences, and that which belongs to sentiment, does it not also have its evidence?"[140] Evidence and emotion, sense and virtue, the epistemological and the moral, had been joined in discussions of the blind since the beginning of the century, threading together through every theoretical and experimental treatment of their thoughts and feelings. Sensation and sentiment, the parts of sensibility, achieved at last a doubly powerful institutional fusion in a school for the blind made temple of the Revolution.

139. On blind and deaf education and the history and travails of the Institut national des jeunes aveugles during the Revolution, see Weiner, "The Blind Man and the French Revolution" (1974); Weiner, *Citizen-Patient* (1993), chap. 8; and Didier-Weygand, *Cécité* (1997–98), part 2, chap. 1. Weiner characterizes the Revolutionary project to educate blind and deaf children as one of socialization, carried out "in the hope that many might progress to the status of citizen-patients." Weiner, *Citizen-Patient* (1993), 225. On Haüy's role in Theophilanthropy, see Didier-Weygand, *Cécité* (1997–98), part 2, chap. 1; and Mathiez, *Théophilanthropie* (1904), 86–117.

140. *Qu'est-ce que la Théophilanthropie, Manuel des Théophilanthropes,* cited in Mathiez, *Théophilanthropie* (1904), 93.

Figure. 3.1. Figure from David Colden, "Remarks on the abbé Nollet's Letters on Electricity," in Franklin, *Experiments and Observations on Electricity* (1751), 135. The figure represents an insulated Colden (the text says he is standing on a cake of wax) discharging two Leyden jars that have previously been charged through their hooks. According to Franklin's theory of perfect symmetries between excesses and surfeits of electrical fire, the jars' insides are positively charged to the same degree as their outsides are electrically negative. Two assistants hold the jars, one by its coating (1) and the other by its hook (2). Colden touches the hook of the one held by its coating and the coating of the one held by its hook. Because the arrangement is perfectly symmetrical, both jars are discharged: according to Franklinist logic, the excess fire inside jar 1 travels through Colden to supply the negativity on the coating of jar 2; at the same time, the negativity on the coating of jar 1 is filled through the hand of the grounded assistant holding it, and the positivity on the hook of jar 2 is dissipated through the hand of the assistant holding it. All of these things must happen simultaneously, so that a balance is preserved and an even distribution of electrical fire restored. Courtesy of the Bancroft Library, University of California, Berkeley.

Chapter Three

POOR RICHARD'S LEYDEN JAR

If a Man casually exceeds, let him fast the
next meal, and all may be well again.
—*Poor Richard's Almanack* (1742)

No gains without pains.
—*Poor Richard's Almanack* (1745)

So wonderfully are these two states of Electricity,
the plus and the minus, combined and balanced
in this miraculous bottle!
—Franklin to Peter Collinson (1 September 1747)

Warnings against the dangers of abstract thinking were not limited to the discussion of blindness, nor was Diderot alone in thinking that a strictly mathematical-mechanist approach to nature on the model of Descartes's physics epitomized a lack of sensibility with baneful consequences, both scientific and moral. Only some fifty years after the "mechanization of the world picture," [1] many of the heirs to that achievement were seeking to undo it. Despite the dramatic successes the mechanist paradigm had recently achieved, around 1740 it came under severe attack.

This widespread rejection of mechanist physics was an expression of sentimental empiricism. To sentimental empiricists, a mechanist approach to nature signaled a closed mind and a misguided self-sufficiency. Many believed, with Diderot, that the architects of rational, mechanist systems of physics ex-

---

1. E. J. Dijksterhuis's 1961 book by that title describes the culmination of the Scientific Revolution in the construction of a single mechanical system embracing the heavens and the earth. Another description of seventeenth-century physics in these terms is Richard Westfall's *The Construction of Modern Science: Mechanisms and Mechanics* (1971). Historians such as Allen Debus (*Man and Nature in the Renaissance* [1978]) and Betty Jo Teeter Dobbs (*The Janus Faces of Genius: The Role of Alchemy in Newton's Thought* [1991]) have complicated this picture by tracing the persistence of mystical, magical, and alchemical philosophies through the seventeenth century. Here I am examining a different complication: the critical reaction against mechanism during the two or three generations following its dramatic success.

hibited the antithesis of sensibility: rather than directing their minds outward through the senses, they tried to capture and internalize the natural world by reducing it to a mechanical model that could be fully contained within a single mind. Making the internal coherence of the model their supreme criterion, mechanists, according to their critics, considered their understanding to be complete and absolute, neither contingent upon experience nor reliant upon the aid of others.

A mechanist explanation of nature also made no reference to any purpose or end, to any agency or act of will. Mechanists banished all appeals to intent, restricting physics to the discussion of moving bits of matter. While they generally assumed the world machine had been constructed by a divine intelligence, they exiled purpose to a place before and beyond nature.[2] In contrast, sensibilists, advocating a sentimental response to natural phenomena, personalized them, making them into things with which one could—and as a good empiricist and a morally sensitive being, would—have sympathy. In the eyes of his critics, the arrogant mechanist, retreating into his own mind and believing he could there encapsulate nature as a machine whose parts acted by constraint alone, was indifferent not only to the world outside himself and to its other inhabitants, but to the agency in nature itself.

The natural world, as sentimental empiricists described it, was imbued with agency, and that agency was more mundane than divine. As empiricists, these critics of mechanist physics understood natural agency in terms of what they saw in action all around them: human agency. They modeled their understanding of the natural world on the domain they knew intimately, the social world. In daily life, they reasoned, when one followed a chain of mechanical causation back far enough, one always arrived at an act of will. Thus, "the only principle that experience shows to be [originally] productive of movement is the will of intelligent beings . . . which is determined, not by motors, but by motives, not by mechanical causes, but by *final* causes."[3]

Sensibilists believed that as in social life, so in nature, the central problem and purpose was the regulation of the dynamic relations among its elements. Whereas the parts of the mechanist's world machine were inanimate and passive, engaging with one another purely by impulsion, the elements of the sentimental empiricist's natural world related to one another actively, by means of behaviors, desires, and intentions. Nature's parts, by acting according to their

2. The question of the role of divine agency in maintaining the world machine was a central point of contention in the epistolary debate between Leibniz and Newton's spokesman, Samuel Clarke. On the Leibniz-Clarke correspondence and the question of God's ongoing intervention in the universe, see Vailati, *Leibniz and Clarke* (1997), chap. 5; Vailati, "Leibniz and Clarke on Miracles" (1995); Shapin, "Of Gods and Kings" (1981); Hall, *Philosophers at War* (1980), 218–23; Kubrin, "Newton and the Cyclical Cosmos" (1967).

3. Turgot to Condorcet, 18 May 1774, in Condorcet, *Correspondance*, 172–73.

inclinations, tended to maintain their relations in a proper state of balance. Here again, in their vision of an overall order emerging from the actions of individual elements, sentimental empiricists projected their understanding of the social world onto the natural world.[4] In contrast, mechanists—who were asocial, according to their critics, in their mode of philosophizing—also held an asocial view of nature, overlooking the importance of the active relations among nature's elements and of the regulation of these relations.

Sentimental-empiricist critics of mechanism considered what they took as its mistakes to be both epistemological and moral. Mechanist imperviousness to the reasons governing the relations in nature meant not only a mistaken physics, but an arrogant and solipsistic self-sufficiency. By contrast, sensitivity to nature's reasons and relations signaled not just a correct physics, but a modest, open, socially collaborative approach to natural science. Knowledge and virtue were inseparable; physics and moral understanding must improve together or not at all, the perfection of each being necessary to the advancement of the other.[5] In reaction against the mechanization of the world picture, in other words, sentimental empiricists sought its moralization.

The Philadelphia moralist Benjamin Franklin provided in his electrical science their favorite example of the good sort of physics, founded on Franklin's assumption of a purposeful natural world. Franklinist physics was essentially teleological, not mechanical, in that its ultimate explanatory causes were final rather than efficient, ends rather than means. Throughout this chapter

---

4. The eighteenth-century notion of an overall social order emerging from the free actions of individuals found its most familiar and powerful expression in Adam Smith's *Wealth of Nations* (1776). However, the idea was older than Smith's application of it and was common in earlier decades in French moral science. For example, the French Physiocrats made it the basis of their own system of political economy around midcentury (the subject of chapter 4). On earlier notions of an "invisible hand" in seventeenth- and eighteenth-century moral and political thought, in particular the writings of Pascal, Montesquieu, Vico, and Mandeville, see Hirschman, *The Passions and the Interests* (1977), 10, 16–17. See also Pack, "Theological Implications" (1995); Ingrao and Israel, *Invisible Hand* (2000), chap. 2; Heilbroner, *Worldly Philosophers* (1980), chap. 3.

5. The view that natural science should make the unveiling of nature's reasons its primary business had the explicit support of Isaac Newton, among others. Despite his authorship of the universal mechanics that unified terrestrial and celestial physics, Newton insisted that the "main Business of natural Philosophy," was not "explaining all things mechanically." Tracing the chain of mechanical influence back toward its source, on pain of infinite regress, one must ultimately arrive at another sort of power. Newton therefore urged his fellows to keep their eyes directed toward "the very first Cause, which certainly is not mechanical." The prime mover in Newton's universe, as in Aristotle's cosmos, was a "Being, incorporeal, living, intelligent, omnipresent." The ultimate questions of natural philosophy concerned this Being's designs: "Whence is it that Nature doth nothing in vain; and whence arises all that Order and Beauty which we see in the World?" The answers to these questions of purpose rather than mechanism would expand moral as well as natural knowledge. The end of natural philosophy would be a full understanding of the first Cause such that "our Duty towards him, as well as towards one another, will appear to us by the Light of Nature." Newton, *Opticks* (1730), query 28, query 31.

and the next, I use "teleological" not in the Christian sense that a well-designed cosmos implies a designer, but in the root Aristotelian sense, to describe causal explanations in terms of purpose. The claim that Franklin's electrical science was teleological will appear odd given his reputation for pragmatism and experimental tinkering. Pragmatic experimentalism and teleology sound like a strange bundle to a modern ear. But they made a coherent package in the context of sentimental empiricism, which cast all-encompassing mechanical systems as the unhappy result of a theorist's imperviousness to living nature. The contrast with strictly mechanical explanations made Franklin's appeals to purposes as physical causes look like a pragmatic independence of mechanist system-building.[6] To locate empirical knowledge not only in sensory experience of nature, but in an open, sympathetic sensitivity to nature's projects, was to make a teleological explanation supremely empirical.

It will no doubt also seem strange that Franklin, who cultivated a reputation for plain speaking and dry humor, should have appealed to people I have been characterizing as sentimentalists. Yet he certainly did. I have been proposing that we must revise our understanding of empiricism, at least during an important part of its history, by recognizing its sentimental and moral dimension. To understand Franklin's appeal to sentimental empiricists requires a complementary revision of our understanding of sentimentalism. If, in the Age of Sensibility, empiricism was sentimental, sentimentalism was also empiricist. This is to say that sentiment was subject to empirical study, and the results, to practical application.[7] In his moralist writing, Franklin took an empiricist's view of sentiment and put his findings to work in redesigning social life.[8] He thus gave sentimental empiricists, seeking a moralized understanding of nature, another side to their coin: an empiricist sentimentalism, a naturalized approach to the moral world. He then carried his naturalist understanding of human desires and tendencies into natural science to describe nature's behaviors, producing exactly the moralized physics that critics of mechanism sought.

6. Franklin himself was therefore able to represent the destruction of philosophical systems as central to his new science: "How many pretty Systems do we build which we soon find ourselves oblig'd to destroy!" Franklin to Peter Collinson, 14 August 1747, in *BFP*, 3:171.

7. Eighteenth-century philosophers' and physiologists' empiricist approach to studying the emotions and the ways in which reformers and revolutionaries then applied the results of these studies in the founding of new social institutions are the subjects of chapter 2.

8. For a classic narrative account of Franklin's empiricist and pragmatic attitude toward human sentiment and morality, see Van Doren, *Benjamin Franklin* (1938), chap. 4. A recent view of Franklin's moralist writing that sets it in the context of eighteenth-century moral psychology and pedagogy is Anderson, *Radical Enlightenments* (1997), chap. 3. For an argument that Franklin approached the living and recording of his own life as the production of an embodied moral and social psychology and philosophy, see Breitweiser, *Cotton Mather and Benjamin Franklin* (1984). On Franklin's paradoxical fusion of moral idealism and moral pragmatism, see Jehlen, "Imitate Jesus and Socrates" (1990).

The resonance of Franklin's teleological empiricism with the new tradition of sensibility in French natural science does much, I think, to explain his extraordinary popularity and influence in France. Many have remarked upon France's love affair with Franklin.[9] I would locate French enthusiasm for Franklinism in the heat of a new struggle against mechanism, with its banishment of purpose, a struggle to institute a morally as well as empirically sensitive physics focusing upon the reasons guiding the relations among nature's parts (§1). This conflict between sensibility and mechanism operated in the electrical debate between Franklinists and followers of the abbé Nollet (§2). In the eyes of sentimental empiricists, Nollet's electrical mechanics came to epitomize a dogmatic imperviousness, and Franklin's electrical teleology, a keen sensitivity to nature's projects.

In the previous chapter, a changing approach to blindness made manifest the eighteenth-century moralization of epistemological questions. The new, moralized epistemology, in which sensations were undetachable from sentiments and ideas inseparable from emotions, engaged philosophical speculation with psychological experimentation and, ultimately, with social reform through civic education. In this chapter, the same moralized epistemology turns electrical science into an example of the contrast between close-minded, self-engrossed mechanism and open, sensitive empiricism, and thereby engages it in an interaction with questions of virtue, good behavior, and the basis of community.

I. POOR RICHARD'S PHYSICS

Franklinist physics rested upon the assumption of an overarching natural purpose: nature's desire to preserve and restore states of balance. In place of a mechanical system, in which each piece moved by constraint alone, Franklin described a dynamic equilibrium of interacting parts, in which the pieces acted willfully but were also constrained by a more general end, the tendency of the whole to maintain its balance (fig. 3.1). The same conservationist teleology and the same interaction between freedom and constraint were at work

9. A perennial question, for example, has been why the results of his electrical experiments were greeted so much more enthusiastically in France than in England. I. Bernard Cohen has suggested that "the American Revolution . . . made Franklin unpopular" in England, and so "it became fashionable to discount his achievement" there. Cohen, *Franklin's Science* (1990), 96. J. L. Heilbron responds that the British reception of Franklin was "perfectly appropriate, neither cool nor hostile, and that it appears ungracious only when contrasted with the French." Heilbron, "Franklin, Haller and Franklinist History" (1977), 546–48. French enthusiasm, not British reticence, was the more peculiar response, and Heilbron traces it to the professional rivalries of Franklin's chief promoter in France. In this chapter I begin with these rivalries, which direct me to an underlying conflict between advocates of mechanism and of sensibility.

in his moralist writing. Franklin advertised America to the French as a place of "happy mediocrity," having neither idle rich nor idle poor, all inhabitants being constrained to earn an active living.[10] The author of *Poor Richard's Almanack* also recommended fasting after feasting and warned, "no gains without pains."[11] To French sentimental empiricists seeking a moralized rather than mechanized natural science, and assuming a natural basis for morality in nature's plan, a physics that balanced will and restraint to maintain proper relations among purposively active parts—a *Poor Richard's Almanack* of electricity—was just the thing.

It was in fact opposition to mechanist physics that motivated the translation of Franklin's letters on electricity into French. The 1752 translation was sponsored by Buffon,[12] who considered mechanist explanations of nature pernicious. He especially disliked the mechanical arrangements of his fellow naturalist Réaumur.[13] Réaumur's protégé, the abbé Nollet, had been fruitfully extending his mentor's methods to the study of electrical action, producing what had become the leading French theory of electricity. Herein lay Franklin's relevance to Buffon's crusade against mechanism: Franklin's account of electrical action contested Nollet's.[14] Against Nollet's electrical mechanics, Buffon counterposed Franklin's electrical teleology.

The translation of Franklin's letters produced a great sensation. Where electrical matters were concerned, before Franklin's intervention, the community of philosophes was fairly peaceful. But in 1753 the *Histoire* of the Academy of Sciences reported that the subject of electricity had recently come to mark an apparently unbreachable rift among *physiciens*.[15] Contemporaries understood the opposing sides of this chasm as Cartesianism, represented by Nollet, and Newtonianism, represented by Franklin. Historians have often adopted this understanding without fully considering the specialized meanings of these contemporary characterizations.[16] Overriding the many similar-

10. Franklin, "Information for Those Who Would Remove to America" (1784), in *Writings*, ed. Lemay, 975–83.

11. Franklin, *Poor Richard, 1742*, in *BFP* (1960), 2:340; *Poor Richard, 1745*, in *BFP* (1961), 3:340.

12. Concerning Buffon's role in the translation of Franklin's letters into French, see Heilbron, *History of Electricity* (1979), 346–52; Roger, *Buffon* (1989), 281–82; and Cohen, *Franklin's Science* (1990), 230n.

13. Buffon, "Discours sur la nature des animaux" (1753), in *Oeuvres philosophiques*, 347; *De la manière d'étudier* (1749), 19–20, 22, 65; Roger, *Buffon* (1989), 122–23, 125, 255–57, 317–20.

14. Heilbron, *History of Electricity* (1979), 346–47.

15. "Sur l'électricité," *MAS* (1753), 6; on the rift caused by Franklin's letters, see Heilbron, *History of Electricity* (1979), 288.

16. "The bitterness of the quarrel between Nollet and Franklin," Cohen writes in *Franklin and Newton* (1956), reflected "...a profound chasm between the metaphysical bases of two theories: the Cartesian and the Newtonian" (387); cf. Heilbron, "Franklin, Haller and Franklinist History" (1977), 546–48. Introducing Franklin's natural science to historians of science in the 1950s, Cohen presented a "Newtonian" Franklin after the Newton of the *Opticks*, the empirical essayer

ities between Descartes's and Newton's systems of philosophy, one difference defined Enlightenment conventions of Newtonianism and Cartesianism.[17] This difference regarded their treatments of natural purpose.[18] Descartes exiled purposes from natural philosophy on principle and restricted himself to mechanism alone; Newton, in declining to give gravity a mechanical cause, reluctantly left a crucial gap where his disciples could fill in the metaphysics of their choosing.

The contrast between Descartes's complete system of mechanical causation and Newton's incomplete one was crucial to Enlightenment Newtonianism. How else can we understand the remarkable tendency, on the part of eulogists of Newton's mechanical system, to cite its breaches? An example is Hume's satisfaction that although Newton "seemed to draw off the veil from some of the mysteries of nature," he also demonstrated "the imperfections of the mechanical philosophy," restoring nature's secrets "to that obscurity in

---

and querist. If Cohen's Franklin was heir to the inaugural tradition of modern science, Heilbron's Franklin is an older-fashioned and more fluid character. He is portrayed by contrast with "the modern scientist" as "the natural philosopher of the Age of Reason." Among his defining characteristics are his "sporadic engagement with experiments and demonstrations, reliance on wide-ranging analogy . . . concern with . . . natural religion, and hope for utility." Heilbron, "Franklin as Enlightened Natural Philosopher" (1993), 2. The Franklin I propose here, the teleologist electrician, had the Newtonian reputation that Cohen describes, but for the Enlightenment reasons listed by Heilbron.

17. On the Enlightenment meaning of "Newtonian," see Buchdahl, *Image of Newton and Locke* (1961), 2; Gay, *Enlightenment* (1969), 2:134; Guerlac, "Where the Statue Stood," in *Essays* (1977), 133; and Hazard, *European Thought* (1954), 137. On the political resonances of Newtonianism and Cartesianism in the Enlightenment, see also Jacobs, *Cultural Meaning* (1980), 45, 54, 61–65, 97, 125. Henry Guerlac has suggested that the Enlightenment struggle between Cartesians and Newtonians "should perhaps be viewed not so much as a contest between rival theories as a confrontation of rival standard-bearers," in "Where the Statue Stood," in *Essays* (1977), 133. Aléxandre Koyré traces this symbolic antagonism to Newton's own insistence upon it in "Newton et Descartes," in *Etudes newtoniennes* (1968), 103, 109. According to Koyré, Newton's representation of himself as Descartes's opposite did not make it so. Newton was the first of a tradition of Newtonians to oversimplify the matter: central elements of his physics were "directly influenced by Descartes" and even "profoundly Cartesian." Guerlac writes similarly that "insofar as there existed throughout the [eighteenth] century such a commonly held 'world view' as . . . 'Newtonian philosophy' . . . it would seem to owe as much to Descartes, and to other influences from the previous century, as it does to Newton." Guerlac, "Newton's Changing Reputation in the Eighteenth Century," in *Essays* (1977), 73. Peter Gay has described how the philosophes then drew upon Newton's representation of Descartes as "his supreme, almost his sole opponent," in order to "construct a Descartes who was the ideal type of the rash metaphysician." Gay, *Enlightenment* (1969), 2:147. See also Home, "Out of the Newtonian Straitjacket: Alternative Approaches to Eighteenth Century Physical Sciences" (1979), in *Electricity*, no. 5:244. Much has been written about the function in Enlightenment philosophy of the symbolic opposition between Newton and Descartes. See, for example, Hazard, *European Thought* (1954), 137.

18. Koyré writes that at the heart of the Enlightenment characterization of Descartes lay the rejection of a "materialism that banished from natural philosophy all teleological considerations." Koyré, "Newton et Descartes," in *Etudes newtoniennes* (1968), 127.

which they ever did and ever will remain."[19] Voltaire, in his *Lettres philosophiques*, popularized for a French audience the contrast between Newton's heroic acceptance and Descartes's dogmatic refusal of obscurity.[20] Newton had set the standard for empiricist openness and modesty by abstaining from assigning a mechanical cause to gravity,[21] and by affirming that the ultimate causes of natural phenomena could not be mechanisms, but must be purposes.[22] Descartes, carried by his materialism to the fantastic plenum of whirlpools and eddies at the heart of his physics, had set the standard for mechanist close-mindedness and anti-empiricist arrogance.[23]

Following this specialized understanding of the difference between Descartes and Newton, which was as much a moral as an epistemological judgment, Franklinists accused Nollet of seeking "to explain everything" and called him vain and opinionated.[24] His insistence on mechanical causes for electrical phenomena—he frequently referred to electricity as the "action of a matter in motion" and the "effect of a mechanical cause"—identified him as a Cartesian, never mind that he explicitly disavowed Cartesianism; Franklin's replacement of mechanical by final causation identified him as a Newtonian, although he went far beyond Newton in his dismissal of mechanical causation.

To begin with Nollet's departures from Cartesianism, he rejected the vortices of electrical matter announced by Cartesian electricians, notably Fontenelle and Nollet's own teacher, Charles-François de Cisternay Dufay. Nollet replaced the vortices with rectilinear currents. He reasoned that since any given electrified body simultaneously attracted some objects and repelled others, electrification must involve two streams of electrical fluid traveling in opposite directions, an "effluent" current carrying repelled objects away from the charged body and an "affluent" current carrying attracted objects toward it. These electrical currents were well demonstrated empirically. In experiments Nollet reported seeing small objects moving to and from an electrified body in radial, diverging and converging lines. When he approached other bodies to the electrified ones, he even saw the electrical matter itself leave the electrified body in little "tufts" of diverging rays[25] (figs. 3.2

19. Hume, *History of England* (1754–62), 6:374.

20. Voltaire, *Letters concerning the English Nation* (1733), 61–66, 74.

21. Gay, *Enlightenment* (1969), 2:133–36.

22. Newton, *Opticks* (1730), query 28, query 31. See supra, n. 5.

23. Koyré, "Newton et Descartes," in *Etudes newtoniennes* (1968), 88, 115, 361–62.

24. Sigaud de la Fond, *Précis historique* (1788), xii, 115. Nolletists, for their part, complained of Franklinists' "misplaced moralizing." Durand, *Le Franklinisme refuté* (1788), 9–10.

25. Nollet, *Leçons* (1759–66), 1:xx–xxi; *Essai* (1746), 67–80. "It is in fact impossible to class Nollet and his fellow French experimental physicists meaningfully as either Cartesian or Newtonian." Home, "The Notion of Experimental Physics in Early Eighteenth Century France" (1985), in *Electricity*, no. 7:111. See also Home, "Nollet and Boerhaave: A Note on Eighteenth Century Ideas about Electricity and Fire" (1979), in *Electricity*, no. 8; and Heilbron, *History of Electricity*

and 3.3). Nevertheless, despite his rejection of Cartesian vortices and his carefully empirical methods, Nollet was destined to play Descartes to Franklin's Newton. By the momentum and impact of the electrical fluid's effluences and affluences, Nollet claimed to be able to explain "all the known facts" about electricity. This emphasis upon matter, motion, and complete mechanical causation sealed his fate.

Meanwhile Nollet's accusations against Franklin helped to identify Franklin as a Newtonian. Nollet's crowning objection to Franklin's theory was that it lacked "truly physical causes."[26] This was because Franklin did not attribute electrical effects primarily to the motion of a fluid. Instead, he referred to the static presence or absence of electrical matter in bodies, to implicit distance forces, and to explicit natural purposes. Franklin assumed the existence of two kinds of matter: common matter, which was mutually attractive, and electrical matter, which was mutually repulsive. These two matters also attracted one another, and in any ordinary object, equal quantities of each were required to balance one another. In the event that too much electricity was present, the extra fluid pooled to form an active electrical "atmosphere." When too little electricity was present, the unbalanced common matter became electrically active. Thus, Franklin explained all signs of electricity by the "wanting" and "abounding" of electrical fluid in bodies and the striving of common and electrical matter to rectify the imbalance.[27] No gain without pain, no feast without a fast. As Newton had demurred from offering an efficient cause for gravitational attraction, Franklin offered none for electrical attraction and repulsion. He never attributed the motion of attracted and repelled objects to the motion of the electrical fluid, or used attraction or repulsion to represent its direction of flow. Even when he used the oscillation of a cork ball between a positive and negative body to demonstrate the equalization of charge, he assumed that the cork ball carried the electrical fire rather than floated in its current.[28] The cause of its motion was its project of rectifying the imbalance.

However, in this preference for purpose over mechanism, Franklin far surpassed Newton. Newton may have left a causal gap at the crux of his system, but Franklin inserted purposes in place of material causes all throughout his. Newton's system, apart from the missing mechanism for gravity, was mechanical,

(1979), 276–77. Empirical observation of the "divergent jets were the foundation, and very likely the immediate inspiration of the système Nollet." Heilbron, *History of Electricity* (1979), 283.

26. Nollet, *Lettres* (1760), 2:13–14; "Conjectures" (1745), 107–8, 110, 139, 151; "Examen" (1753), 492.

27. Franklin, "Opinions and Conjectures" (1750), in *Franklin's Experiments*, 213–18; Franklin to Collinson, 29 April 1749, in ibid., 188.

28. Franklin to Collinson, 1 September 1747 and 29 April 1749, in ibid., 185, 189. See Home, "Franklin's Electrical Atmospheres" (1972), in *Electricity*, no. 11:133; Heilbron, *History of Electricity* (1979), 336–37, 366–67.

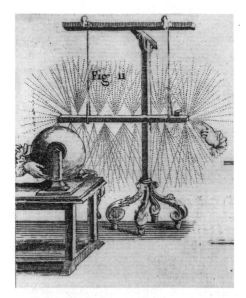

Figures 3.2 and 3.3. Nollet's effluences and affluences, as pictured in his *Essai sur l'électricité des corps*, 2nd ed. (1750). The second plate also shows, at bottom left, a gentleman preparing to shock himself with a Leyden jar, while at the bottom right, a lady generates electricity.

a study of the action of forces upon objects. Franklin's electrical science was not essentially mechanical: he was not primarily concerned with the dynamics of electrified objects acted upon by electrical forces. He was interested in attraction and repulsion only insofar as they manifested the desire of positive bodies to give, and of negative bodies to gain, electricity.

Thus Franklin's physics was not so Newtonian, nor Nollet's very Cartesian, except in the specialized sense dictated by sentimental empiricism: Nollet was a "Cartesian" in that he accepted only efficient causes as causes, while Franklin was a "Newtonian" because he arranged his physics around final causes.

The electrical world, like the social world, Franklin believed, was driven by its own overarching tendency to maintain a state of balance. Traveling between natural science and moral philosophy, Franklin always took along his accustomed ontological and moral scheme of things. The concept of electrical charge as a surplus or deficit of electricity, measured against a neutral, balanced condition, as J. L. Heilbron has observed, parallels the argument of Franklin's "Dissertation on Liberty and Necessity" of 1725. Here, Franklin had supposed that when "*a Creature is form'd and endu'd with Life, 'tis suppos'd to receive a Capacity of the Sensation of* Uneasiness *or* Pain." The creature's desire to be freed from pain was the "Spring and Cause of all Action"; life was a succession of escapes from discomfort. These escapes caused pleasure, and since "the *Desire* of being freed from Uneasiness is equal to the *Uneasiness,* and the *Pleasure* of satisfying that desire equal to the *Desire,* the *Pleasure* thereby produc'd must necessarily be equal to the *Uneasiness* or *Pain* which produces it."[29]

Though Franklin "later repudiated this doctrine, as useless or mischievous . . . he by no means abandoned its form," but applied the same sort of bookkeeping to myriad other subjects. "Franklin's difficulty in conceiving of accumulation without compensatory deficits," writes Heilbron, "indicates the strength of the habitual mode of thought that created the concepts of plus and minus electricity."[30] It also indicates Franklin's inclination to view both the natural and the social worlds as ordered by an interaction between individual and cosmic purpose: the end of the whole to maintain balanced relations among its parts, and the volitional behavior of the parts, which realized the balance.

Being primarily interested in the desires and tendencies of the electrical fluid, Franklin departed from the early study of electricity, which focused upon attraction and repulsion. These phenomena lent themselves to mechanist effluvial theories, with effluent and affluent fluids carrying attracted and repelled bodies in the required directions. Franklin, unconcerned with electrified objects' means of transportation, dismissed effluvia and used attracted and repelled objects merely as signs of electrification. Even then, he preferred the more fiery variety of signs, such as sparks and glows. Joseph Priestley, translating Franklinist practice into doctrine, said that attraction and repulsion constituted poor empirical evidence. They appealed to the mechanist fondness for reducing the world to moving chunks of matter. Priestley predicted that the days of the mechanists' effluvial theories were numbered now that electricity had begun "to make itself sensible to the smell, the sight, the touch and the hearing: when bodies were not only attracted and repelled, but made to emit strong sparks of fire, . . . a considerable noise, a painful sensation, and a strong phosphoreal smell."[31]

Franklin directed his Philadelphia experiments precisely at coaxing the electrical fire to make itself sensible. Showing the "wonderful effect of pointed bodies," drawing off and throwing off electrical fire, Franklin noted that in the dark, one could see the fire "continually running out silently at the point."

---

29. Franklin, "Dissertation on Liberty and Necessity" (1725), in *BFP*, 1:57–71.
30. Heilbron, "Franklin as an Enlightened Natural Philosopher" (1993), 13.
31. Priestley, *History* (1775), 15, 16, 18.

Likewise smoke from dry rosin on hot iron was attracted to form atmospheres around positively electrified objects, "making them look beautifully, somewhat like some of the figures of *Burnet's* or *Whiston's Theory of the Earth.*"[32] The Franklinist reliance upon sparks and glows, so much at odds with all previous assumptions about the importance of attraction and repulsion, looked to Nollet like deliberate provocation. He responded with exasperated disbelief that there was a certain "affectation in rejecting attraction and repulsion as equivocal signs."[33]

In Nollet's view, Franklin pushed aside what should have been at the very center of his science, electrical motions, focusing instead upon flashes and winds and vapors. Moreover, Franklin attributed these effects to static surpluses and deficits of fire. Nollet protested that electrical matter could explain nothing on its own, just as "the air is not the wind." Matter itself did not constitute a physical cause; it had to be set in motion. Franklin confused "the subject with its modifications." Electrical effects were caused, Nollet insisted, not by the fluid itself, but by the fluid's "movements."[34] Moreover, attention to these movements, he maintained, revealed fatal flaws in Franklin's theory. How could a body repel things by receiving matter, as Franklin's theory implied a negatively charged body must do? How could a body attract things by giving forth fluid, as a Franklinist positive body must? More generally, how could a single current "cause movements opposite to its own?" Nollet pronounced Franklinism conceptually incoherent: it confused effects with causes and accidents with necessities, and, above all, it conflated matter and motion.[35]

Franklin and his supporters, as Nollet was quick to note, recognized the inconsistencies and missing mechanisms in their theory.[36] But in their most successful responses to mechanist critics, they did not try to resolve these difficulties in mechanist terms. Instead, they responded on their own terms, devising ingenious experimental displays of electricity's purposeful behavior.

The effects of negative electricity posed the gravest conceptual problem for Franklinist theory.[37] True, Newton had declined to give gravity a cause, but he had never hinted at so mysterious a cause as a negativity. In fact, Franklinist explanations in terms of negative and positive electricity were not even internally consistent. F. U. T. Aepinus, an ambivalent Franklinist at the

32. Franklin to Collinson, 11 July 1747, in Franklin, *Franklin's Experiments*, 173. Later Franklin used the same means to demonstrate that electrical atmospheres followed the shape of the bodies they surrounded. Franklin, "Opinions and Conjectures" (1750), 216.

33. Nollet, "Examen" (1753), 487.

34. Nollet, "Réponse au supplément" (1753), 511.

35. Nollet, "Examen" (1753), 486; Nollet, *Lettres* (1767), 3:159–60, 243.

36. Nollet, "Examen" (1753), 481.

37. Heilbron, *History of Electricity* (1979), 337.

Berlin Academy of Sciences, pointed out that if ordinary matter attracts electricity, and electricity repels itself, then neutrally charged bodies, containing a balance of normal and electrical matter, should attract one another. The electrical matter in each would attract the common matter in the other, overwhelming the mutual repulsion of the electrical matter in each—unless, Aepinus speculated, common matter was mutually electrically repulsive as well. That would have the advantage of explaining minus-minus repulsion, though it would have the disadvantage of reversing Newton.[38]

More generally, it was "difficult to imagine" how negative electricity could cause the various phenomena of electrification, that is, how a lack of something could cause anything. This was especially problematic in the case of the mutual repulsion of two negative bodies, for how could two absences repel one another?[39] The Philadelphia Franklinist Cadwallader Colden was troubled by this difficulty, and Franklin's "Meteorological Observations," addressed as a response to Colden, opens with the confession "that it seems absurd to suppose that something can act where it is not." Ebenezer Kinnersley, another Philadelphian, wrote to Franklin proposing to do away with repulsion altogether, and to explain all apparent instances of it by "the mutual attraction of the natural quantity [of electricity] in the air, and that which is denser or rarer" in electrified bodies. Franklin agreed with equanimity that minus-minus repulsion was "difficult to be explained," and indeed that "attraction seems in itself as unintelligible as repulsion."[40] But he did not labor to resolve this unintelligibility in the mechanics of his electrical theory. Instead, he relied upon its teleology, which was perfectly intelligible. Negative bodies were electrically active because they needed electrical fire. They repelled one another because they had nothing to offer each other.

Franklin's followers too, for the most part, argued on behalf of negative electricity in other ways than by trying to make it intelligible to a mechanist. A notable example is Jean-Baptiste Le Roy, the leading French Franklinist electrician, who showed great versatility in elaborating upon Franklin's own original logic to respond to Nolletist challenges. Franklin had initially established the existence of symmetrical electrical states by the appearance of two intensities of sparks, stronger ones between negatively and positively charged bodies, and weaker ones between either and a neutral body.[41] Le Roy improved upon these initial displays to show that the glass globe of an insulated generator became

38. See ibid., 388–89; Heilbron, "Franklin, Haller and Franklinist History" (1977), 548–49.

39. Le Roy, "Mémoire sur l'électricité" (1753), 449–50.

40. Franklin to C. Colden, 6 December 1753, in *BFP*, 5:145; Ebenezer Kinnersley to Franklin, 12 March 1761, in Franklin, *Franklin's Experiments*, 349–50; Franklin to Kinnersley, 20 February 1762, in ibid., 365–66.

41. Franklin to Collinson, 11 July 1747, in ibid., 175.

electrified symmetrically to its framework, the globe positively and the framework negatively: the sparks that the globe and framework gave one another were twice as great as the sparks they gave to a person on the floor.[42] Furthermore, if either the globe or the framework were grounded, sparks from the other to a person on the floor were greater, which supported Franklin's balance hypothesis that more electricity could be fed into the positive side of the apparatus as more could be drawn from the negative side, and vice versa.

Next Le Roy went beyond spark intensities to reveal symmetry by means of glows and brushes of fire. Rods presented to the conductor and framework by one end, and to one another by the other, showed a brush of fire on the conductor side and a glowing point on the framework side. Le Roy surmised that the electrical fire converged to a glowing point on its way into a rod and diverged in a brush as it exited. [43] Thus the brush showed fire moving out of the overfull conductor, and the glow marked its entrance into the depleted framework.

To explain minus-minus repulsion, Le Roy's tactic was to argue that repulsion never happened by means of a fluid in motion, even in the more intelligible cases that involved positive charges. He pointed out that one could feel no breath of wind, nor see any glow or flow of electrical matter between two mutually repulsive bodies. On the contrary, when two positive bodies were opposed to one another, they would make one another's visible brushes of electrical fire disappear "like snuffing out a candle." Le Roy argued that this absence of any sign of effluence between positively charged, mutually repulsive bodies cast doubt on the notion of a fluid cause for repulsion in general.[44] However, wind and sparks were exactly the sorts of evidence that Nollet distrusted as variable, unreliable impressions, because they indicated no directional flow of fluid. In response to Le Roy's argument, he pointed out that there was indeed an undeniable sign of effluence between two electrified bodies: mutual repulsion itself.[45]

In short, Nollet and Le Roy were talking right past one another. The quandaries of mechanical causation that Nollet pointed out were of relatively little interest to Franklinists. Of primary importance was simply that electrically

42. This observation had ambiguous implications, however. In principle, the amount of electricity exchanged between the positive globe and the negative framework should be the same as the amount exchanged between the positive globe and a grounded person: just the globe's excess electricity. Still, the greater effects exhibited between the two electrically charged parts of the apparatus did seem to suggest contrary states of electrification.

43. Nollet wrote, "Mr Leroy me prouve par des raisons qu'il appelle luy meme des raisons d'analogie, . . . quil passe de la matiere Electrique dans les pointes ou l'on voit paroitre un point lumineux." Dossier Nollet, AS.

44. Le Roy, "Mémoire sur l'électricité" (1753), 455–66.

45. Nollet, "Examen" (1753), 500, 487.

charged objects moved from one place to another in order to restore an equal distribution, never mind what carried them, affording an "occasion of adoring that wisdom which has made all things by weight and measure!"[46] Nollet deemed this teleology a perversion of good Newtonian philosophy. When the Italian Franklinist Giambattista Beccaria invoked Newton's stricture against hypotheses to justify a physics of motions without mechanical causes, Nollet wrote: "I find the rule of Newton very good, Reverend Father, as long as one doesn't abuse it." For Nollet, Beccaria's proposal to explain electrical attraction by the need of full bodies to give, and of empty bodies to receive, electrical matter represented a clear abuse of Newton's example. "If I saw an inanimate object move itself toward another, for the sole reason that it lacked what the other could give it, I would believe I had seen a miracle."[47]

To Franklin, however, it was no miracle but an observable fact that the electrical fire acted purposefully. He discerned in its willful behavior a source of answers to problems posed by a standard mechanics of matter in motion. This departure from mechanism was bound to please Buffon. In the first volumes of his *Histoire naturelle, générale et particulière*, written with the collaboration of Louis Daubenton, Buffon had criticized the mechanist theory of life put forth by Nollet's mentor, Réamur. In particular, Buffon disliked Réaumur's Cartesian assumption that life was imposed as a rational scheme upon brute matter and must be understood in terms of "simple inorganic parts." Buffon deplored this reduction of life to lifelessness. He viewed each part of an organic being as homologous to the whole, so that the whole defied mechanistic analysis. To the sentimental-empiricist natural historian, life was no mechanical arrangement, but the palpable result of organic matter's tendency to live.[48]

Living matter, according to Buffon, was imbued down to its smallest parts with vital force, a "penetrating" and "active" power. In the late 1740s, he and another collaborator, the abbé Needham, had tested their idea that organic matter was permeated with soul, consciousness, and a vegetative or reproductive tendency. In a series of experiments involving the microscopic study of mixtures including almond seeds, crushed wheat, and meat juices, they claimed to have triggered spontaneous generation.[49] Réaumur responded in kind. He spent the holidays of 1751 with his cabinet keeper, Mathurin-Jacques

46. Franklin, "Opinions and Conjectures" (1750), 214–15.

47. Nollet, *Lettres* (1760), 2:161–62.

48. On eighteenth-century physiologists' turn against mechanism and toward "forces in some way immanent in matter and acting perhaps . . . according to their own teleological principles," see Moravia, "From 'Homme machine' to 'Homme sensible'" (1978) (quote from 49.) See also Kiernan, *Science and Enlightenment* (1968).

49. Buffon, *Histoire naturelle* (1750), 2:7–9, 15–19, 28, 154–56; Needham, *Nouvelles observations* (1747), xi, 190–98.

Brisson, and the Jesuit naturalist Lelarge de Lignac, refuting Buffon and Need-ham's experiments. Lignac, their chosen spokesman, brought out a three-volume attack on Buffon's natural history entitled *Lettres à un Américain sur l'Histoire générale et particulière de M. de Buffon.*[50] The disagreements between the two teams were basic. Buffon and his collaborators subscribed to the view that nature was no rational system but "a state of sublime disorder" to whose irreducible behaviors the naturalist must be receptive. The rationalist Réau-mur responded by calling their primordial stews "chaos."[51] As for Buffon and Needham's animating "vegetative force," Lignac rejected it for violating New-tonian force laws.[52]

By the end of 1751, the two sides were at loggerheads: Buffon, Daubenton, and Needham faced Réaumur, Brisson, and Lignac; three volumes of *Lettres à un Américain* answered three volumes of *Histoire naturelle*; a sublime, irreduc-ible, vitalist ontology confronted a rationally ordered, mechanist system of nature. Meanwhile Buffon had been preparing to reveal his own American letter writer: Franklin, who, like Buffon, believed in a monist suffusion of soul in matter, an irreducibly animate material world. Several years earlier, Frank-lin had taken on Andrew Baxter's *An Enquiry into the Nature of the Human Soul*, in which Baxter described a Malebranchian dualist universe of passive mat-ter and active, immaterial substance.[53] Franklin argued against any such divi-sion, maintaining that matter was itself essentially capable of thought, and that "if any part of Matter does not at present act and think, 'tis not from an In-capacity of its Nature but from positive Restraint."[54]

While Buffon had been mixing his vital concoctions, Franklin had been working on his first electrical experiments. The "extremely subtle" matter described in Franklin's letters, like the force that infused living matter in Buf-

---

50. Torlais, *Esprit* (1936), 238; Roger, *Buffon* (1989), 256; Lignac, *Lettres* (1751).

51. "Cabinet" (1751),in Diderot and d'Alembert, eds., *Encyclopédie*, 2:490; Réaumur to Charles Bonnet, 14 March 1754; cited in Torlais, *Esprit* (1936), 243. Louis Daubenton's anatomical descrip-tions of the king's cabinet accompanied Buffon's natural history, and the article "Cabinet" was drawn largely from Daubenton's text. In a similar phrase, Buffon wrote that nature's "disorders" excited his "whole admiration." Buffon, *De la manière d'étudier* (1749), 17.

52. Needham claimed the vegetative force arose from "two simple and contrary forces, a force of resistance and a force of expansion." He thereby implied, as Lignac pointed out, that equal and opposite forces summed to a positive result rather than counterbalancing one another in proper Newtonian fashion. Next, against Buffon's distrust of mathematical abstractions, Lignac countered that to perceive was to abstract: each sense selected its qualities; moreover, vision geometrized an irregular world into regular shapes so that, for example, one saw a die as a perfect cube. Finally Lignac's patience was fully exhausted by Buffon's belief that parts of an organism were homologous to the whole, such that a horse was made of infinite tiny horses. Lignac, *Lettres* (1751), letter 8:43–44; letter 10:13; letter 11:12–14; letter 12:107; letter 6:6.

53. Baxter, *Enquiry* (1737), 79–210. See Yolton, *Locke and French Materialism* (1991), 163–64.

54. Franklin to Thomas Hopkinson?, 16 October 1746, in *BFP*, 3:84–89.

fon's natural ontology, permeated common matter "with such ease and freedom as not to receive any perceptible resistance."[55] Franklin's electrical fire was particularly apt to Buffon, who had recently claimed that experimental physics had been misconstrued as the demonstration of mechanical effects. Repudiating such an approach to experimental physics, he maintained that its true aim should be to explain "all of those things which we cannot measure by calculating."[56] This domain for experimental physics would be abandoned in the next century, when physicists, notably William Thomson, Lord Kelvin, would lay claim precisely to its complement. But Buffon's recommendation was borne out by Franklin's electrical science. Franklin's electrical matter did not act by a standard mechanics of matter in motion, by momentum and impact. Instead, it influenced the world it saturated by means of its own idiosyncratic preferences and tendencies. Franklin invoked his electrical fluid to explain things that, he thought, defied mechanical explanation. An example was the perforations in a church weathervane that had been struck by lightning. These showed that the lightning had followed a crooked path, as it "most easily turned to follow the Direction of good Conductors."[57] This ability to choose its path implied that lightning was unlike a solid projectile: it seemed not to have momentum. Standard mechanics could not explain its trajectory, which was explicable, instead, by the electric fluid's preference for certain materials over others.

Franklin's expansive electrical fluid and self-compressing common matter behaved in ways that resembled the expansions and resistances of Buffon and Needham's organic matter.[58] Both were governed more by appetites than by forces, requiring one another, like the pairs of contrary elements in an Aristotelian compound, as hunger requires food.[59] Buffon, who approved of Aris-

---

55. Buffon, *Histoire naturelle* (1750), 2:17, 156; Franklin, "Opinions and Conjectures" (1750), 213–15.

56. Buffon, *De la manière d'étudier* (1749), 18, 72–73. See also Diderot, "Art" (1751), in Pons, ed., *Encyclopédie*, 1:252–53.

57. Franklin to Jan Ingenhousz, 21 June 1782 (dated by the Benjamin Franklin Papers Archive, Yale University), in Franklin, *Writings*, ed. Smyth, 7:88–97. Similarly, Franklin proposed that light was a "subtle elastic Fluid," since if it consisted of material particles, their very great speed would give them the momentum of cannonballs. Also, the sun would shrink as it gave off light, weakening its hold on the planets. Franklin to Cadwallader Colden, 14 May 1752, in *BFP*, 4:310–12.

58. Heilbron notes the nonmechanical mode of action of electrical atmospheres: "The static character of the atmospheres implied a step away from mechanism; only effluvia in motion can cause action by impact." Heilbron, *History of Electricity* (1979), 336–37. Of the Franklinist avoidance of mechanistic explanations, he writes that Franklinists "differed from their opponents in eschewing mechanical analogies to attraction and repulsion" (367).

59. Aristotle, *On Generation and Corruption*, II.8, 335a5–16. Aristotle attributed "appetites" only to mind, and not to matter, but he did say that "food is akin to the matter, that which is fed is the 'figure'—i.e. the 'form.'"

totle as a "better Physicist" than Plato,[60] had reason to admire Franklin's physics and to find it familiar. A depletion of electrical fluid in common matter, like a scarcity of Buffon's nutritive material in a living body, resulted in a violent "hunger" to be filled. Common matter, with its voracious appetite for electrical fluid, was "a kind of spunge to the electrical fluid."[61]

The active electrical "atmospheres" in Franklin's theory, the pooling of electrical overflow upon a body's surface resulting from an excess of fire within it,[62] also had an analogue in Buffon's theory of reproduction. When Buffon's animating fluid was present in greater abundance than was necessary for a body's nutrition, the excess was pooled in "reservoirs." The particles of animating fluid in these reservoirs served the purpose of reproduction.[63] Franklin explained electrical activity as Buffon explained nutrition and reproduction, in terms of lack and excess, and a tendency toward balance and repletion. Electrical and organic matter acted according to their inclinations; at the same time, these very inclinations worked to regulate their activities and to conserve a dynamic order, whose preservation was nature's guiding purpose.

These appetites, desires, and tendencies were no more directly observable than the cosmic vortices of Cartesian physics, and yet they were acceptable to sentimental empiricists. Franklin, who frequently declared his abhorrence of hypotheses,[64] relied heavily upon assumed natural purposes. Electrical positivity and negativity themselves, and the direction of electrical flow between the two states, are prime examples. "We know that electrical fluid is *in* common matter," Franklin reasoned, "because we can pump it *out*" by using a spinning globe generator or by rubbing a glass tube.[65] Sparks caused by contact between an emptied and an overfilled body showed the restoration of equilibrium: they were the electrical fire traveling from the replete to the depleted body.

Yet sparks do not show direction. Only Franklin's theory of electrical transfer led him to surmise that a spinning globe was provoked, by the rubbing of the cushion, to take in electrical fire and was therefore overfilled, or positively charged.[66] Franklin acknowledged this lack of direct empirical support for his notions of electrical positivity and negativity in an account of his research into the charge on thunderclouds. Having charged one glass phial from

60. Buffon, "Systèmes sur la génération" (1749), in *Oeuvres philosophiques*, 258.

61. Franklin to Collinson, 1 September 1747, in Franklin, *Franklin's Experiments*, 181; Franklin, "Opinions and Conjectures" (1750), 213–14.

62. Franklin, "Opinions and Conjectures" (1750), 214.

63. Buffon, *Histoire naturelle* (1750), 2:7, 154.

64. Franklin to Perkins, 13 August 1752, in *Writings*, ed. Smyth, 3:96; Franklin to abbé Soulavie, 22 September 1782, in ibid., 8:601; Franklin to unknown correspondent, 14 June 1783, in ibid., 9:52–53.

65. Franklin, "Opinions and Conjectures" (1750), 214.

66. Franklin to Collinson, 11 July 1747, in Franklin, *Franklin's Experiments*, 175.

a spinning globe generator and the other through a lightning rod, he hung a cork ball between them and by its oscillation concluded that they were differently charged. On the assumption that the globe was positive, he concluded that thunderclouds must be negative. This meant that "in thunder-strokes, 'tis the earth that strikes into the clouds, and not the clouds that strike into the earth." This piece of counterintuition, Franklin said, would not trouble the experienced electrician, who would understand that whether the clouds were positive and the earth negative or vice versa, there would ensue "the same explosion, and the same flash between one cloud and another, and between the clouds and mountains, &c., and same rending of trees, walls, &c., which the electric fluid meets in its passage, and the same fatal shock to animal bodies." [67]

That is to say, all appearances would be identical, whether the earth was positive and the clouds negative or the reverse. These states of emptiness and fullness, the very crux of Franklin's electrical physics, were simply not observable. But this did not disturb the empiricist Franklin and his supporters. Why were electrical negativity and positivity not "hypotheses" in the pejorative sense? Only because they represented a teleology rather than a mechanics. Cartesian-style mechanism had banished teleological considerations from natural philosophy. Sentimental empiricists, condemning mechanist speculation, welcomed teleology back in. Wherever they disallowed hypothetical efficient causes, in their place they attributed speculative motives to nature.

Franklin's dramatization of the tendency of different electrical states to move toward equilibrium, when they were allowed communication through a conductor, offers a nice, final example of the essentially teleological nature of his physics. He made a cork ball oscillate between two wires, one attached to the positively charged top of a condenser, and the other to its negatively charged bottom (fig. 3.4). Rather than attributing the oscillation to the motions of the electrical fluid as it carried the cork along in its path, Franklin did the reverse. He attributed the motion of the fluid to the motion of the cork, investing the cork with the purpose of equalizing the two electrical states, "fetch[ing] and carry[ing] fire from the top to the bottom" of the bottle. (Indeed, the apparently lifelike behavior of the cork inspired Franklin to give it legs made of linen and fashion an "artificial spider." [68]) In place of a mechanical cause, Franklin gave the cork a project, a role to play in furthering nature's guiding plan.

67. Franklin to Collinson, September 1753, in ibid., 271.

68. Franklin to Collinson, 1 September 1747, in ibid., 183; Franklin to Collinson, 11 July 1747, in ibid., 177.

II. POOR RICHARD'S LEYDEN JAR

Franklin's teleological empiricism proved peculiarly effective when applied to the central apparatus of his physics, the Leyden jar.[69] This was a glass jar filled with water into which (according to the prevailing theory) electrical matter was fed through a wire.[70] After the jar was charged, if one touched both the wire and the bottle at the same time, one received a massive shock. One could also draw sparks off the outer surface of the jar during charging. The mystery was that the jar had to be grounded, not insulated, during charging, in order to get a major explosion by discharging it. This seemed counterintuitive to those who reasoned like Nollet: If the jar was grounded, why did the electricity not all run out the bottom, since, as attraction across glass screens suggested, glass was permeable to electrical matter? Why should the jar charge at all? Why should there be any electrical effects on the outer surface of the glass, let alone the violent commotion during the discharging?

Crucial to Franklin's account of the Leyden jar were his willingness to assign qualities to electrical and other substances with no (or with indifferent) mechanical justifications and his assumption that charge was conserved. The operative qualities were an alleged "springiness" of electrical matter and an inability to penetrate through glass. These caused the electrical fire to compress inside the bottle. Conservation of charge dictated that the surfeit of electricity inside it correspond with a symmetrical negativity on its outer

Figure 3.4. Franklin's industrious cork ball, suspended between wires extending from the positively charged inside of a Leyden jar and its negatively charged outside, fetches and carries the electrical fire back and forth until equilibrium is restored. From Franklin, *Experiments and Observations on Electricity* (1751), fig. 2. Courtesy of the Bancroft Library, University of California, Berkeley.

69. The Leyden jar, the first capacitor, was named for the city in which it was inadvertently invented when a visitor to the laboratory of Pieter van Musschenbroek picked it up from its insulating stand, thereby grounding it. Previously, electrical experimenters had assumed that in order to collect electrical fire in a jar, they needed to insulate the jar in order to prevent the electricity from running out the bottom. See Heilbron, *History of Electricity* (1979), 312–16.

70. On the design of the Leyden jar, see Hackman, *Electricity from Glass* (1978), chap. 5.

surface: as much fluid could be fed into the inside as could be forced from the outside. Franklin explained the Leyden shock provocatively as the rush of the "abounding of fire" within the jar, which "presses violently" to ease the "hungry vacuum" without, where the "wanting" of fire "seems to attract as violently in order to be filled."[71]

Nollet objected that springiness was a "vexatious" supposition, which violated sound philosophical practice by multiplying hypotheses beyond, or even counter to, empirical evidence. A flexible electrical matter belied the "extreme speed" with which it made its effects felt.[72] Furthermore, for electricity to be compressed in electrified bodies, the bodies would have to be "vessels incapable of letting [the electrical matter] escape," which they clearly were not, since electrical effects could be transmitted through or across them.[73] Even allowing the electrical fluid its vexatious property of elasticity, the bouncing back of the fluid would not suffice to explain the Leyden effect, Nollet protested. Such a brusque, sharp sensation could only be caused by a collision, the impact of two streams moving through him in opposite directions, one from the wire and one from the bottle: "The spark bursts with a sort of precision," he insisted, "the inflammation, the noise and the pain ... do not proceed by degrees.... [T]he effect is all that it can be in the instant it appears."[74] A unidirectional flow could never cause a sudden, sharp commotion. Nollet had no good answer to the central problem of how, if the jar was grounded and the glass permeable, there could be electrical effects on the sides of the bottle. But he maintained that Franklin's replete and hungry surfaces provided no mechanism for an instantaneous effect.

As for the "alleged impermeability of glass," Nollet likewise rejected it "out of repugnance for useless objects and forced hypotheses." Impermeability was contradicted by observations of electrical effects between charged bodies separated by glass. It had the further problem of implying that all of the "great quantity of electrical emanations" from the Leyden jar's exterior surface originated there, rather than in the generator. How could an uncharged surface be made to produce such a great quantity of electrical matter?[75] Finally, there was

71. Franklin to Collinson, 29 April 1749 and 1 September 1747, in Franklin, *Franklin's Experiments*, 188, 181. The modern definition of a capacitor is two conductors separated by an insulator. In the case of the Leyden jar, the conductors are the water and the jar's conductive stand. According to the modern explanation of capacitance, it works by electrostatic induction: the negative charge on one conductor (the water, in the case of the Leyden jar) induces a positive charge on the other (the stand), or vice versa, setting up an electrostatic field in the insulator (the glass), which stores energy.

72. Nollet, *Lettres* (1760), 2:57; "Examen" (1753), 494–95.

73. Nollet, *Lettres* (1760), 2:214–15.

74. Nollet, "Examen" (1753), 489–90.

75. Nollet, *Lettres* (1767), 3:172–73, 1:71.

no plausible reason why electrical fire should be able to move into and within glass but not through it. Franklin attempted an explanation of this difficulty, suggesting that the "texture [of glass] becomes closest in the middle, and forms a kind of partition, in which the pores are so narrow, that the particles of the electrical fluid, which enter both surfaces at the same time, cannot go through... yet their repellency can." [76] Even Franklin later rejected this hypothesis, having ground away more than half the thickness of a piece of glass and found it as good as before for making a condenser, a test also suggested by a Nolletist. [77]

Never mind. The outer surface of the Leyden jar emptied itself of electricity in order to maintain a balance with the inner surface, however the particles of glass might be configured: "So wonderfully are these two states of electricity, the plus and the minus, combined and balanced in this miraculous bottle! situated and related to each other in a manner that I can by no means comprehend!" [78] The miraculous balance was experimentally indicated. Cadwallader Colden's son David, also a Franklinist, showed that the bottle could be charged only to a certain point. This limit implied that the jar could achieve a state of fullness such as Franklin described. (The frontispiece to this chapter, fig. 3.1, illustrates another of Colden's demonstrations of the Leyden jar's continual preservation of a balance.) But Colden did not seek a mechanism to explain the jar's effects. At the end of his rejoinder to Nollet, he admitted that the Leyden jar, even as explained by Franklin, was "still a mystery not to be accounted for." [79] Sigaud de la Fond, sympathetic to the Franklinist program, likewise confessed "in good faith that [Franklin's theory] seems at first glance so paradoxical, that it can repel the mind of the reader." But he, too, embraced the contradictions of the Leyden jar. Though the idea of conservation seemed a paradox, "it is upon this paradox that Doctor Franklin's entire theory is founded . . . and it is this paradox that we propose . . . to establish as an incontestable truth." [80]

The establishment of what appeared to be a paradox as an incontestable truth expressed an acceptance of the limits of mechanical explanation and was thus a rebuke to Cartesian arrogance. So, too, was the unexpected behavior of electricity; propagandists of modest, sensitive philosophy placed a premium on accidents. Nollet was critical of the contemporary tendency to celebrate the unforeseen: "It is to chance, they say, that we owe a large part of our discoveries. I admit that this is true up to a point," but "chance presents itself indifferently to everyone," so its results must depend ultimately upon the philoso-

76. Franklin, "Opinions and Conjectures" (1750), 230.

77. Franklin to Dr. L., 18 March 1755, in Franklin, *Franklin's Experiments*, 333; Romas, *Oeuvres* (1911), 138.

78. Franklin to Collinson, 1 September 1747, in Franklin, *Franklin's Experiments*, 181.

79. David Colden, "Remarks on the Abbé Nollet's Letters," in ibid., 285–86, 290.

80. Sigaud de la Fond, *Précis historique* (1788), x, 302.

pher's exploitation of it.[81] To Priestley, however, the exploitation of chance meant simply a readiness to be humbled by it. He found that electrical research had been cursed by the virulent "species of vanity" of its students, but complementarily blessed by its inclination to baffling, theory-defying accidents.

The electrical shock was a shock in all senses of the word, as "surprising" as any discovery of Newton's. And the Leyden phial, which had been invented entirely by mistake (its inventors unintentionally grounded their bottle, though they assumed it ought to be isolated)[82] represented the pinnacle. It had defeated all pre-Franklinist theories "by exhibiting an astonishing appearance, which no electricians, with the help of any theory, could have foreseen."[83] Priestley admired the Leyden jar and its celebrated interpreter for their humbling and humble tendency to privilege inscrutable purpose over intelligible mechanism. By defying logic they were true to sense, and so to virtue.

III. POOR RICHARD'S SCIENTIFIC METHOD

The sentimental empiricist's natural world, imbued with life, will, and purpose, implied a scientific method. This method included a particular kind of mathematics, one whose goal was "moral" rather than mathematical certainty; a distinctive mode of reasoning, by "analogy"; and an ethic of what a number of historians have characterized as philosophical modesty.[84]

To begin with mathematics, a natural world governed by purpose rather than mechanism demanded a calculus of relative rather than absolute values. Buffon argued that in nature, as in social life, a numerical value had no instrumental meaning in isolation, but became meaningful only in relation to other values. Thus he urged natural philosophers to renounce their quest for mathematically certain knowledge and to be content with a lesser kind of certainty, but one better suited to understanding nature and human behavior alike: moral certainty.[85] While mathematical certainty was absolute and com-

---

81. Nollet, *Leçons* (1759–66), 1:lxxiii.

82. Heilbron, *History of Electricity* (1979), 312–16. See n. 69.

83. Priestley, *History* (1775), 2:20; 1:xv, 192–93.

84. Keith Baker describes a great "chorus in denunciation of the 'spirit of system' and in praise of... epistemological modesty." Baker, *Condorcet* (1975), 87 (see 87–95). Gay similarly characterizes the "Enlightenment's campaign against system-building and in behalf of 'philosophical modesty.'" Gay, *Enlightenment* (1969), 2:138–39; cf. 2:134–35. On the importance of modesty in the rhetoric of early modern science, see Shapin and Schaffer, "The Modesty of the Experimental Narrative," in *Leviathan* (1985), 65ff.; and Shapin, *A Social History of Truth* (1994), chap. 4.

85. During the second half of the preceding century, rational theologians and natural philosophers seeking a happy medium between the twin dogmas of radical skepticism and scholastic certainty had developed a tripartite division of probabilities into mathematical, physical, and moral. Accordingly, mathematical certainty was absolute and compelled, but unattainable in matters of natural philosophy, religion, and society. Here, inferior forms of certainty—physical

pelled, it was unattainable with regard to events in the real world. "The absolute, of whatever kind," Buffon wrote in his essay "Moral Arithmetic," "is of the domain neither of nature nor of the human mind." Moral certainty was relative, but precisely for that reason, it was appropriate to nature's purposes, to those of human beings, and especially to the human purpose of studying nature's purposes: "All our knowledge is founded upon proportions and comparisons." It took a most abstract, insensible thinker to cling to the ideal of mathematical certainty. Buffon wrote that absolute certainty resembled other abstract mathematical concepts like infinity. These were "privative ideas," arrived at by imagining a real object and then removing its sensible qualities. The manufacture of absolutes by means of this process of abstraction was responsible for the gravest philosophical mistakes, which would be remedied by a calculus oriented toward moral rather than mathematical certainty, relative rather than absolute values.[86] Diderot agreed: he announced in his *Encyclopédie* article "Probabilité" that he would "speak here only of moral certainty," because that was sufficient for "a wise man."[87]

Adopting a calculus of moral certainties was not only wise but good: it had moral benefits, while an insistence on absolute, mathematical certainty was morally detrimental. A person who subordinated quantitative evaluation to concrete purposes, who understood all values as meaningful only in relation to other values, was thereby integrated into a world of interacting agents: electrically overcharged and undercharged bodies, market exchanges of money for goods. In contrast, one who saw the world in terms of discrete, absolute values was blind to the interaction of its parts and carried out his reckoning in abstracted isolation. "The Miser is like the Mathematician," Buffon wrote, since "both estimate money by its numerical quantity." In contrast, the "sensible man" measured money only comparatively, in relation to its uses.[88] For moral and not only epistemological reasons, mathematics should be always subordinated to purpose and related thereby to a world of physical and social interaction.

Franklinist electrical science, as we have seen, rested upon relations rather than absolutes: emptier and fuller, less and more. In general, Franklin's uses of mathematics were in keeping with Buffon's recommendations. In his *Autobiography*, Franklin reports having "twice failed" at learning arithmetic in

for natural philosophy, and moral for religion and society—were sufficient. See Daston, *Classical Probability* (1988), 56–57, 60; Shapiro, *Probability and Certainty* (1983), 105, 32, 84, 4, 16, 31; Hacking, *Emergence of Probability* (1975), 146.

86. Buffon, "Essai d'arithmétique morale" (1777), in *Oeuvres philosophiques*, 475–76, 484–86.
87. Diderot, "Probabilité" (1765), in Diderot and d'Alembert, eds., *Encyclopédie*, 3:105.
88. Buffon, "Essai d'arithmétique morale" (1777), in *Oeuvres philosophiques*, 469.

school and having "never proceeded far" with geometry. He also recalls having read at an early age Antoine Arnauld and Pierre Nicole's *La Logique, ou l'Art de penser* (1662), generally known as the Port Royal *Logique*.[89] This was among the first accounts of quantitative probabilities and was important in establishing the epistemological category of moral certainty. The Port Royal authors sought to codify prudent risk taking by using a conception of "expectation" that combined the likelihood of an outcome with its value. They judged it prudent to take a risk if the unlikelihood of the favorable outcome was balanced by its favorability. The benefits of winning a lottery, for example, balanced the unlikelihood of drawing the winning ticket, and so justified the risk of losing the ticket price. In this scheme, probability was relative to purpose: one's expectation of an event depended upon how earnestly one desired or feared it.[90]

Franklin knew this model of prudence and recommended something similar in a letter of friendly advice to Priestley. The letter proposes a "Moral or Prudential Algebra" to govern choices in the face of uncertainty. This algebra involved drawing up a list of pros and cons, "endeavor[ing] to estimate their respective Weights," and then using them to cancel one another until they revealed "where the Ballance lies." Franklin confessed that the "Weight of Reasons cannot be taken with the Precision of Algebraic Quantities." The value of the exercise lay not in the individual weights, but in their comparison. A choice could be right or wrong only in relation to the rest of one's projects and possibilities.[91]

Buffon, similarly yoking probabilistic calculations to human emotion and purpose, restricted his realm of moral probabilities to events within the purview of expectation: any likelihood that could be sensed by the emotions fell within the domain of moral probability, while likelihoods too small to affect the emotions were excluded. The calculus thus included a scale of moral probabilities calibrated by intensities of anticipation and fear, and bounded at either end by extremes of conviction and indifference. Since "of all moral probabilities, man is most affected by the fear of death," Buffon defined his scale according to this fear. A healthy man in the prime of life feels no fear of dying the next day. Buffon therefore defined the likelihood of such an event

89. Franklin, *Autobiography* (1771–88), in *Writings*, ed. Smyth, 1:243.

90. Indeed the Port Royal authors cautioned against an undue devotion to mathematical precision. Like Buffon later would, they identified the mistaking of mathematical probabilities for realities as a leading cause of philosophical errors. Daston writes that the Port Royal discussion was "as remarkable for its restrictions on quantitative reasoning as for its use of it." Daston, *Classical Probability* (1988), 17–18, 39, 243.

91. Franklin to Joseph Priestley, 19 September 1772, in *BFP*, 19:299–300.

as the zero point of his scale. With the help of mortality tables, he defined all events with a likelihood below .0001 as morally impossible, and all events with a likelihood above .9999 as morally certain.[92] As it was miserly and overly mathematical to count money by its numerical value rather than according to its uses, so it was misguided to reckon probability and certainty in absolute terms. They too had meaning only insofar as they could be felt and acted upon.

In his electrical theory, as in his "Moral Algebra," Franklin aspired to achieve a kind of certainty that might well be called moral, in Buffon's sense of the word. Professing not to "clothe" his "Nonsense . . . in algebra, or adorn it with fluxions,"[93] Franklin used quantities in just the way Buffon preferred. They described sensible qualities, never mathematical abstractions. They presented relations rather than absolute quantities. And these relations manifested an underlying scheme of natural purposes. Often they represented states of balance and imbalance. Describing the conservation of charge in a condenser, for example, Franklin said its top could be electrified positively only in exact proportion as its bottom was electrified negatively. Suppose, he explained, that "the common quantity of electricity in each part of the bottle . . . is equal to twenty." In that case, one could electrify the top only as far as forty, since the bottom would then be at zero.[94]

Franklin's use of geometry was also "moral" in this specialized sense of the word: he used figures not for their geometrical properties, but to portray sensible qualities, like wetness or electrification, and to sketch the tendencies that governed them. Franklin frequently surmised that mutually repulsive particles—either of electricity or of air—formed equilateral triangles. Air particles were held in triangles by gravity, and electrical particles by their attraction to common matter. The smaller the triangles of electrical fire, the higher the degree of electrification. The wetter the air, the smaller the triangles. Rain and electrical discharge occurred when the triangles shrank enough for the attractive forces to fully overcome the repulsive ones. The release of the water or electrical fire reestablished equilibrium.[95] The geometrical properties of triangles were irrelevant to this process; Franklin used triangles merely to represent greater and lesser concentrations. For this moral approximation of geometrical reasoning, Franklin was rebuked by the anonymous author of a pamphlet on "his Pretensions to the title of Natural Philosopher":

92. Buffon, "Essai d'arithmétique morale" (1777), in *Oeuvres philosophiques*, 457, 459–69.
93. Franklin to John Perkins, 4 February 1753, in *BFP*, 4:442.
94. Franklin to Collinson, 1 September 1747, in Franklin, *Franklin's Experiments*, 189.
95. Franklin, "Thunder-gusts" (1749), in ibid., 203; "Opinions and Conjectures" (1750), 215; "Meteorological Observations" (1751?) in *BFP*, 4:235–36.

[I] shall just take a little notice of your method of introducing Mathematics. You certainly are not so ignorant as to imagine that it was in this manner that Newton applied Mathematics to explain Natural Philosophy? . . . He introduces no figures from which he does not deduce consequences; and what is more, the very figure he introduces is the only one which could have answered his purpose; whereas, for any use you make of your *triangles*, any other figure would have done your business equally well.[96]

To Buffon, in contrast, Franklin's casually descriptive use of quantities and geometry to sketch the relations among nature's parts, and the goals that oriented them, surely proved that he was an *homme sensé*, neither mathematician nor miser. Never turned inward upon abstract mathematical absolutes, his eye was always directed outward toward the relations of more and less, give and take, whose regulation was the guiding purpose of nature, as of social life.

Like a mathematics of relative rather than absolute values, an analogical mode of reasoning represented each piece of the world in terms of its relations to other parts. The components of an analogical system interacted not by mechanical constraint, but in such a way as to preserve a more general pattern of relations. Analogy was thus the form of explanation that sentimental empiricists recommended as the best suited to illuminating a cosmos animated by the purpose of maintaining certain relations among its parts.

Analogies carried a natural philosopher beyond his immediate sensory experiences, but nonetheless kept him faithful to their implicit message. Therefore Locke had identified analogy as "the only help we have" for all matters "falling not under the reach of our senses,"[97] and sentimental empiricists embraced this view. Buffon judged that if "experience is the foundation of all our physical and moral knowledge, analogy is its first instrument."[98] Diderot, in his article "Encyclopédie," explained that his chief encyclopedic purpose, to "change the common manner of thought," resided in the "analogical" referrals

96. "A Letter to Benjamin Franklin" (1777), in Franklin, *Franklin's Experiments*, 435. Shape, as opposed to geometry, was also instrumental in Franklin's account of how electrical fire could more easily be drawn from the pointed than from the smooth parts of bodies. To explain the "power of points" to attract and dispel electrical fire, Franklin invoked the smaller surface area in communication with a point on a pointed protrusion than with a point on a flat surface. The reduction in proximate surface area meant a local minimum of force between common matter and electricity. Once again, although he referred to a triangle with labeled vertices, Franklin called upon no geometrical properties beyond the sensible property of pointiness. Franklin, "Opinions and Conjectures" (1750), 215–19.

97. Locke, *Essay* (1690), 665.

98. Buffon, "Essai d'arithmétique morale" (1777), in *Oeuvres philosophiques*, 457. See also Buffon, *De la manière d'étudier* (1749), 62.

at the ends of the articles. These became emblematic of the encyclopedic project, not least because its critics so deplored them. They were, one wrote, "eternal and oppressive . . . drag[ging] the reader alphabetically, from letter to letter, from page to page, from column to column."[99] D'Alembert, having resigned his co-editorship of the *Encyclopédie*, parodied the examples Diderot had offered of the utility of analogical "proofs." Analogies were indifferent to physical causation, d'Alembert pointed out, such that one might use analogical reasoning to "prove" that a barometer rises in order to announce rain.[100] But Diderot stood by his referrals, insisting that they sent the reader "to places one would never be guided except by analogy," by relating the articles to one another, "interlacing the branches [of the tree of knowledge] with the trunk."[101]

Franklin's chief propagandist, Priestley, agreed that "analogy is our best guide in all philosophical investigations" and hoped his *History and Present State of Electricity* would persuade electricians to follow Franklin's example in part by "deducing one thing from another by means of analogy."[102] Franklin was indeed a bold analogist.[103] The lightning rod is the most dramatic and successful example: thundercloud was to iron rod, Franklin surmised, as static electricity generator to bodkin.[104] An analogical bent also led Franklin to link the cushion and globe of his electrical generator to the water and salt in the ocean, which ostensibly rubbed together to generate lightning and electricity. When a simple experiment (the shaking of a vial of salt water) failed to confirm this hypothesis, he extended the analogy to the friction of the wind against hills and trees.[105]

Often analogies filled the gaps left by missing mechanisms in Franklin's physics. For example, he relied upon analogy to explain the baffling "power of points," his claim that pointed conductors could both attract and dispel electricity more easily than other shapes. To accept this assertion meant assigning opposite effects to the same material cause, as a Nolletist electrician, M.-J. Brisson, pointed out. He took Franklin to task for his claim that points could

99. Linguet, *Annales* II (1784), 370.

100. D'Alembert, *Essai sur les éléments de philosophie* (1759), 234. See Pappas, "L'Esprit de finesse" (1972), 1229–53.

101. Diderot, "Encyclopédie" (1755), in Pons, ed., *Encyclopédie*, 2:54–56.

102. Priestley, *History* (1775), 2:13, 54–56.

103. Heilbron has written of Franklin that "the spring of his creativity lay in the play of his great power of analogy across all the arts and sciences." "Franklin as an Enlightened Natural Philosopher" (1993), 2.

104. Franklin, "Opinions and Conjectures" (1750), in Franklin, *Franklin's Experiments*, 213–36; quote on 221.

105. Franklin, "Thunder-gusts" (1749), in ibid., 202–3; Franklin to Collinson, September 1753, in ibid., 267–68.

both "draw off" and "throw off" electrical fire, wondering how if "the point of an electrized body has less force for attracting and retaining its atmosphere than has one of [its] sides . . . it [can] be that the point of a non-electrized body has more force than one of [its] sides?" [106] To explain the apparent contradiction, Franklin argued by analogy. It requires less force to pluck out the hairs of a horse's tail one by one than all at once. Likewise, pointed conductors plucked small bits of electricity rather than large clumps; they could pick bits of electrical matter either from themselves, when throwing off fire, or from other objects when drawing off fire. [107]

Le Roy, too, in one of his defenses of negative electricity against mechanist critics, argued by analogy. He suggested replacing the words "negative" and "positive" with "condensed" and "rarefied." Imagine a container covered by a piece of parchment. One could tear the parchment by rarefying the air inside the container as well as by condensing the air outside. So a rarefaction could indeed have the same effect as its opposite, a condensation of substance. [108] Nollet responded that the analogy failed precisely because the parchment example was made intelligible by a mechanical cause that the phenomena of negative electricity obviously lacked. The parchment was torn, whether through condensation or rarefaction, by the pressure of the air outside the container. In many of the phenomena of negative electricity, in contrast, all effects had to be attributed to the depletion of electrical fire in an electrified body. Le Roy had been misled, Nollet reckoned, because he considered "electricity . . . as a virtue in itself, in abstraction from all mechanism." [109]

It was precisely because they abstracted from mechanism that analogies appealed to sentimental empiricists. Analogies indicated nature's tendency to preserve certain relations among her parts, never mind how she accomplished the task. Consider Franklin's use of an analogy to defend his account of the Leyden jar. To critics of that account, its three most mysterious elements were, first, Franklin's insistence that a discharge of "inconceivable quickness and violence" involved no impact of opposing fluids; second, his assertion of a strict symmetry between the positivity inside the jar and the negativity outside; and third, his contention that, owing to this symmetry, a charged jar had no more electricity in it than an uncharged one. Franklin asked his readers to imagine the jar as a bent spring. In order to restore itself to its natural configuration, a spring must symmetrically "contract that side which in the bending

---

106. Brisson, *Dictionnaire raisonné* (1799–1800), 87.

107. Even Franklin immediately admitted the dubiousness of this account, yet contended: "I do not cross [it] out: for even a bad solution read, and its faults discovered, has often given rise to a good one." Franklin, "Opinions and Conjectures" (1750), 219.

108. Le Roy, "Mémoire sur l'électricité" (1753), 450.

109. Nollet, "Examen" (1753), 492.

was extended, and extend that which was contracted." It must perform both operations simultaneously in order to perform either one. And despite the violence of the spring's snap, it involves no collision, and one would never dream of claiming that it had gained elasticity in the bending, or released it in the restoration.[110] The Leyden jar, like the bent spring, acted to restore an equilibrium. This purpose, and not a story of colliding fluids, rendered it intelligible.

Shortly after the translation into French of Franklin's letters on electricity, Diderot promoted Franklinism as a model of how to avoid doubling nature's veil with abstract speculation: "Open the works of Franklin . . . and you will see how the experimental art demands sight, imagination, wisdom, resourcefulness." Invoking Franklinist electrical science as his leading example, Diderot argued that only a sensitive "instinct"—"looking, tasting, touching, listening"—could penetrate beyond the confines of rationalist solipsism.[111] According to sentimental empiricists such as Buffon and Diderot, a tendency on the part of a natural philosopher to reason by analogy and to reckon by moral certainties showed his instinctual sensitivity to the ends that guided the relations and interactions of nature's parts.

This sensitivity was an epistemological advantage, enabling the natural philosopher to understand, for example, the transfer of electricity from an overcharged to an undercharged body. But I have been suggesting that sentimental empiricists also viewed this approach to studying nature as a moral matter, expressing humility and openness to the outside world. "If there is no Use discover'd of Electricity," Franklin wrote, "it may *help to make a vain man humble*."[112] The observation manages to be wryly self-effacing, while at the same time calling attention to its author's willingness to be chastised by his volatile subject.

In the previous chapter, the sentimentalization and moralization of the epistemological discussion of blindness culminated in a social project, the civic education of the blind. Franklinist physics provides another example of epistemology moralized, with the result, once again, that scientific concerns blended smoothly into social ones. Along with the epistemological and moral virtues they discerned in Franklinist physics, sentimental empiricists associated a social virtue, an openness to the influence of others. Sensibility was empirical, humble and sociable by the same token: it focused the philosopher's eyes outward and "preserv[ed] the ears open to instruction."[113] The

---

110. Franklin to Collinson, 29 April 1749, in Franklin, *Franklin's Experiments*, 190–91.

111. Diderot, *De l'interprétation de la nature* (1753), 185, 198–214, 217. The abbé de la Roche praised Franklin for preferring "a well-observed fact to all abstract reasoning not founded in nature or truths of experience." Abbé de la Roche, "Fragments d'une vie de Franklin," BIF, MSS 2222, 72–73.

112. Franklin to Collinson, 14 August 1747, in Franklin, *Franklin's Experiments*, 171.

113. Hume, *Enquiries* (1748–51), 263, 7–8.

three sorts of virtue—epistemological, moral, and social—comprised a single amalgam that can best be described as philosophical modesty.[114]

The opposite of philosophical modesty was rationalist vanity, the conceit that one could capture the natural world, reduced to a system of mechanical causes and effects, fully within one's own mind, with no help from others and no need for correction by experience. Such a conceit was not only vain but solipsistic: the rationalist lived "remote from communication with mankind, and ... wrapped up in principles."[115] Priestley condemned the rationalist authors of mechanist systems for their "vain-glory, self-conceit, arrogance." Franklin, by contrast, was unequaled in his "diffidence" and in the "modesty with which ... [he] proposes every hypothesis."[116] That Franklin came to personify philosophical modesty was thanks partly to his European publicists, especially Priestley, but was also due to Franklin's own efforts. He was ostentatiously diffident.[117] Frequently, he drew the same connection as Priestley between philosophical method and social virtue. A "confident *Self-Sufficiency*," Franklin judged, was among not only "the greatest Vices," but also the worst "Weaknesses, that the human Mind is capable of." Not merely a scoundrel, the "self-sufficient Man, who proudly arrogates all Knowledge and Science to himself," was also a "*Fool*."[118] Arrogance was stupidity and stupidity, arrogance: both resided in insensibility.

114. The influence of sociability on early modern scientific practice has recently focused much attention in the history of science. See, for example, Biagioli, *Galileo* (1993), 112–20; Shapin and Schaffer, *Leviathan* (1985), 332–44; Shapin, *A Social History of Truth* (1994), chap. 4. For Shapin's and others' discussions of philosophical modesty in early modern science, see above, n. 84.

115. Hume, *Enquiries* (1748–51), 263, 7–8.

116. Priestley, *History* (1775), 1:xxvi (here Priestley is citing David Hartley), 192–93; cf. 2:15–16, 39–40, 56.

117. Peter Gay remarks upon a similar irony: "The philosophes' modesty was expressed in the most confident of tones." Gay, *Enlightenment* (1969), 22:143–44. The philosophes' attitudes toward modesty were often contradictory. Hume approved of the modest philosopher who "purposes only to represent the common sense of mankind in more beautiful and engaging colors," but he also hated humility, calling it a "monkish virtue" serving "no manner of purpose," and so really a vice. Hume, *Enquiries* (1748–51), 7–8, 270. Franklin likewise began his life's story by excusing the intrinsic vanity of the autobiographical project: "Most people dislike vanity in others, whatever share they have of it themselves," he wrote, "but I give it fair quarter wherever I meet with it, being persuaded that it is often productive of good to the possessor, and to others within his sphere of action.... [I]t would not be altogether absurd if a man were to thank God for his vanity." However, further along, Franklin entered "Humility" as the last of a list of thirteen virtues comprising moral perfection. Franklin, *Autobiography* (1771–88), in *Writings*, ed. Smyth, 1:328, 227. Instrumentalism makes sense of these apparent contradictions: vanity and modesty were each to be recommended commensurately with their uses. Montesquieu, for example, in a strikingly Franklinesque phrase, gave an instrumentalist assessment of the importance of modesty to philosophy. "A fund of modesty," he wrote, "yields a very great interest." Montesquieu, *Mes pensées* (1720–55), 988.

118. Franklin, "On a Pertinacious Obstinacy in Opinion" (1735), in *Writings*, ed. Lemay, 253.

In his *Autobiography*, Franklin reported having retained from Xenophon's *Memorabilia* of Socrates the lesson that self-improvement requires social intercourse and therefore adopted "the habit of expressing myself in terms of modest diffidence." This meant avoiding words like "certainly, undoubtedly," which provoke "modest men ... to leave you in the possession" of your errors. Franklin professed to sprinkle his discourse with disclaimers like "*I should think it so*" or "*I imagine it to be so*" and cleaved to the pronoun "we" in his accounts of the Philadelphian experiments.[119] In his moral writings, too, Franklin had argued for a link between social solidarity and philosophical modesty. An interlocutor in his 1735 "Dialogue between Two Presbyterians," having denied papal infallibility, concludes that he cannot "modestly claim Infallibility" for his own interpretations of Scripture and must choose unity over orthodoxy.[120]

A social impetus, then, propelled the rejection of mechanist natural science. The impossibility of capturing nature's ways in a rational system, containable in a single head, would keep the sensible philosopher sociable. Rather than looking to universal principles of equality or sameness for a source of social unity, sentimental empiricists did the opposite, finding a guarantee of sociability in nature's diversity. This principle operated at the heart of a major philosophical and political development of the second half of the century: encyclopedism. Diderot devised his *Encyclopédie* to be the embodiment of empiricist modesty tied to social collaboration. Its very massiveness signaled its aim: to unite, never to reduce. The irreducible diversity of nature meant that it was not "given to a single man to know all that can be known." Therefore an encyclopedia could never be "the work of a single man," nor yet of a single learned academy, nor even of the combined efforts of all the world's academic societies. It required authors "spread throughout the different classes," the artisan *homme du peuple* and the *savant* alike. Philosophical modesty founded an encyclopedic collaborative community permeating all ranks and corners of human society.[121]

Franklin, like Diderot, put to work his association of empiricist modesty with social collaboration. The Junto, his Philadelphian club for "mutual improvement" in the pursuit of "Morals, Politics or Natural Philosophy," was a gathering of artisans worthy of the *Encyclopédie*: a copier, a surveyor, a shoe-

119. Franklin, *Autobiography* (1771–88), in *Writings*, ed. Smyth, 1:244–45; Franklin to Collinson, 28 March 1747 and 11 July 1747, in Franklin, *Franklin's Experiments*, 169, 171. Franklin also carefully segregated his scientific writing into "experiments and observations," on the one hand, and "opinions," "conjectures," "suppositions," or "loose thoughts," on the other, professing to treat facts as facts and fantasies as fantasies. See Franklin to Perkins, 13 August 1752, in *BFP*, 4:340–41.

120. Franklin, "Dialogue between Two Presbyterians" (1735), in *BFP*, 2:27–33.

121. Diderot, "Encyclopédie" (1755), in Pons, ed., *Encyclopédie*, 2:51, 44–45, 64, 49. On Diderot's preoccupation with "sociability," see Gordon, *Citizens* (1994), 28–30.

maker, a joiner, several printers, a clerk, and one gentleman, an expert in the art of punning.[122] Also like Diderot and Buffon, Franklin associated unsociability with mathematics. There was only one mathematical member of the Junto, Thomas Godfrey, whom Franklin described as "not a pleasing companion; as, like most mathematicians I have met with, he was forever denying or distinguishing upon trifles, to the disturbance of all conversation. He soon left us."[123]

Sentimental empiricism, with its fusion of social, moral, and epistemological ideals, provided the fertile French ground in which Franklin's moralized electrical science took root. True, his reputation as an electrician benefited more by the lightning rod experiment at Marly-la-Ville in May 1752 than by any other single event. Yet the analogy between lightning and electricity neither relied upon, nor especially supported, Franklin's account of electrical action. Indeed, as Nollet protested, he himself had proposed this analogy several years earlier.[124] As for seizing "the lightning from the heavens,"[125] Jacques de Romas, a Nolletist electrician in Bordeaux, obtained certificates of priority for the electrical kite experiment from the Bordeaux Academy of Sciences and also ultimately from the Paris Academy. Romas may even have conceived of a sort of lightning rod and submitted the idea to a notable member of the Bordeaux Academy, the baron de Montesquieu, in the year before the Marly experiment. Romas explained the lightning rod with effluvia, claiming that the effluences from the electrified rod forced back the effluences from the thunderclouds.[126] This account was certainly no less satisfactory than Franklin's. Franklinist advocates of lightning rods remained unable to explain how long, thin objects both attracted and dispelled electrical fire, and to decide whether one or the other (or both) of these powers was instrumental in protecting civilization from lightning.[127]

122. The fact that the Junto was a secret society does, however, limit the scope of its members' socially collaborative vision.

123. Franklin, *Autobiography* (1771–88), in *Writings*, ed. Smyth, 1:298–99.

124. Torlais, *Un Physicien* (1954), 112–13; "Au reste, mon amy, s'il cela confirme et il soit prouvé par là que le tonnerre et l'électricité sont au fond la meme chose, ouvrez mes lecons de physique p. 374 et vous verez que je l'ay dit 4 ans avant l'homme à qui l'on s'efforce d'attribuer cette idée." Nollet to Jean Jallabert, 2 June 1752, in *Correspondance*, 216. See Heilbron, *History of Electricity* (1979), 351.

125. The epigram inscribed on Jean-Antoine Houdon's bust of Franklin, ascribed to Turgot, is "Eripuit coelo fulmen sceptrumque tyrannis": "He seized the lightning from heaven and the scepter from tyrants."

126. Years later Montesquieu kindly reported to Romas that Franklin had toasted Romas's health at dinner one evening. This was the best gratification Romas would receive, and he duly recounted it to his wife. "Nothing pleases me more," she responded, "than this enthusiasm of our English *savants*." Romas, *Oeuvres* (1911), 289, 73, 183–85. See also dossier Le Roy, AS.

127. "Thus the pointed rod either prevents a stroke from the cloud, or, if a stroke is made, conducts it to the earth." Franklin, "Of Lightning, and the Method (now used in America) of se-

It was not the lightning rod experiment, therefore, that made the reputation of Franklin's theory in France.[128] Priestley gleefully suggested a different factor: "Dr. Franklin's . . . reputation was greatly increased by the opposition which the Abbé Nollet made to his theory."[129] This opposition lent Franklin's experiments a significance that transcended the electrical debate. They came to represent not just a theory of electricity, but a kind of natural science. In the year of Franklin's first visit to France, Priestley noted that in many philosophical journals, "the terms Franklinism, Franklinist and the Franklinian system occur[red] on almost every page."[130] A couple of years later, urging Franklin to return to France for a second visit, one devoted sectarian promised him that he would be "amid franklinists." This disciple signed his letter "Bertier Frankliniste" and added piously: "I was a franklinist without knowing it, and now that I know I never fail to cite the author of my sect, Sir."[131]

What could it mean to be a "franklinist without knowing it"? Might Bertier have unconsciously believed in negative and positive electricity? In the conservation of charge in a condenser? Unlikely. He must have been referring not to the particulars of Franklin's account of electricity, but to an approach to natural science that he took them to embody: the scientific method of the author of *Poor Richard*. In this sense, Bertier's sentiments were widely shared. Nollet observed sarcastically that Franklin was the "Evangelist of the day."[132] Franklin's French readers immediately perceived the relevance of his electrical theory to an urgent conflict in progress.

On one side was a mechanist natural science that banished reference to natural purpose, rendering nature as a machine whose parts moved by constraint alone. The mechanist believed he could describe each discrete motion of the machine in absolute terms and could fully capture in his mind's eye the entire rational system of mechanical parts. Thus, he was vulnerable to the accusation that he arrogantly viewed himself as complete and independent in his understanding, with no need of any agency outside himself, natural or human. On the other side was a sensibilist natural philosophy depicting a natural world suffused with agency. This sensibilist physics did not seek absolute knowledge of nature's parts but focused instead upon their dynamic relations.

curing Buildings and Persons from its mischievous Effects" (September 1767), in Franklin, *Franklin's Experiments*, 391.

128. Home, "Electricity in France in the Post-Franklin Era" (1974), in *Electricity*, no. 12:2.

129. Priestley, *History* (1775), 1:193. See also Heilbron, "Franklin, Haller and Franklinist History" (1977), 548; Cohen, *Franklin and Newton* (1956), 511.

130. Priestley, *History* (1775), 1:193.

131. Joseph-Etienne Bertier to Franklin, 27 February 1769, in *BFP*, 16:56.

132. Nollet, *Lettres* (1753), 1:10.

It worked by relating each part to the others—each emptier to a fuller, each more to a less—treating the relations themselves as primary, and their regulation as the guiding purpose in nature. Mechanists understandably found this vision of nature to be deplorably animist and mystical. But a degree of animism suited the sentimental empiricists. Because their world was irreducible to any rational, mechanical system, it could never be fully captured in the mind's eye. Therefore, its advocates purported to keep their eyes and ears modestly open to instruction and, that way, to remain responsibly connected to a world of harmonious natural and social interaction.

ERIPUIT COELO FULMEN SCEPTRUMQUE TIRANNIS

A. GENIE D. FRANKLIN

Figure 4.1. Marguerite Gérard's 1779 etching *To the Genius of Franklin*, after Jean-Honoré
Fragonard, with Turgot's epigram of Franklin beneath it: "He seized the lightning from heaven
and the scepter from tyrants." Davison Art Center, Wesleyan University.

Chapter Four

FROM ELECTRICITY TO ECONOMY

If there is a field where we have been warned
against systems, it is politics.
—Condillac, *Traité des systèmes* (1749)

. . . the physics of Descartes applied to politics.
—d'Alembert (disparagingly, in reference to
Montesquieu's *Esprit des lois* [1748]) to Gabriel
Cramer (21 September 1749)

The only principle that experience shows to be productive
of movement is the will of intelligent beings . . . which
is determined, not by motors, but by motives, not by
mechanical causes, but by *final* causes.
—Turgot to Condorcet (18 May 1774)

The multiplicity of systematic writings in a
nation is almost always a sign of decadence . . .
and . . . of revolution.
—Parlement de Paris, *Remontrances* (1776)

Cobblestones were the first things Franklin found to admire in France
when he landed there in the summer of 1767: he liked the smoothness
under the wagon wheels that bore him gently from Calais to Paris. He praised
also the rationality of the cobbles' Cartesian construction: because they were
cubes, they could be turned when one side wore down and made as good as new.
Only one cloud threatened the sunny view that Franklin's comfortable journey
afforded him of French efficiency: "The poor peasants complained to us griev-
ously that they were obliged to work upon the roads full two months in the year,
without being paid for their labour."[1] Thus, the cobblestones served Franklin
well in more ways than one. During his first hours on French soil, they introduced him to a primary goal of the Economists' program of reform, the elimination of the feudal *corvée* that exacted unpaid labor from landless subjects.

---

1. Franklin to Polly Stevenson, 14 September 1767, in *BFP*, 14:251.

Franklin's early encounter with the *corvée* coincided with an emerging interest in political economy. Reciprocally, the French Economists—or "Physiocrats," as they named themselves in the year of Franklin's first visit—began to take a keen interest in Franklin. Before his travels to France, Franklin's relations with the philosophes had been based entirely upon his electrical research,[2] and it was his physics that first caught the Physiocrats' fancy. "Physiocracy" invoked the methodological union of natural and moral science: the term meant "rule of nature." Its partisans argued that "nature did not limit her physical laws" to the traditional fields of natural science but extended them to "society" as well.[3] The Physiocrats set out to harness the methods of natural philosophy to their new moral science. Franklin, having applied the methods of moral reasoning to his new natural science, met them halfway. Their easy extrapolation from Franklinist physics to Physiocratic politics is implicit in the epigram of Franklin coined by A.-R.-J. Turgot, Physiocratic fellow traveler and reformist finance minister, and illustrated by Jean-Honoré Fragonard (fig. 4.1): "He seized the lightning from heaven and the scepter from tyrants."

Franklin's utility to the Physiocrats derived from the particular sort of natural science his electrical theory represented. Historians writing on the Physiocrats have generally emphasized the naturalism of their economic theory.[4] But it is not enough to say that the Physiocrats drew upon the methods of natural science, for, as we have seen, these methods were themselves subject to heated controversy. One must therefore specify that a certain kind of natural science served as the model for Physiocracy, the kind of natural science of which Franklinist physics had become the leading exemplar. This was an anti-mechanist science whose practitioners assumed that nature was propelled not by matter in motion, but by benign purpose, and that their business was to remain sensitive to nature's purposeful behavior. Naturalism was common to the economic theories that arose during the latter part of the eighteenth century, particularly in France and Britain, a cluster of theories of which Physiocracy was an early and influential member.[5] But naturalism in the sentimental-empiricist mode was specific to Physiocracy and crucially shaped the theory.

The Physiocrats adopted Franklin's teleological physics as a model for their economic science during the 1760s and '70s, and they developed their Franklinist natural-moral science in the language of sensibility (§1). However, this language was as versatile as it was powerful: it served not only the

---

2. Aldridge, *Franklin* (1957), 23.

3. Dupont, "Science nouvelle" (1768), in *Oeuvres politiques*, 1:536; Fox-Genovese, *Origins of Physiocracy* (1976), 9.

4. See, for example, Fox-Genovese, *Origins of Physiocracy* (1976), 9–10.

5. On the naturalism of classical economics in general and Adam Smith's economic thought in particular, see Roll, *History of Economic Thought* (1954), 147–48; and Blaug, *Economic Theory* (1968), 30–32.

Physiocrats, but also their opponents, who wielded it against them during Turgot's tenure as inspector-general of finances from 1774–76 (§2). The leading charge against Turgot and the Physiocrats during Turgot's administration, in which he tried unsuccessfully to enact Physiocratic reforms, was that he and they were guilty of "the spirit of system," a favorite epithet among sentimental empiricists denoting the arrogant and dogmatic antithesis of sensibility. Later, however, the Physiocrats successfully reappropriated this language and applied it, along with the mantle of Franklinist science, to policymaking during the Revolution (§3).

Thus, sensibilist science shaped not only Physiocracy but the surrounding political debates, and from all sides. Economists, reformist administrators, members of the Parlement, the king, and ultimately the National Assembly, all came to speak the language of sentimental empiricism.[6]

I. THE PHYSIOCRATS' NATURALIZATION OF VALUE

While debating the epistemological limits of rational mechanics, Turgot—philosophe, Physiocratic sympathizer, and finance minister—observed to the marquis de Condorcet that impulsion could not account for the existence of movement in the universe because it presumed a prior source of motion. Moreover, daily experience established that matter in motion was always causally preceded by an act of will. Both to avoid an infinite regress and to be properly empirical, therefore, physics must ultimately arrive at teleological explanations. It was Turgot, in the course of his epistolary conversation with Condorcet, who penned the passage quoted in chapter 3: "The only principle that experience shows to be productive of movement is the will of intelligent beings ... which is determined, not by motors, but by motives, not by mechanical causes, but by *final* causes."[7]

Political economy was Turgot's primary concern, as a provincial intendant and then a minister of finance, but, especially through his articles for the *Encyclopédie*, he was also involved in the natural sciences and drew upon them for models.[8] A believer in the limits of mechanism, he particularly admired Franklinist physics and compared a detractor of Physiocracy to "Nollet disputing

---

6. For an examination of the Enlightenment view that "economic life [was] ... a matter of sentiment" and its implications for eighteenth-century economic theory, focusing on Condorcet and Smith, see Rothschild, *Economic Sentiments* (2001).

7. Turgot to Condorcet, 18 May 1774, in Condorcet, *Correspondance*, 172–73. See chap. 3, p. 70 n. 3.

8. For example, Turgot wrote the article "Expansibilité" for the *Encyclopédie*, which figured prominently in imponderable fluid theories, including Lavoisier's theory of caloric. See infra, chap. 7.

against Franklin about electricity."[9] As in the debate between Franklinists and Nolletists, in the Physiocrats' program and the controversy surrounding Turgot's reforms, the alleged evils of Cartesian mechanism were central. The Physiocrats professed to be anti-Cartesian, overthrowing a model of causation they deemed artificial and bringing about a common recognition that the economy's complexity eluded mathematical reduction.[10] They opposed the calculations and mechanical manipulations of mercantilism, which, they liked to insinuate, amounted to Cartesian economics.[11]

The Physiocrats argued that tariffs, complex tax laws, industrial regulations, and currency manipulations were attempts to impose a human, rational order on an economic domain to which it was profoundly unsuited.[12] Using sensibilist science's nastiest word, they called mercantilist policy a "system." The phrase "spirit of system," denoting the evil opposite of sensibility, was drawn from the internecine battles of natural philosophers beginning around mid-century. In these battles, system-building—the construction of mathematical and mechanical systems to explain natural phenomena—had emerged as the evil opposite of a philosophy of sensitive responsiveness to nature's ways. Systems were attempts to impose one's own scheme on nature. The Physiocrats pronounced mercantilism an "artificial system," a system with "no limits," a system "impossible to execute," a "system . . . [in which] all the errors touch and support one another." The abbé Baudeau, founder and editor of one of the main Physiocratic journals, proclaimed that if such mercantilist "systems had never interfered, the ancient and primitive liberty . . . would still subsist; for it is evidently the natural state."[13]

Sensibilists argued that the spirit of system must give way, in the sciences of society as in natural science, to an empiricist sensitivity to nature's intentions. François Quesnay, the Physiocrats' founder, traced the origins of social life to primitive people's "*involuntary sensibility* to physical pleasure and pain," which

9. Turgot to Lespinasse, 26 January 1770, in Turgot, *Oeuvres*, 3:420. Turgot referred to Ferdinando Galiani's *Dialogues sur le commerce des blés* (1768).

10. Dupont, "Science nouvelle" (1768), in *Oeuvres politiques*, 1:566, 577.

11. On the way in which the Physiocrats and other critics of mercantilist policies projected an image of mercantilist "theoretical systematicity," see Larrère, *L'Invention de l'Economie* (1992), chap. 3.

12. Anon., "De l'état actuel des Sciences et des Arts," in *Les Ephémérides du citoyen, Ou bibliothèque raisonnée des sciences morales et politiques* (Paris, 1765), vol. 2, bk. 2, 220–21, 238; "Sur la prosperité du commerce," in ibid., vol. 3, bk. 2, 266; "Réponse à l'Oeuvre intitulé Principes de tout Gouvernement," in ibid. (1767), 7:143.

13. Mercier de la Rivière, *L'Ordre naturel* (1767), 527, 581, 575, 577, 560; Baudeau, *Introduction à la philosophie économique* (1771), in Daire, ed., *Physiocrates*, 2:716. See also Quesnay, "Mémoire sur les avantages de l'industrie et du commerce" (1765), in *François Quesnay*, 2:744: "If each one reasons from his own particular point of view, the result is so many discordant systems harmful to the advancement of the science." The essay is written as a dialogue, and in this passage Quesnay adopted the voice of his adversaries.

induced them to provide for their own subsistence. To assure themselves of the "peaceful possession of the fruits of their labors," they needed "to live in society." Nature was therefore "the first teacher of social man." It was she who "wanted the reunion of men in society; it was she who dictated the essential conditions of this reunion; it is she finally who makes *sensible* to them the necessity of society, and that of the conditions to which they must submit in order for society to form and perpetuate itself."[14] Thus, state policy should be suited not to any man-made system, but to nature's own governance.

The Science, as the Physiocrats dubbed their program of reform, taught that nature, through agriculture, was the only truly productive part of the economy, meaning that it alone created more than it consumed. All other areas of human endeavor conserved value but did not create it. In manufacture and trade, the value of the inputs equaled the value of the outputs. The worth of the raw materials and labor, in industrial production, added up to the same amount as the value of the resulting products. These products were then exchanged, in commercial transactions, for money or for other products of equal worth. Thus, neither industry nor commerce generated value; they merely transformed it, and agriculture was the sole source of all the wealth in the economy.[15] This meant that taxes on commerce and industry were ultimately taxes on agriculture. Only nature herself could be taxed, and, moreover, nature herself determined the true tax each year by creating an agricultural surplus, or what Physiocrats termed the "net product." In order to suit taxation to the annual increase in wealth, following nature's own rule, all taxes should be drawn directly from the agricultural net product. Taxes imposed upon any other part of the economy were arbitrary and destructive.[16]

The practical implications of these principles consisted largely in the lesson that what was good for the landed proprietors was good for everyone: the wealth of the landowners must be maximized in order to maximize the kingdom's wealth. This chiefly meant freeing the grain trade, which the Physiocrats expected to bring about the "good price" of grain (a high price).[17] Until the middle of the eighteenth century, the Crown imposed an elaborate array of restrictions on the market in grain: it prohibited export of grain to foreign countries, restricted its movement within the kingdom by means of permits and certificates, and disallowed its measurement, sale, or purchase outside of

14. Quesnay, in Mercier de la Rivière, *L'Ordre naturel* (1767), 610.

15. "Here is the essential point, and the one most ignored or at least neglected in France: we have not even recognized the difference between the product of work that returns nothing but the price of the labor, and that of work that pays for the labor and procures revenues. As a result of this inattention we have preferred industry to agriculture, and commerce in manufactured goods to commerce in raw materials." Quesnay, "Grains" (1757), in *François Quesnay*, 2:472.

16. Dupont, "Science nouvelle" (1768), in *Oeuvres politiques*, 1:550, 575–76.

17. Quesnay, "Grains" (1757), in *François Quesnay*, 2:509.

stipulated areas within each province. The Crown, towns, and provinces also levied tolls on the transit and sale of grain. The Physiocrats proposed doing away with this whole structure of restrictions and tolls.

More generally, they advocated replacing the entire baroque tax structure of the Old Regime with an *impôt unique*, a single tax on land rent. A policy of taxing only land rent might seem disadvantageous to the proprietors, but, in fact, advocates of the *impôt unique* explained that the policy would benefit landowners first and foremost, as it would maximize their profits from the sales of their agricultural products.[18] Finally, the Physiocrats advocated a mode of government by "legal despotism," in which the monarch and his administrators empirically derived and enacted the law as dictated by nature.[19] This doctrine was intended as a rebuke to Montesquieu's parliamentary theory of government, in which its functions were divided among mutually balancing bodies. The Physiocrats judged such an arrangement of counterweights to be, like mercantilist economic policy, a "system"—"mechanical," artificial, and counter to the order of nature.[20]

To be sure, the Physiocrats' policy recommendations did not spring fully formed from the brow of a philosophical movement. They responded to certain features of the Old Regime economy: first, that France was a predominantly agricultural country; second, that French agriculture was dominated by small-scale, impoverished subsistence farming; and third, that the tax structure of the Old Regime had become increasingly chaotic, due largely to the nobility's successes at securing fiscal privileges and tax exemptions, and imposed an ever-more crippling burden on peasants and on a new class whose interests the Physiocrats sought to advance, reform-minded agricultural entrepreneurs. The Physiocratic program accordingly had two purposes: to simplify fiscal policy and to channel resources into agriculture.[21] However, if economic problems inspired Physiocratic policy, it was the language and arguments of sentimental empiricism that provided the medium in which the Physiocrats developed and defended their proposals. The ideals of sensibility shaped discussion of Physiocratic reforms not only among the philosophes, but in court and in Parlement, and defined the terms in which these reforms were both instituted and opposed.

The Physiocrats understood their injunction that nature be allowed to determine taxation as the crux of an empiricist economic philosophy. They pro-

18. Fox-Genovese, *Origins of Physiocracy* (1976), 58; Meek, *Economics of Physiocracy* (1962), 25; Blaug, *Economic Theory* (1968), 26–27; Dakin, *Turgot* (1939), 92–93.

19. See Mercier de la Rivière, *L'Ordre naturel* (1767), chaps. XVIII, XIX, XX, XXI.

20. Quesnay, "Maximes générales du gouvernement économique d'un royaume agricoles," November 1767, in *François Quesnay*, 2:949; Turgot to Dupont, 10 May 1771, in Turgot, *Oeuvres*, 3:486–87. See infra, §3, and Weulersse, *Le Mouvement physiocratique* (1910), 2:52–55.

21. Meek, *Economics of Physiocracy* (1962), 23–27.

fessed to replace abstractions with evidence. The economic historian Georges Weulersse observed that "there was hardly a word the Physiocrats used with greater frequency than the noun 'evidence,' the adjective 'evident,' and the adverb 'evidently.' Their whole moral philosophy, their whole politics rested upon the notion of evidence."[22] In a similar characterization of Economic empiricism, Steven Kaplan has described the Physiocratic program as the demand for "nothing less than a *tabula rasa* in subsistence affairs. The police apparatus designed to impose subsistence conditions was to be abandoned, and provisioning left solely to 'nature.'"[23]

At the heart of Physiocratic doctrine was a naturalist theory of value. The wealth of the land was "annually reborn"; it was "renascent." Money was "sterile." It could not constitute the "true wealth of a nation" because "money does not breed money." Value could be bred, generated, born, but never made. This accounted for its unique origin in agriculture: "industry and mercantile trade" constituted "only an artificial realm."[24] In their naturalization of value, the Physiocrats differed from their contemporaries, who located value in conventional, man-made bases: money, population, and—particularly after the 1776 publication of Adam Smith's *Wealth of Nations*—labor. Rejecting the labor theory of value, Quesnay warned that it was artificial, abstract, a "system," a "vain and frivolous prejudice," and potentially "a dangerous and perfidious error."[25]

Smith, for his part, called Physiocracy "the nearest approximation to the truth" so far, and may even have thought of dedicating *The Wealth of Nations* to Quesnay.[26] Nevertheless, Smith rejected the Physiocrats' central theoretical claim that wealth originated in nature. For him, economic value was manmade. The naturalism of Smith's economy was thus of a more limited sort than the Physiocrats' naturalism. Smith assumed an analogy between nature, governed by natural laws, and economic life, propelled by laws of human nature. But the Physiocrats made their economy an integral part of the natural world, its naturalism literal and absolute.[27]

Crucial to the development of this rigorously naturalist moral science was a congenial model of nature and of natural science: one that ascribed moral

22. Weulersse, *Le Mouvement physiocratique* (1910), 2:120.

23. Kaplan, *Bread, Politics* (1976), 2:682.

24. Quesnay, "Maximes générales du gouvernement économique d'un royaume agricole" (November 1767), in *François Quesnay*, 2:967–69. For a thorough examination of Quesnay's and the Physiocrats' theory of value and an argument that it encompassed notions of the exchange value of commodities despite its naturalism, see Vaggi, *Economics* (1987).

25. Quesnay, "Sur les travaux des artisans" (1766), in *François Quesnay*, 2:912.

26. Smith, *Wealth* (1776), bk. 4, chap. IX, ¶39; Stewart, "Account" (1793), §3, ¶12.

27. "To see that labor, not nature, was the source of 'value' was one of Smith's greatest insights." Ibid., 47. On the naturalism of Smith's economy and his theory of value, see Roll, *History of Economic Thought* (1954), 146–48, 154–60.

purposes to nature and privileged these purposes in scientific explanation. The Physiocrats needed something like Franklinist physics upon which to model their moral physics of wealth. Quesnay drew up a *Tableau économique* to show the flow of the natural fluid of wealth through the economy, originating in the agricultural surplus and conserved through every subsequent transaction. The *Tableau* made a great impression among reformist philosophes and became emblematic of Physiocracy.[28] Observing that Quesnay was a doctor and surgeon who had, moreover, written several treatises on bloodletting, historians have often identified the circulation of the blood as the model for his theory of the circulation of money.[29] In fact, in the preceding century, even before William Harvey's demonstration that blood circulated, political economists had already liked to call money the blood of the body politic.[30] Physiocratic authors did revisit the old analogy, but only in passing.[31] Harvey's hydraulics of the circulatory system were not very useful to them. Their wealth was a fluid more like Franklin's electricity, which did not act mechanically, but instead pursued its own goals and tendencies.

Once born from the land, wealth spread itself through the economy "by an influence as rapid as electrical fire."[32] The natural purposes guiding the flow of wealth through Quesnay's *Tableau* were the same as those directing Franklin's electrical economy: conservation and the maintenance of balance. Pierre-Samuel Dupont de Nemours, another leading Physiocrat, emphasized the necessary symmetry of commercial exchanges. "We will repeat incessantly," Dupont wrote, "that all trade assumes equilibrium, a balance of sales and purchases." Those who failed to understand this fact, and tried to buy without selling, were lucky "that the thing is impossible." When Physiocratic authors did call upon hydraulic and mechanical analogies to explain such principles, as when Franklin did so, it was not to reduce the economy to a machine, but, on the contrary, to invoke the hidden operation of natural purpose even in mechanical arrangements. For example, using an analogy that recalled Franklin's account of symmetry in the Leyden jar, Dupont argued that buying and selling were inextricably connected as two sides of a single spring,

28. Quesnay, *Tableau économique* (1758), in *François Quesnay*, 2:667–73; Heilbroner, *Worldly Philosophers* (1980), 47; Fox-Genovese, *Origins of Physiocracy* (1976), 246.

29. See, for example, Heilbroner, *Worldly Philosophers* (1980), 47; and Barre, *Economie politique* (1966), 1:34–35. For a full examination of the physiological and chemical models for Quesnay's theory of circulation, see Christensen, "Fire, Motion and Productivity" (1994).

30. See Larrère, *L'Invention de l'Economie* (1992), 107–8.

31. See, for example, Turgot, *Réflexions sur la formation . . . des richesses* (1766), §68.

32. Dupont, "De l'exportation et de l'importation des grains, mémoire lû à la Société royale d'agriculture de soissons" (1764), in *Oeuvres politiques*, 1:120–21. Christensen emphasizes the importance of "a highly subtle and active ether—the matter of fire" as a model for Quesnay's understanding of the movement of wealth. Christensen, "Fire, Motion and Productivity" (1994), 277.

which cannot restore itself to its proper state without simultaneously contracting the expanded side and expanding the contracted side.[33]

Similarly insisting upon the tendency of the economy to seek a balance, Turgot wrote that although "the different uses of capital result in very unequal products," this inequality did not prevent the products from "establishing a kind of equilibrium, as between two fluids of unequal weight that occupy two branches of an inverted siphon." The fluids' appearance would be asymmetrical, but "the height of one could not augment without the other climbing as well."[34] Behind the operation of springs, siphons, and market exchanges lay nature's eternal tendency toward balance.

Central to the teleological naturalism of the *Tableau*, and of the Physiocrats' theory of value, was the familiar injunction to remain sensitive to natural purpose and the accompanying stricture against arrogant attempts to surmount nature's tendencies with rational designs. Quesnay assured his readers in *Philosophie rurale* (1764), upon which he worked with his closest collaborator, the marquis de Mirabeau, "If it has taken so much work to dissect the body politic, that does not mean we will need the scalpel in hand to maintain its health." The "spirit of regulation" had inflicted countless evils upon humanity, as administrators refused to recognize that "the world goes on its own." When once the world was left alone, humanity would recognize the Physiocrats' "principles executed in virtue of the innate order of things."[35]

The *Tableau* represented the "basis of moral order on earth," an "immutable" order from which "Sovereigns and Subjects can only stray . . . to their disadvantage." As Quesnay and Mirabeau explained, "Perfect government is not of human institution."[36] Governments must simply allow the natural flow represented in the *Tableau* to take place unimpeded. They ought "have no care but to pave the way, remove the rocks from the path."[37] Wealth would then follow its own course from agriculture through industry and commerce, and back to agriculture to renew the cycle. Like electrical fire in Franklinist physics, it would maintain a dynamic equilibrium, rather than a clockwork regularity, subordinating physical principles to moral purposes in its eternal return to balance and harmony.

The Physiocrats were among the *franklinistes* who had arisen in the wake of Buffon's efforts on behalf of Franklin's electrical writings. If Franklin's effusive

33. Dupont, "De l'exportation et de l'importation des grains, mémoire lû à la Société royale d'agriculture de soissons" (1764), in *Oeuvres politiques*, 1:162.

34. Turgot, *Réflexions sur la formation . . . des richesses* (1766), §87.

35. Mirabeau and Quesnay, *Philosophie rurale* (1764), chap. VII, in Quesnay, *François Quesnay*, 2:727. *Philosophie rurale* was signed by Mirabeau but was written in collaboration with Quesnay, who wrote all of chapter 7. See Quesnay, *François Quesnay*, 2:687, n. 1.

36. Mirabeau and Quesnay, *Philosophie rurale* (1764), xviii.

37. Ibid., chap. VII, in Quesnay, *François Quesnay*, 2:727.

thank-you notes of 1768 are to be trusted, his French followers awaited him with open arms when he arrived in Paris for the first time in 1767.[38] They also provided a service beyond enthusiastic hospitality: they offered a social and intellectual bridge from Franklin's older interest in electricity to his more recent interest in political economy. That is, they began to recruit the moralist natural philosopher into the project of founding a naturalist moral science.

The Economists were not naive in their fascination with Franklin's electrical economy; they were trained in natural philosophy and experienced in electrical research. The Physiocrat Jacques Barbeu-Dubourg, for instance, was also a physician, medical reformer, botanist, and mathematician, as well as an electrical experimenter. In this last capacity, Barbeu-Dubourg had presented his compliments to Franklin as early as 1754 through Thomas François Dalibard, Franklin's original French translator. Sometime after Franklin's visit, Dubourg was inspired to add translation and editing to his list of vocations, volunteering himself to become Franklin's French editor. He published an edition of Franklin's works that included both electrical and political economic writings.[39] Dupont soon introduced himself to Franklin in a letter that reflected Dubourg's success as Franklin's publicist: "I had known you as . . . the Physicist, the man whom nature allows to unveil her secrets. . . . My friend Monsieur le Docteur Barbeu du Bourg has since communicated to me several of your writings concerning the affairs of your country."[40]

Thus the electricians and Economists among Franklin's correspondents were friends, colleagues, and, often, the very same people. When Franklin sent Dalibard a copy of Priestley's *History and Present State of Electricity*, Dalibard responded by presenting Barbeu-Dubourg's compliments and lamenting that Le Roy had made off with the *History* immediately upon its arrival.[41] Le Roy, meanwhile, began to include economic problems along with electrical ones in his correspondence. He asked Franklin about the free exportation of grain and the advantages to be gleaned from competition. Franklin was initially reluctant to offer concrete answers to such queries, although he went so far as to agree with Le Roy that the "Principles of Commerce are yet but little

38. See Franklin to Bertier, 31 January 1768, in *BFP*, 15:33; Franklin to Dalibard, 31 January 1768, in *BFP*, 15:35.

39. Franklin, *Oeuvres de M. Franklin* (1773); Aldridge, *Franklin* (1957), 22; see headnote in *BFP*, 15:112–13; Dalibard to Franklin, 31 March 1754, in *BFP*, 5:253–54.

40. Dupont to Franklin, 10 May 1768, in *BFP*, 15:118–19. Interest in Franklinist science on the part of political economists was not limited to Physiocrats. The anti-Physiocratic finance minister Jacques Necker, who also rested his economic program upon a notion of social balance, performed a series of Leyden jar experiments and presented a memoir on them to the Académie des sciences in 1761. Procès-verbaux, 1761, AS.

41. Dalibard to Franklin, 14 June 1768, in *BFP*, 15:140.

Understood."[42] But when Priestley's *Essay on the First Principles of Government* was published, Franklin forwarded it to Barbeu-Dubourg.[43]

In their eagerness to enlist Franklin, the Physiocrats prematurely announced his support for their cause in their journal, the *Ephémérides*.[44] This announcement was followed by an account of Franklin's examination before the House of Commons during the Stamp Act controversy, in which he expressed no Physiocratic principles whatsoever.[45] His argument concerned only the political and not the natural or economic basis for taxation. Dupont, who had edited the volume in question, was scolded for name-dropping by Turgot: "To announce to the public the opinions of a man like Franklin, you must either be charged to do so, or be very sure of your facts. You are not yet cured of the sectarian spirit." A more interesting article, Turgot admonished, would have discussed "in detail the question of the colonies, carefully presenting the opinions of M. Franklin, which are not at all in accord with the true principles" of Physiocracy.[46]

Franklin had recently written two economic essays, but these were not especially Physiocratic either. In "On the Price of Corn and the Management of the Poor," he argued that it was discriminatory against farmers to allow the exportation of manufactured goods but not raw materials. The ban had been publicly justified as a measure to keep domestic prices down; Franklin interpreted this as "a tax for the maintenance of the poor" and objected that poverty was better eased by inducement to work than by charity.[47] In "On the Labouring Poor," Franklin pursued this theme by condemning the Poor Law. This was an Elizabethan body of doctrine that continued to impose on English parishes an obligation to collect funds for the relief of their poor, by housing the old and sick, apprenticing the young, and employing the able-bodied.[48] Franklin disapproved of it and disparaged "the malignant censure some writers have bestowed upon the rich for their luxury and expensive living while the poor are starving, etc." After all, Franklin explained, "the rich do not work for one another." They hire the poor and buy the products

42. Franklin to Le Roy, 31 January 1769, in *BFP*, 16:34–35.

43. Franklin to Barbeu-Dubourg, 22 September 1769, in *BFP*, 16:205.

44. *Les Ephémérides du citoyen* (1768), 7:32.

45. Franklin argued that taxes could not fairly be imposed upon a population by an assembly in which it was not represented. He was so un-Physiocratic as to allow import taxes on the principle that these were payments for the service of maintaining the sea for safe travel. See "The examination of Doctor Benjamin Franklin &c., in the British House of Commons, Relative to the Repeal of the American Stamp Act, in 1766," in *BFP*, 13:129–59.

46. Turgot to Dupont, 5 August 1768, in Turgot, *Oeuvres*, 3:13.

47. Franklin, "On the Price of Corn and the Management of the Poor" (29 November 1766), in *BFP*, 13:510–16.

48. See De Vries, *Economy of Europe* (1976), 7.

of their labor. Thus, all that the rich spend goes straight to the poor. It was Franklin's standard conservationist teleology. He concluded with the observation: "Our labouring poor receive annually the whole revenue of the nation, and from us they can have no more."[49]

So although Franklin opposed various forms of taxation and regulation, as did the Physiocrats, his economic views were surely not sufficiently Physiocratic to justify the Physiocrats' claiming him as their own. But in 1769 Franklin finally made good the Physiocrats' claim, and the promise they had gleaned from his electrical philosophy, in his "Positions to Be Examined," where he argued that a nation could honestly acquire wealth only by "a continual Miracle wrought by the Hand of God," that is, agriculture. Manufacture could transform but never create value. Fair commerce, the exchange of equal values, was also transformative rather than productive.[50]

This argument explicitly advanced the central tenets of Physiocracy: that all economic value was natural in origin; that once born from the land, wealth was conserved throughout the economy; that economic value was subject to nature's own process of self-governance; and that human manipulation of the economy was thus either futile or pernicious. The social and economic orders must be allowed to regulate themselves freely, so that the inscrutable but beneficent motives governing them could be realized and nature's balance maintained. At last Dupont had what he wanted, a fully Physiocratic declaration signed by the interpreter of the Leyden phial. The Physiocrats had succeeded in extending Franklinism to political economy, and wealth was a Franklinist fluid.

The Physiocrats' naturalization of economic value provoked criticism even among their political allies. Condillac, for example, supported the same economic policies as they did; in particular, he favored freedom of commerce, but for opposite reasons. The Physiocrats, as we have seen, argued for freedom of commerce on the ground that nature, through agriculture, was the source of all new wealth and must accordingly be allowed to determine the annual tax, which must therefore be drawn directly from the agricultural surplus. Condillac, in contrast, rooted value not in nature, but in social convention.[51] He opposed the tendency to "regard value as an absolute quality, in-

49. Franklin, "On the Labouring Poor" (April 1768), in *BFP*, 15:103–7.

50. Franklin, "Positions to Be Examined" (4 April 1769), in *BFP*, 16:107–9. Another prominent example of the Physiocrats' influence upon American ideas about the natural origins of value is Thomas Jefferson's *Notes on the State of Virginia* (1784–85), which was written in response to a questionnaire submitted by Quesnay. On Jefferson and the Physiocrats, see Chinard, ed., *Correspondence* (1931); Kaplan, *Jefferson and France* (1967); O'Brien, *Long Affair* (1996).

51. On the importance of "sociability" in Condillac's economic theory, see Larrère, *L'Invention de l'Economie* (1992), 66–69. On the subjectivity of Condillac's theory of value, see Knight, *Geometric Spirit* (1968), 238–39. Others among the Physiocrats' opponents, including the abbé Galiani

herent in things independently of the judgments we make." The value of a thing, Condillac said, arose primarily from assessments of its "utility," the needs and uses people had for it. He therefore argued for free trade on the basis of the social, rather than natural, origins of value. Condillac's conventional origin of value implied that commerce was not sterile, despite the claims of the Physiocrats; exchanges between people who valued what they received over what they traded always maximized value. Taxes would inhibit such exchanges, and it was for this reason, and not the sterility of commerce, that Condillac opposed them.[52]

Because judgments were at the source of all wealth, a proper general understanding of the economy would bring about a properly functioning economy. In Condillac's view, such an understanding was best achieved through a careful use of language. He insisted, for example, upon the distinction between "values" and "sums," the confusion of which allowed bankers to make their profits: equal sums had different values in different places, thus one could make a profit in value through an exchange of equal sums. Condillac surmised that "bankers affect an obscure language" in order to prevent "people from seeing their operations clearly."[53] Economic health lay in clarity of expression and public instruction, not, as the Physiocrats would have it, in allowing nature to govern.

For similar reasons, David Hume disapproved of the Physiocrats, rather vehemently. To the abbé Morellet, a Physiocratic sympathizer though not an orthodox believer, he urged, "Thunder [the Physiocrats], and crush them, and pound them, and reduce them to dust and ashes! They are, indeed, the set of men the most chimerical and arrogant that now exist."[54] The Physiocrats so trusted nature over human judgment, Hume charged, that they were willing to sacrifice political to economic freedom with their doctrine of "legal despotism." In contrast, he associated a healthy economy with benevolent politics, not unhampered nature. Individual autonomy far outweighed natural facts like soil and climate. In fact, he maintained that barren soil and

and J. J. L. Graslin, gave similarly subjectivist definitions of value. Graslin wrote, "Wealth is the thing to which value is attributed; the thing is not wealth except in proportion to this attribution. But the attribution of value is foreign to the nature of the thing; its principle is in man uniquely, it grows and shrinks with the needs of man and disappears with him." Graslin, *Essai* (1767), 51. See Airiau, *Opposition aux Physiocrates* (1965), 96–101.

52. Condillac, *Commerce* (1776), 254–59, 249, 367, 320–21, 369–70. On this difference between Condillac and the Physiocrats and for an argument that the Physiocrats missed an opportunity in categorically rejecting Condillac's views, see Eltis, "L'Abbé de Condillac and the Physiocrats" (1995), esp. 231–34.

53. Ibid., 302–4, 308, 301. On John Horne Tooke's similar association of clarity of expression with political health, see Smith, *Politics* (1984), 110–53.

54. Hume to Morellet (1769); cited in Andrew S. Skinner, "David Hume: Principles of Political Economy," in Norton, *Hume* (1993), 247.

inclement weather had actually benefited English peasants because difficult conditions gave rise to an advantageous social system. Having a more complicated task than their French or Italian counterparts, English peasants received larger funds and longer leases. Their independence bred industry and inventiveness.[55]

Like Condillac, who traced economic prosperity to clear language, and Hume, who located it in a social system that fostered autonomy, Jacques Necker, the Genevan banker who served as controller-general of finances during the 1770s and '80s, also assigned a cultural rather than natural origin to economic value. In keeping with Physiocracy, Necker denied that the accumulation of money was the measure of a state's strength (money was not a good in itself, but only a "sign of the good truly useful and agreeable to men"), but he then parted company with the Physiocrats. He made the size of a state's population the index of its success and traced population growth to happiness, arguing that a population increased when there was "a happy harmony among the different classes in society." Lifting export barriers might maximize revenue, but not happiness. If wine were consumed at home rather than exported abroad, Necker surmised, the "French nation would not be less happy." To foster happiness and an increased population, administrators should focus their efforts on perfecting domestic "political relations" through a harmonious distribution of goods.[56] In particular, their chief duty was to insure subsistence: "The subsistence of the people is the most essential object that must occupy the administration."[57] During his three terms as finance minister, Necker introduced various social and administrative reforms—lifting the mortmain on royal domains, reducing the number of tax farmers,[58] introducing new provincial assemblies—but he did not relinquish his commitment to the principle that provisioning was the job of administration, not nature.[59]

Finally Antoine Lavoisier, a tax farmer and an economic theorist among his other callings, also settled upon conventional rather than natural sources of prosperity. He had started off in search of natural ones. But after eight years of experimentation on his farm at Freschines, Lavoisier reported political

55. Hume, "Of Commerce" (1752), 297–98.

56. Necker, *Grains* (1775), 216, 219, 357–58.

57. Necker to Sartine, 14 February 1778, AN FII*1, fol. 258; cited in Kaplan, *Provisioning Paris* (1984), 23. Kaplan writes that except "for two brief interludes, this commitment to the consumer-people of France was the uncontested tenet of public policy during the old regime. It was founded on the conviction that social stability could be guaranteed only by guaranteeing the food supply" (23).

58. Tax farmers (*fermiers généraux*) were members of the Tax Farm, the corporate body to whom the Crown contracted tax collection under the Old Regime.

59. On Necker's ministries, see Hardman, *French Politics* (1995), 133–43, 162–67; and Harris, *Necker, Reform Statesman* (1979) and *Necker and the Revolution* (1986).

rather than strictly agronomic findings to the Paris Society of Agriculture. He had begun with a principle informed by the new "agronomy," an experimental science of agriculture founded around midcentury by the chemist and botanist Henri-Louis Duhamel du Monceau:[60] successful farming required fertilization by manure, which meant animals. Lavoisier argued that French farmers had too few animals; the greater success of English agriculture was due to the English emphasis on raising and selling livestock rather than wheat. This allowed the English to plant clover and turnips every third year, which they could use profitably as nourishment for the animals, rather than leaving their fields to lie fallow. The English system had the further advantage of supplying farmers with plenty of manure to fertilize their crops. To increase the numbers of animals on his farm, in keeping with the recommendations of the new agronomy, Lavoisier had planted "artificial prairies" with grass for grazing. But his improvements produced few results, and he now said that he "recognized with pain" that no matter "what care, what economies" cultivators made, they could never hope for a return of more than 5 percent.

The problem was not insufficient food for the animals but "a moral obstacle, more difficult to vanquish than most physical obstacles." As he would repeat continually during the early years of the Revolution, Lavoisier now concluded that it was "principally from our institutions and our laws that arise the obstacles to the progress of agriculture." These obstacles included the arbitrary and fluctuating *taille* and the *corvée*, which "humiliat[ed] the tax-payer" and "punish[ed] industry" (see fig. 4.2); feudal tithes that carried off more than half, and sometimes all, of the net yield of a region; the excise taxes on salt and tobacco, which involved "inhuman and indecent inspections that . . . tend to render odious the authority" of the Crown; monopolies on milling, which led to poor quality and loss of grain from faulty practices and subjected the people to the greed of the millers; and the prohibition on the export of grain, which "limit[ed] the industry of the cultivator and prohibit[ed] him, in a sense, from harvesting more wheat than the nation could consume."[61] Although Lavoisier identified the same impediments to prosperity as the Physiocrats, he differed from them in that he assessed these obstacles as political not only in their origins, but in their effects. The Physiocrats held that taxation and regulation

60. The seminal texts were Henri-Louis Duhamel du Monceau, *Traité de la culture des terres* (1750–56) and *Eléments d'agriculture* (1762). On Duhamel du Monceau, the origins of agronomy, and the new agriculture, see Bourde, *The Influence of England* (1953); Viel, "Duhamel du Monceau" (1985); and Dinechin, *Duhamel du Monceau* (1999).

61. Lavoisier, "Assemblée d'Orléans" (1788), in *Oeuvres*, 6:256–58, 260–61; "Instruction" (1787), in ibid., 6:205, 248; "Expériences" (1788), in ibid., 2:814–16, 819–21; "Encouragements" (1787) in ibid., 6:218–19.

Figure 4.2. Anonymous Revolution-era engraving representing the peasantry crushed under the weight of the *taille*, the *corvée*, and other taxes pressed down by the clergy and the nobility.
© Photothèque des musées de la ville de Paris, negative: Habouzit.

impeded the operations of nature, but Lavoisier argued that these institutions distorted the social conditions necessary for the production of wealth.[62]

There were many, then, even among those who advocated some of the same reformist policies as the Physiocrats, who did not harness their recommendations to an economic philosophy that naturalized wealth. This was the Physiocrats' innovation. In making wealth a natural fluid like Franklin's electricity, they armed themselves with the arguments of sentimental empiricism: that the economy was not a mechanical arrangement to be rationally grasped and deliberately manipulated, but an organic, balance-seeking process whose flow must be left unimpeded; that good administration was a matter of sensitivity to this organic process. These arguments shaped the meaning of Physiocratic policies such as freedom of commerce. In Condillac's hands, freedom of commerce expressed something quite different: a conviction that value was not an objectively measurable quantity but the product of subjective human needs, uses, and judgments. With Smith, freedom of commerce became

62. "What is interesting is that it should have been a scientist who came to these conclusions. For what could science do about any of these, the real problems?" Gillispie, *Science and Polity* (1980), 387. Cf. Bensaude-Vincent, "Balance" (1992), 230.

something else again: a response to central tendencies of human nature. To the Physiocrats, however—among the earliest advocates of a policy of free trade, and promulgators of what has remained that policy's slogan for two and a half centuries, "*laissez-passer . . . laissez-faire*" [63]—freedom of commerce was the policy of an administration sensitive to a purpose-driven natural world.

Extending the language and principles of sentimental empiricism to political economy, the Physiocrats armed themselves with a powerful set of arguments. They also made themselves vulnerable to the same arguments in the hands of their enemies.

II. THE SPIRIT OF SYSTEM AND THE FALL OF TURGOT

As finance minister from 1774 to 1776, Turgot tried assiduously to enact Physiocratic economic reforms but was hampered by an increasingly fervent opposition. The most common and damning accusation it leveled against him and the Physiocrats was that they were guilty of the "spirit of system." As we have just seen, the Physiocrats themselves had borrowed "system" as a term of abuse from sensibilist natural science. But their opponents appropriated the language of sentimental empiricism in the mid-1770s, making "system" their own catchword. To understand its meaning at that point, we must begin with its emergence as a term of censure around midcentury.

In the autumn of 1748, the Cartesian former secretary of the Académie des sciences, Dortous de Mairan, addressed an indignant paper to the assembly. He had had his fill of the sudden vogue of system-bashing: "System and chimera seem to be synonymous today. 'It's a system,' is often the whole criticism of a book, and to declare oneself against systems, and emphasize that what one offers the public is not one, has become a commonplace of prefaces." Mairan identified a certain false modesty in protestations that the human intelligence was too feeble to achieve a rational understanding of nature: "Those who condemn us to an eternal ignorance of first principles, have they so perfectly seen the heart of things? . . . [O]ne must know a lot to decide thus the extent of human knowledge, present and future." [64] The initial targets of the epithet "spirit of system" had been the mechanical systems of Cartesian physics, but the field grew to embrace all branches of natural and moral science. The year

---

63. Quesnay, "Lettre de M. Alpha" (1767), in *François Quesnay*, 2:940. Quesnay's usage of the phrase was not the first, nor was Vincent de Gournay's (to whom Quesnay himself attributed it). But it was the Physiocrats who transformed *laisser faire* into a slogan. On the history of the phrase and the Physiocrats' role in promulgating it, see Beer, *Inquiry* (1939), 89, 122; Oncken, *Die Maxime Laissez faire* (1886); and Quesnay, *Oeuvres économiques*, ed. Oncken (1888), 671–72 n. 1.

64. Mairan, *Dissertation* (1749), v, vii.

after Mairan's speech, Condillac wrote, "If there is a field where we have been warned against systems, it is politics."[65] When, in the same year, d'Alembert described Montesquieu's *De l'esprit des lois* (1748) as "the physics of Descartes applied to politics," he clearly intended no compliment.[66]

Although he himself occasionally used "Descartes" as a term of opprobrium, d'Alembert was worried about the vogue against systematic philosophy. In 1751 he urged the members of the Académie that "having learned to be suspicious of our industry, we must not carry our suspicions to an excess." Like Mairan, d'Alembert suggested that dogmatic skepticism was no better than other forms of dogmatism. It was "often as dangerous to pronounce upon what [the mind] . . . cannot do, as upon what it can."[67] The first volume of Diderot and d'Alembert's *Encyclopédie* appeared in the midst of these polemics, and d'Alembert had the delicate task of writing its methodological preliminary discourse. He reached for compromise. The "spirit of system" invoked arbitrary metaphysical hypotheses, he conceded, and that was bad. But this pernicious tendency must be distinguished from the good "systematic spirit," which rigorously reduced as many phenomena to as few principles as possible. D'Alembert's attempts at conciliation often resulted in confusion. On the one hand, he wrote that "vain" geometry consisted of "mind games," its axioms empty tautologies. On the other hand, the mathematical sciences were the only source of certainty about nature. They alone were destined "always to perfect themselves."[68]

D'Alembert cited Condillac, who, like Mairan and d'Alembert himself, warned those who denounced system-building against falling "to the other extreme, and assert[ing] that there is no knowledge at all to which we may aspire." Condillac had proposed pure mathematics and technical languages as good systems, and d'Alembert followed suit, arguing that safety lay in rigorous calculation, which was "seldom found in those frivolous conjectures we honor with the name of systems."[69] Throughout his articles in the *Encyclopédie*, d'Alembert sprinkled similarly mixed messages. In "Geometry" he

65. Condillac, *Traité des systèmes* (1749), chap. 15, 249.

66. D'Alembert to Gabriel Cramer, 21 September 1749; cited in Hankins, *D'Alembert*, 81.

67. D'Alembert, *Réflexions* (1751), 6, 18–19.

68. D'Alembert, "Discours préliminaire" (1751), in Pons, ed., *Encyclopédie*, 1:89, 92–93, 95, 91, 155–56. D'Alembert's insistence on the status of the mathematical sciences was polemical since the debate over system-building was marked by an exodus away from mathematics. Buffon, Voltaire, and Diderot, all of whom had had an early interest in the discipline, became disenchanted with it in the years leading up to midcentury, and Diderot reported "a general movement towards natural history, chemistry, and experimental physics." Diderot, "Encyclopédie" (1755), in Pons, ed., *Encyclopédie*, 2:49. See Hankins, *D'Alembert* (1970), 99.

69. Condillac, *Traité des systèmes* (1749), 27–28, 196, 267; Hankins, *D'Alembert* (1970), 107–8; D'Alembert, "Discours préliminaire" (1751), in Pons, ed., *Encyclopédie*, 1:155–56.

wrote that although this science had "all kinds of detractors among us," its un-deniable truths perhaps provided the only means for victims of despotism to "[shake] off the yoke of oppression." [70] And in "Expérimental" he surmised that "this would be the place to make some observations on the abuse of cal-culations and hypotheses in Physics, if that object had not already been filled" amply by other writers. Perhaps some natural philosophers did misuse math-ematical and mechanical reasoning, but "finally, what do men not abuse?" [71]

Yet Mairan's, d'Alembert's and Condillac's protestations fell upon deaf ears. Indeed, although Condillac's complaint in his *Traité des systèmes* was that most supposed systems were unworthy of the name, his contemporaries took the *Traité* as their rallying cry against the spirit of system. [72] Diderot, for one, was not pleased by d'Alembert's ambivalent and often approving attitude toward system-building. [73] *De l'interprétation de la nature* (1753) was Diderot's published rebuttal to d'Alembert's "Discours préliminaire." [74] It began with a poetic cele-bration of raw experience: "It is about nature that I am going to write," Diderot announced, and to do so properly, "I will let the thoughts flow from my pen, in the same order in which the objects present themselves to my reflection." He identified his subject as "that phenomenon that seems to occupy all of our philosophers, and to divide them into two classes," the phenomenon of sys-tematic speculation. Citing Buffon, he affirmed that mathematical truths were empty tautologies. He called mathematics a game, an affair of mere conven-tion, and, notoriously, predicted the imminent demise of geometry.

Diderot exempted from eternal obscurity those "happy geometers" (d'Alembert) who maintained an interest in the beaux arts. [75] This was a con-cession to friendship, an attempt to find redemption in d'Alembert's mostly systematic spirit. Some years later, in *Le Rêve de d'Alembert*, Diderot made a fic-tional d'Alembert, lost in a philosophic dream, recognize the unsystematic vi-tality of a swarm of bees as the basis of all things in nature, animate and inan-imate. In the lucidity of his dream, the d'Alembert character accepts Diderot's principle of "sensibility" as a "general and essential quality of matter," with its implication that even a "stone feels." [76] The systematic mind was closed to this vibrant truth, Diderot implied: the capacity of all things in nature to feel.

70. D'Alembert, "Géométrie" (1757), in Diderot and d'Alembert, eds., *Encyclopédie*, 7:628.

71. D'Alembert, "Expérimental" (1756), in Pons, ed., *Encyclopédie*, 1:92–93.

72. See Anon., "Système," in Diderot and d'Alembert, eds., *Encyclopédie*, 15:777; Knight, *Geo-metric Spirit* (1968).

73. Pappas, "L'Esprit" (1972), 1235; Hankins, *D'Alembert* (1970), chap. 4. Buffon disapproved of d'Alembert's discourse as well. See Roger, *Buffon* (1989), 264–65.

74. Hankins, *D'Alembert* (1970), 89–90; Pappas, "L'Esprit" (1972), 1235.

75. Diderot, *De l'interprétation de la nature* (1753), 177–80.

76. Diderot, *Le Rêve de d'Alembert* (1767), 291–92; "Entretien entre d'Alembert et Diderot" (1767), 257–58.

It is an old theme of Enlightenment historiography that the philosophes used natural philosophical words, especially the words "nature" and "reason," as coded promotions of social and political reform.[77] But this observation does not do justice to the versatility of political rhetoric drawn from the language of natural science. For one thing, the arguments of natural science served both sides of the political spectrum, reformers and conservatives alike.[78] For another, this language included negative slogans that were at least as powerful as the positive ones; witness the word "system." Natural philosophers' terms of opprobrium for themselves and one another were effective and spread quickly from philosophy to politics. Peter Gay has called the philosophes a bickering family whose criticisms of one another provided the counter-Enlightenment with its most persuasive claims.[79] The central family quarrel of the second half of the century, the quarrel over sensibility and system-building, provided political slogans that were both potent and versatile, serving the needs of reformist administrators and of their critics as well.

Turgot's tenure as controller-general of finances from 1774 to 1776, the first years of Louis XVI's reign, was the philosophes' moment of closest involvement with the court and the height of their political power. Turgot had served as intendant of the Limousin, one of the poorest regions of France, for the previous thirteen years. There he had established a laboratory of enlightened political economy and administration, successfully enacting the essentials of the Physiocrats' program.[80] He had reformed the tax system, compiling a land register and lifting feudal taxes, the *taille*, from which the clergy and nobility had been exempt, and the *corvée*, a tax of labor on unlanded peasants for the maintenance of roads. He had also maintained a local free trade in grain through difficult conditions with good results.[81]

In 1763 and 1764, successive finance ministers had issued decrees freeing first domestic and then foreign commerce in grain. These were initially well received, but beginning in 1765, there was a sequence of bad harvests, and

77. Cassirer, *Philosophy of the Enlightenment* (1951), 248; Gay, *Enlightenment* (1966), vol. 2, chap. 3; and Baker, *Inventing* (1990), 159. For a contrasting view of the relation between scientific explanation and political authority, see Gillispie, *Science and Polity* (1980), 549.

78. Kaplan writes that in the political history of grain regulation, "parlementary politics" were "much more nuanced than is generally supposed." Members of the Parlement treated "political, social and economic questions without reference to the narrow range of motives usually ascribed to them." Kaplan, *Bread, Politics* (1976), 2:696.

79. Gay, *Enlightenment* (1966), 1:4–5.

80. Concerning Turgot's intendance in the Limousin, see Dakin, *Turgot* (1939); Maurepas and Boulant, *Ministres* (1996), 345–46.

81. For a discussion of these enlightened administrative measures, see Dakin, *Turgot* (1939), chap. 7.

grain prices began to rise. High prices brought extensive popular resistance to the free commerce laws, disobedience on the part of municipal authorities, and opposition by the Parlement of Paris and by provincial parlements, all of whom demanded a return to the old laws. This opposition to a free market in grain culminated in 1770 in a resumption of traditional trade restrictions. But Turgot received a special dispensation from the court to continue enforcing the free commerce laws of 1763 and 1764 in the Limousin, where the bad harvests had brought famine. In order to bring grain into the region, Turgot offered incentives to the merchants: he insured their losses, guaranteed their capital advances, and in addition promised them bounties. The result was that 800,000 livres worth of grain were brought into the Limousin during the famine years of 1769 and 1770 and sold at a loss of 5 percent as compared with a 30 percent loss on grain transactions in Paris.[82] Thus Turgot had supplied the Physiocrats and other reformist philosophes with a modest but significant track record. In 1774, with the death of Louis XV and the accession of his adolescent grandson to the throne, reformist administrators such as the comte de Maurepas, then minister of state, saw the chance for Turgot to apply their program on a grander scale.

One of Turgot's first acts as finance minister, in 1774, was to issue an edict freeing domestic commerce in grain. The edict left in place all restrictions on foreign trade and on trade in Paris and was essentially a reenactment of the 1763 law.[83] Like the 1763 law, Turgot's edict was initially quietly received in Parlement and in the provinces. But once again, in 1774, a poor harvest sent prices up, and as had been the case a decade earlier, high prices brought widespread opposition to free commerce, which culminated in the spring of 1775 in the bread riots known as the *guerre des farines*. Turgot responded in part by offering bounties to merchants who agreed to import grain, and by announcing an amnesty for those rioters who returned home. However, he also threatened to arrest those who did not—almost six hundred people were arrested and two were publicly executed—and quartered troops in the towns to enforce his edict. Where inducement failed, he fell back on force and used it unflinchingly. When the *guerre des farines* was over, he proceeded to expand the free

82. See Dakin, *Turgot* (1939), 104–10; Poirier, *Turgot* (1999), 125–32. Poirier writes that "in spite of his doctrinal liberal positions, Turgot planned *dirigiste* measures" in the Limousin because he believed that liberal reforms would take awhile to make their good effects felt (126). He also established a system of social welfare in his region. He issued a decree requiring each parish to convene an assembly to organize poor relief. These assemblies were to set up voluntary subscription lists or taxes to provide the poor with food, money, and work. Turgot himself set up soup kitchens and public works projects including a roadworks that produced three hundred miles of roadway. Dakin, *Turgot* (1939), 112–17; Poirier, *Turgot* (1999), 127–28.

83. See Morilhat, *Prise de conscience* (1988), 39–40; and Dakin, *Turgot* (1939), 177–79.

trade in grain with twenty-three new regulations eliminating taxes and suppressing the privileges that created monopolies.[84]

In the meantime, however, the Parlement of Paris had begun to oppose Turgot and his policies, demanding that the king reduce and fix the price of grain.[85] The members of the Parlement had abundant reason to dislike Turgot, with his repeated insistence upon free commerce against their opposition and his affiliation with the antiparlementarian Physiocrats. In the heated atmosphere of the *guerre des farines*, they began a campaign against him. Others, too, went on the offensive. Turgot's opponents, both within and outside the Parlement, shared a common language: they accused Turgot of being "systematic," lost in a dogmatic dream of how the economy should operate, insensitive to the nuance and complexity of the real world. By common consensus among his contemporaries, Turgot was a system-builder. This verdict pursued the Physiocrats and Turgot throughout his ministry and into his fall from power, becoming the slogan of the campaign against him.[86]

As early as the summer of 1774, when Maurepas first proposed Turgot to the new king for the position of controller-general of finances, Louis XVI replied doubtfully that he feared Turgot was "quite systematic." Maurepas responded with irritation that "none of those you might ask would be exempt from this criticism. . . . You will perhaps see that his systems reduce to ideas you find correct." Then, during the period before his official appointment, Turgot's close friend and adviser, the abbé Véri, promoted him by seeking occasions for him to "purge himself of the coloring of a man of systems."[87]

These accusations took on new force in the wake of a treatise by Necker, criticizing the Physiocrats' economic philosophy in general, and Turgot's (limited) implementation of it in particular, entitled *Sur la législation et le commerce des grains*. Necker's attack was published with Turgot's overconfident blessing in 1775, at the height of the *guerre des farines*.[88] In the work Necker decried "men who meditate in their study" and contended that truth "refuses

84. Dakin, *Turgot* (1939), 181–90; Baker, *Condorcet* (1975), 61; Rudé, *The Crowd* (1964), 30.

85. Dakin, *Turgot* (1939), 185–88.

86. "The opposition," according to Weulersse, incessantly "reproached the Economists for their spirit of system." Weulersse, *Physiocratie sous . . . Turgot et Necker* (1950), 223; *Physiocratie à l'aube* (1950), 349. Poirier similarly observes that by the end of 1775, a profusion of "songs, pamphlets and epigrams described [Turgot] as an encyclopedist, a man of systems, a stubborn type capable of driving the State to its demise rather than renouncing his ideas." Poirier, *Turgot* (1999), 280–81. Ironically, while the Physiocrats' allies were often practitioners, proprietors, or grain traders, rather than theoreticians, the Physiocrats' enemies, those who so deplored the spirit of system, tended to be philosophes. Diderot, for example, was a stern critic of the Physiocrats and their friends, accusing them of "abstract" reasoning. Kaplan, *Bread, Politics* (1976), 2:687–88, 697, 608.

87. Véri, *Journal*, 9 August 1774, 13 August 1774, 1:160, 173–74.

88. Ibid., 10 May 1775, 1:283.

any simple or general notion, surrounding itself with exceptions, reservations and modifications."[89] Morellet summed up Necker's work as a tissue of "middling and moderate opinions, and some declamations against the spirit of system, that is the whole work." Another critic concurred: "M. Necker is like all Writers who combat systems, destroying much more easily than he builds."[90] But others found the treatise persuasive, and it became enormously popular.[91] "Wherever our steps carry us in a rioting city," Mirabeau wrote during the *guerre des farines*, "everywhere there is bitterness ... against people of system."[92] Véri observed that a "furious cry" had arisen among "a class of people in the capital" labeling the freedom of commerce a "*dangerous system, a famishing system,*" and that the same people were calling to Maurepas to save "the State against a systematic head."[93]

Finally Baudeau, a friend of Turgot, having tired of all the talk of systems, published the following anecdote in his *Chronique secrète*: when a lady in the court accused Turgot of system-building, Baudeau retorted, "Yes, Madam, he is systematic, that is to say his ideas are well examined and joined by principles; as that is what the word *systematic* means. Ah! Do you believe, then, that to govern a realm like France, one must have ideas that are desultory?"[94] His words had been foreshadowed over a decade earlier by Turgot's own lament that "any man who thinks has a system," and yet "this name, 'man of systems,' has become a sort of weapon."[95] "Man of systems" would become Turgot's political epitaph. Shortly after his disgrace, a memoir writer reported that "Turgot's enemies do not stop letting loose against him" and insisting that his "views" were "too systematic."[96]

Historians have echoed contemporary analyses of Turgot's disgrace.[97] But this is worth reexamining. It was Turgot, after all, who recommended that

89. Necker, *Grains* (1775), 212–13.

90. Morellet, *Mémoires* (for the year 1775), 202; Bachaumont, *Mémoires*, 10 April 1775, 8:19.

91. Baker calls Necker's treatise "a patiently undogmatic appeal for a more pragmatic approach to matters of social legislation than Turgot's rational convictions would allow." Baker, *Condorcet* (1975), 63.

92. Mirabeau, "Discours de la rentrée," 1776; cited in Weulersse, *Physiocratie sous ... Turgot et Necker* (1950), 35.

93. Véri, *Journal*, 26 May 1775, 20 February 1776, 1:286, 412.

94. Baudeau, *Chronique secrète* (7 July 1774); cited in Weulersse, *Physiocratie sous ... Turgot et Necker* (1950), 35.

95. Turgot, "Eloge de Gournay," 1759, in *Oeuvres*, 1:618–19. In Dupont's edition of Turgot's works, the second part of the quote is replaced by "a man without a system or any linkage in his ideas could only be an imbecile or a lunatic" (619n).

96. Bachaumont, *Mémoires*, 16 April 1776, 9:86.

97. See, for example, Gillispie, *Science and Polity* (1980), 16–17; Cornette, "Turgot" (1995), 64. Cf. Dakin, *Turgot* (1939), 282–85. Dakin defended Turgot against charges of system-building, writing that Turgot was a sensationist who succeeded in giving "to sensational philosophy a richer and ... less ambiguous meaning than that with which Locke and many of his followers had endowed

natural philosophers look beyond nature's "motors" to her "motives." It was he again who wrote, in another letter to Condorcet, that geometry would never capture the atomic structure of matter, just as mechanism could never explain motion.[98] Nor did Turgot ever entirely accept the Physiocrats' system, although he was sympathetic to the essentials of their program.[99] He was often severe with his friend Dupont for his orthodoxy, criticizing Dupont's *La Physio-cratie* as "too systematic, too tight, too abridged by essential omissions."[100]

Turgot also mocked Dupont's loyalty to the Physiocrats' "Master," Quesnay. He wrote irreverently that "criticism based upon facts is not the strength of the Master" and deplored the publication of one treatise by Quesnay as "truly the scandal of scandals, it's the Sun crusting over!"[101] Turgot reviled the Physiocratic doctrine of "legal despotism" to ensure proper taxation, which "incessantly dirtie[d] the works of the economists."[102] Although he advocated a single tax on land, he did not admit the exclusive productivity of agriculture. He supported the liberty of commerce and believed the economy would seek its own balance, but not as a consequence of the agricultural origins of wealth. He also distrusted Quesnay's *Tableau économique*.[103]

Neither as a mechanist, nor as a mathematician, nor as a Physiocrat, therefore, was Turgot systematic in his commitment. What then was the spirit of his system-building? Charles Gillispie has diagnosed a certain dogmatic rigidity

---

it." Lockeans, seeking to make the thin database of "immediate experience" support the full weight of their philosophy, inevitably fell back upon theoretical presuppositions in their interpretations of experience. According to Dakin, Turgot's broader conception of "experience," embracing the experience of the past as well as the immediate experience of the senses, was better able to found philosophical understanding without recourse to dogma. Another, more recent defense of Turgot is to be found in Poirier, *Turgot* (1999). Poirier describes Turgot as "driven by a concern for the public good and social justice," as a "technocrat" and liberal reformer who nonetheless made concessions to the contemporary political and social realities by incorporating *dirigiste* measures into his reforms, and as the "most brilliant advocate" of a set of sweeping social reforms ultimately put in place by the Constituants in 1789 (371, 368–69, 374).

98. Turgot to Condorcet, 27 October 1772 and 13 November 1772, in Condorcet, *Correspondance*, 101, 106.

99. Dakin, *Turgot* (1939), 302. Claude Morilhat writes that Turgot's central difference from the Physiocrats was his distaste for their systems, arising from his sensationist-empiricist impulses: "Sensationist empiricism was the principle of Turgot's theoretical position in regard to the physiocratic school." Morilhat, *Prise de conscience* (1988), 209.

100. Turgot to Dupont, 18 November 1767, in Turgot, *Oeuvres*, 2:677.

101. Turgot to Dupont, 26 December 1769, in ibid., 3:78. Turgot referred to Quesnay's *Recherches philosophiques sur l'évidence des vérités géometriques*, which was published in 1773.

102. Turgot to Dupont, 10 May 1771, in ibid., 3:486.

103. See Turgot, "Eloge de Gournay," 1759, in ibid., 1:595; Fox-Genovese, *Origins of Physiocracy* (1976), 67. Concerning Turgot's emphasis, distinguishing him from the Physiocrats, on "moral causes over physical causes" in political economy, see Fontaine, "Turgot's 'Institutional Individualism'" (1997), quote on 3.

that rendered Turgot unable to reconcile the demands of efficient administration with those of diplomacy.[104] Yet consider Turgot's response to the cattle plague that began during his first year as finance minister. Henri Bertin, secretary of state for agriculture, had recommended the slaughter of infected animals as the safest method for containing and ending the plague. But Turgot hoped for a less drastic approach and requested a commission from the Academy of Sciences to look into the matter. In his correspondence with the commissioners and local notables in the plague-stricken areas, Turgot continually cited the limits of human control and of his own knowledge and the importance of remaining sensitive to the farmers' suffering.

In one letter he explained his recommendation of slaughtering only the first eight or ten sick animals. If more than that number were infected, he reasoned, the whole parish must be written off, "in which case it could be humane not to take from the Proprietors the feeble glimmer of hope that they might save those still alive." Moreover, Turgot added, "I pray you to remember that I am very far from the center of the evil, and . . . believe I can do no better than to yield my judgment to those who are better established to see the effects of the harms and remedies."[105]

Turgot was authoritarian in the enforcement of his policy of limited free commerce in grain. But authoritarianism is not the same thing as system-building. Keith Baker has noted that Turgot found nothing "more to be feared and repressed than the irrational disposition of the inflamed and ignorant mob."[106] It was this fear, which Turgot shared with many of his enlightened contemporaries, and not an attachment to abstract principles or an imperviousness to real-world complexities, that drove his response to the popular resistance against his policy of free trade. Answering a worried subordinate who wrote to say that the people of Dijon continued to oppose the liberty of commerce, Turgot replied that it was "not the people that should guide you, but the Law." He authorized severity in enforcing it "to make an example that intimidates the people and serves to contain them." True, Turgot's belief in free commerce seemed unshakeable. He insisted that the local inspectors, by visiting bakeries to investigate the hoarding of grain, had themselves caused all the "tumult." Urging them not to manage the market but to allow it to right itself, Turgot insisted that free trade would make the prices "as low as they can be." In any event, he maintained, even if the prices did remain high, "they

104. Gillispie, *Science and Polity* (1980), 16–17.

105. Turgot to M. l'Archevêque de Narbonne, 1 January 1775, AN, F12 151. Charles Gillispie has called it "characteristic" of the results of liberal reform during that period that Turgot and his commission "ended by ordering slaughter far more widely than Bertin had imagined." Gillispie, *Science and Polity* (1980), 28.

106. See Baker, *Condorcet* (1975), 61.

themselves [would] calm the alarms they cause" by attracting wheat from other regions to ensure provisioning. On the other hand, as we saw earlier, Turgot helped the forces of "nature" with a scheme of incentives for merchants and, in the Limousin, with an extensive new apparatus for poor relief. He also noted in his correspondence with local authorities that his instructions were "not a rule to follow scrupulously in all its particulars; but an indication of what can be done and how one can proceed; and I yield to your prudence for whatever you believe needs to be changed." [107]

Free commerce in grain was a policy that Turgot was neither the first nor the last to institute, and his version of it was comparatively moderate.[108] He made important concessions. The "Marseillaise du blé," Turgot's 1774 edict freeing the commerce in grain, was significantly limited, applying only to the domestic trade and only outside of Paris.[109] Over the opposition of Hue de Miromesnil, Louis XVI's keeper of the seals, Turgot also acceded to demands that the clergy be exempted from the general tax with which he proposed to replace the *corvée*.[110] On the evidence, then, just as Turgot was not especially rigid in his commitment to theoretical systems—quite the contrary—he was also not unusually "systematic" (although he could be authoritarian) in his implementation of policies. Certainly he was not utterly flexible, but neither was he hopelessly doctrinaire. Like most administrators, he was somewhere in between.

Turgot's problems may have been exacerbated by a diplomatic and social incompetence. He was described as shy, awkward, and brusque.[111] "There is currently much noise on the subject of M. Turgot," wrote one observer of the court in the spring of 1775, who professed to offer his readers "the exact truth of what one must think of this Minister." He reported that Turgot's fellow courtiers found him "inflexible" and even his friends judged him to be "too severe." [112] Véri frequently cautioned Turgot about his overconfident temperament. "Without regard for persons, without consideration of the ignorance in which you might be of a thousand details," Véri scolded, "you pronounce your judgment ... you never pronounce a word that might signify the slightest hesitation." Sometime later Véri demanded, "Will you never learn

107. Turgot, "Commerce des grains en Bourgogne," AN, H 187, #108, 110, 111, 119.

108. Three finance ministers under Louis XV—Jean-Baptiste de Machault d'Arnouville (1745–54), Henri-Léonard-Jean-Baptiste Bertin (1759–63), and Clément-Charles-François de Laverdy (1763–68)—had pursued policies of liberalizing the grain trade. Turgot's immediate predecessor, Joseph-Marie Terray (1769–74), had reinstituted taxes and restrictions. See Maurepas and Boulant, *Ministres* (1996), 346–37.

109. See Morilhat, *Prise de conscience* (1988), 39–40. The quote is from Jules Michelet.

110. Véri, *Journal,* March 1776, 1:423.

111. See Faure, *Disgrâce* (1961); Hardman, *French Politics* (1995), 50–51.

112. Métra, *Correspondance* (1787), 1:268.

to condescend to an evil" for the sake of diplomacy? Turgot's response was all good-natured humility. "I agree with you on all the points of your sermon," he teased. Still, after Turgot's fall from power, Véri judged that his "character ... had more to do with his dismissal than any other cause." His "considerable" faults all bore upon his "qualities as a courtier and a colleague."[113] Louis XVI's ultimate decision to dismiss Turgot in disgrace in 1776 seems to have resulted mainly from his alienation of many people at court, especially Marie-Antoinette and her entourage, who resented the economies Turgot tried to impose on the royal household.[114]

That system-building was the loudest, commonest, and longest-lived charge against Turgot says as much about the charge as it does about Turgot. Here was the epithet of the moment, expressing the terms in which people had come to understand political as well as philosophical wrong-headedness. The Physiocrats themselves had brought the language of sensibility from natural philosophy into political economy in their invention of a natural-moral science, and their critics responded in kind. The philosophes had long used "spirit of system" to associate qualities of character—solipsism, arrogance, insensitivity, authoritarianism—with a quality of intellect, extreme rationalism. Turgot's critics reversed the relation, associating the flaws of character they discerned in Turgot—arrogance and insensitivity—with a type of intellect, the system-builder. Their insistence that Turgot was a man of systems was the converse of Diderot's argument that geometers were inhumane. Rationalism was insensitivity, and insensitivity rationalism. In the same way, historians have understood Turgot's authoritarian enforcement of free trade as system-building, eliding the political evil with an intellectual one.

Toward the end of Turgot's tenure as finance minister, he became embroiled along with Louis XVI in a major struggle with the Parlement of Paris concerning the registration of the so-called Six Edicts with which Turgot intended to enact a new set of reforms. The most important of the edicts replaced the *corvée* with a tax on all landowners except the clergy; another abolished guilds; and the remaining four eliminated various dues and offices. The Parlement received the Six Edicts from the king in February 1776 and refused to register them. Instead, it issued two remonstrances, one against the edict elimi-

113. Véri, *Journal*, June 1775–May 1776, 1:319, 328, 331, 392, 446.
114. Hardman, *French Politics* (1995), 50–51; Maurepas and Boulant, *Ministres* (1996), 348–49; Cornette, "Turgot" (1995), 69; Poirier, *Turgot* (1999), 252–63. Even Turgot's enemies at court accused him of system-building. One of Marie-Antoinette's allies against Turgot, the baron de Besenval, wrote in 1776 that Turgot was an "*homme d'esprit*, but systematic from having long worked in a study, through which he arrived at speculations mostly false or impracticable, the usual pitfall for people given to a metaphysical theory who always go astray in administration." Besenval, *Mémoires* (1805–7), 187. See also Poirier, *Turgot* (1999), 256.

nating the *corvée*, and the other against the other edicts. Upon receiving these the following month, Louis XVI forced the Parlement to register all six edicts in a *lit de justice*.[115]

The remonstrances were written in a language that was, by then, familiar. Against the proposed tax to replace the *corvée*, the Parlement cited the "natural right" to property, a "law of the Universe that, despite the efforts of the human mind, maintains itself in each empire." The remonstrances continually referred to free commerce as a "system" and relied upon the following arguments: Generalities belied the diversity of nature and human society; Paris was not like the provinces, the grain trade was not like other commerce, and these distinctions must be reflected in fiscal policy; the Economists had only speculative "knowledge without practice"; their indifference to practical concerns had transformed mere opinions into systems; whereas an isolated opinion might easily be forgotten, systems, by their very nature, were evil.[116] When the Parlement referred to pernicious "systematic opinions," Edgar Faure, chronicler of Turgot's disgrace, has noted "the word 'systematic' indicated Turgot from a mile away." [117]

It is a poignant detail that the Parlement appropriated these arguments not just from the writings of the philosophes in general, but from Turgot himself. Searching for weapons to use against him, researchers in the Parlement studied his early discourses at the Sorbonne and circulated a passage from his *Discours sur les avantages que l'établissement du christianisme a procurés au genre humain* (1750). The passage begins: "Sorrow to those nations in which the spirit of system has directed the legislators." [118]

In the midst of the struggle between the Crown and Parlement, a supporter of Turgot, Pierre-François Boncerf, added fuel to the fire by publishing a pamphlet entitled *Les Inconvénients des droits féodaux* (1776), in which he advocated the elimination of seigneurial privileges.[119] The Prince de Conti, a supporter of the queen and enemy of Turgot, called the Parlement's attention

---

115. Parlement de Paris, *Remontrances* (1888–98), 3:275–77, 326–56; Cornette, "Turgot" (1995), 68. A *lit de justice* was a special session of Parlement in which the monarch arrived in person to reclaim his supreme legislative authority. The Parlement in turn expressed its opposition to the law by recording it as registered "at the express command of the king." See Hardman, *French Politics* (1995), 263.

116. Parlement de Paris, *Remontrances* (1888–98), 3:277–324. The Parlement also neatly inverted the Physiocrats' argument against regulation, that it gave the state undue control over a natural process; the Parlement claimed that *freeing* the grain trade made the state responsible, in the eyes of the people, for fluctuations in price and supply that had previously been blamed upon seasonal variation alone. See Kaplan, *Bread, Politics* (1976), 2:446–47.

117. Faure, *Disgrâce* (1961), 486.

118. Métra, *Correspondance* (1787), 3:26–27.

119. Boncerf (under the pseudonym Francaleu), *Les Inconvénients des droits féodaux* (1776).

to the pamphlet, and the Parlement promptly ordered it shredded on the great staircase of the Palais de Justice. After the ceremonial shredding, the Parlement intended to interrogate Boncerf and Pidansat de Mairobert, the censor who had permitted publication of the pamphlet, but Louis XVI put his foot down, forbidding any further hostilities. Here was an occasion for a third Turgot-related remonstrance.[120]

This one was in the full flower of sentimental empiricism, with warnings against the hubris of rationalism and the dangerous detachment of mind from natural experience. Boncerf's pamphlet was—what else?—"purely systematic." The Parlement warned the monarch, "There are limits that nature prescribes to man; there is, so to speak, a wall of separation that she forbids him to breach." Having arrived at this limit, with "nothing more to discover," the "simple progress of spirit" gave way to "subtleties of mind." Man began to "blame that which exists" and to substitute for it "systems built upon sophisms." Thus, the Parlement concluded, "the multiplicity of systematic writings in a nation is almost always a sign of decadence . . . and . . . of revolution."[121] In a final remonstrance in response to the *lit de justice* at which the Six Edicts were registered, the Parlement warned against "purely speculative" principles and wrote, "Your Majesty will doubtless see in the totality of all these edicts, the branches of a system that destroys ceaselessly." When Louis XVI received the remonstrance in May, he had already fired his ill-starred finance minister.[122] J. E. B. Clugny, Turgot's successor, repealed his reforms. The *corvée*, the guilds, and the old restrictions on the grain trade would now survive until the Revolution.[123]

That the Parlement had successfully appropriated its targets' own sources of uncertainty and ambivalence is reflected in Mirabeau's mournful and self-accusatory assessment of the Physiocrats' position in 1776: "Always," he wrote, "the greatest number, even while following our principles and profiting from our work, will say that the Economists were . . . people of imagination and of systems, who dazzled and caused illusions because they suffered from them themselves. . . . They'll say that, Messieurs, and they won't be wrong."[124] The philosophes' self-directed admonitions against system-building had spread during Turgot's tenure as finance minister to permeate the language of the

120. Parlement de Paris, *Remontrances* (1888–98), 3:356–64.

121. "Remontrances sur l'interdiction des poursuites, dirigée contre l'auteur de la brochure sur les inconvéniens des droits féodaux" (March–April 1776), in Parlement de Paris, *Remontrances*, 3:362–63.

122. Parlement de Paris, *Remontrances* (1888–98), 3:374, 384, 368.

123. Dakin, *Turgot* (1939), 264.

124. Mirabeau, "Discours de la rentrée," 1776; cited in Weulersse, *Physiocratie sous ... Turgot et Necker* (1950), 46.

court, the king, and the Parlement.[125] Just over a decade later, it would be heard on the floor of the National Assembly.

III. THE NATIONAL ASSEMBLY
VERSUS THE SPIRIT OF SYSTEM

In 1776, six months after Turgot's disgrace, Franklin arrived in Paris for the third time. He had then just taken a revolutionary stand on the question of colonial taxation and was as much an Economist as an electrician. This new collaboration, like the older one, was mutually beneficial. Franklin had endorsed the Physiocrats' economic program, and they now supported his political arguments. These were principally against checks and balances and in support of a unicameral legislature. The Physiocrats took these campaigns to be further applications of Franklin's sensible skepticism about mechanical arrangements. La Rochefoucauld approved of the unicameral legislature as the political expression of Franklin's avoidance of systematic, mechanist complexity. Bicameralism, with its "mechanical" system of checks and balances, was the spirit of system applied to governing. Quesnay deplored the "system of counterweights in government" as a "fatal opinion," and so did Turgot.[126] Condorcet hoped that in light of the example set by the "naive expression of good sense" in American government, people might "stop promoting these complicated machines, in which a multitude of springs make the operations violent." [127]

The Physiocrats were more successful at campaigning against "mechanist" politics in Franklin's name than they had been on their own behalf—and, indeed, than was Franklin himself. Though the unicameralists lost the American debate, they remained strong in France. A little over a decade later, during their own revolution, French Physiocrats and others continued to adhere to what they saw as a Franklinist scientific method applied to politics. Franklin's support of a unicameral legislature was, according to Alfred Owen Aldridge, even more influential in France than his condemnation of a hereditary aristocracy, because of its measurable effect upon "particular leaders in the French

125. There are many examples of Parlement and royal appropriations of Enlightenment vocabulary. In one, Antoine-Louis Séguier, *avocat du roi*, told the king, "You have all around you ministers whose . . . enlightenment inspired hope for the end of these disgraces and the restoration of the old magistrature." "Lit de justice pour le rétablissement du Parlement" (12 November 1774), in Parlement de Paris, *Remontrances*, 3:238–39.

126. Aldridge, *Franklin* (1957), 86–87; Quesnay, "Maximes générales du gouvernement économique d'un royaume agricola" (November 1767), in *François Quesnay* 2:949; Turgot to Dupont, 10 May 1771, in Turgot, *Oeuvres*, 3:486–87.

127. Condorcet, "On the Influence of the American Revolution on Europe" (1786), in *Selected Writings* (1976), 80.

Revolution" and because, although "many other writers and statesmen condemned a hereditary aristocracy, . . . none but Franklin and his personal disciples championed a single legislature." [128] When Franklin died, La Rochefoucauld and Condorcet capitalized on his death by arguing for unicameralism in their eulogies. Condorcet lamented that Franklin had "witnessed with pain," during his own country's constitutional debate, "the majority decision to give a complicated form to an assembly that, by the nature of its functions, seemed to require the most simple." And La Rochefoucauld proclaimed that "Franklin alone" had "freed the political machine of these numerous gears, and these much-admired counter-balances that complicated it." [129]

Franklin's death in 1790 came at a moment of maximal uncertainty, volatility, and anxiety in French politics about the extent to which one should trust abstract theories of government over the experience of centuries. Most agreed, despite the general confusion, that Franklinism was the model of wisdom and safety. Prompted by Mirabeau, the National Assembly voted to declare three days of national mourning. Journalists reported that the vote for an official period of mourning had been unanimous, though the proceedings of the Assembly record that there were a number of pointed abstentions. One editorialist insinuated that Franklin would not have supported the new constitution; another demanded why, as "free men, we subject ourselves to these puerile and dishonorable observances?" But a third, who wrote that Franklin's life should interest all who could recognize "virtue and genius," was more typical. Sentimental empiricism, with its particular fusion of the moral and the epistemological, dominated the great majority of the editorials that celebrated Franklin and Mirabeau's motion for mourning. Gilbert Chinard has summed up the eulogists' message thus:

> This great *savant* . . . applied to political problems the method he followed in scientific research. In a century dominated by the love of systems, he devoted himself directly to attacking difficulties, without preoccupying himself with any system or any theory.[130]

The posthumous political celebrations of Franklin announced a general conviction that Franklin's moral and political qualities were tied to his scientific sensibility. La Rochefoucauld mentioned the "observational genius" that Franklin brought to bear "in the sciences and in politics." Le Roy praised Franklin's avoidance of systematic complexity: "In his philosophical and political views, [he] always grasped the simplest side of a question." Vicq-d'Azyr

128. Aldridge, *Franklin* (1957), 86–87.

129. Condorcet, "Eloge de Franklin" (13 November 1790), in Chinard, *L'Apothéose*, 137; La Rochefoucauld, "Discours" (1790), in ibid., 96.

130. Chinard, *L'Apothéose* (1955), 171–72.

cited the lightning rod as the supreme example of this talent for simplicity.[131] Brissot celebrated as supremely modest the man who had fashioned modesty into an epistemological tool, writing that "Franklin had genius, but still he had virtues, he was simple, good, modest above all. Ah! what talent can get by without modesty?"[132] And Condorcet lauded Franklin's forbearance from forming "a general system in politics," or trying to "give once and for all the highest degree of perfection to human institutions." The eulogists identified in Franklinism a tendency to follow "the order of events" and to "await the passage of time."[133] This tendency, they believed, reflected a continual appeal to a diachronic order governing natural and social phenomena, revealed in the dynamic equilibrium of exchanges (whether economic or electrical), rather than to what they saw as the artificial and timeless stability of a clock-work mechanism. An unchanging mechanism would be belied by experience. Historical experience, like sensory experience, could be understood only as it was felt.

The eulogists came not to bury Franklin but to praise him, and thereby to keep his influence alive for their own political purposes. After Mirabeau's eloquent announcement to the National Assembly beginning simply "Franklin est mort," tears allegedly "poured from every eye."[134] Weeping, the Assembly members translated sentimental empiricism into Revolutionary policy. The fusion of scientific method and moral theory, having been developed by natural philosophers over the preceding half-century, had been nascent policy for several decades. The king, his ministers, and the Parlement all spoke the language of sensibility.

Farmers as well as philosophes had considered the connection between Franklinist physics and Physiocratic politics; provincials as well as Parisians had debated it. The connection had been sharply apparent in 1764, for example, when Dupont addressed the Société royale d'agriculture de soissons. Dupont told the Société: "All is linked, all connects to the land, all is joined by secret chains, tokens of divine goodness and by an influence as rapid as electrical fire." He concluded that when "wealth spreads over a branch of cultivation, all the others feel the commotion."[135] A sensitive empiricism was the way to know the benevolent dictates of "divine goodness" and the force of electricity alike. "The most exquisite Folly is made of Wisdom spun too fine,"

131. La Rochefoucauld, "Discours" (1790), in Chinard, L'Apothéose, 93; Le Roy, "Note" (1790), in ibid., 121–22; Félix Vicq-d'Azyr, "Eloge" (1790), in ibid., 159.

132. Chinard, L'Apothéose (1955), 27. (The author is Jacques Pierre Brissot de Warville.)

133. Condorcet, "Eloge de Franklin" (1790), 140–41.

134. Chinard, L'Apothéose (1955), 171–72.

135. Dupont, "De l'exportation et de l'importation des grains, mémoire lû à la Société royale d'agriculture de soissons" (1764), in Oeuvres politiques, 1:120–21.

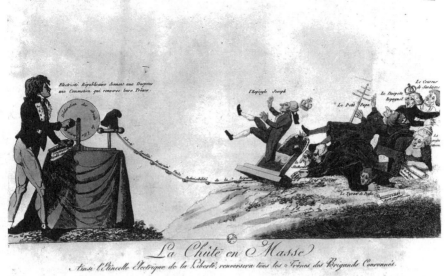

Figure 4.3. A later example of the electrical analogy applied to politics: "La Chute en Masse,
Ainsi L'Etincelle Electrique de la Liberté, renversera tous les Trônes des Brigands Couronnés"
("The mass-collapse: thus will the electric spark of liberty overturn all the thrones of
the crowned brigands"). Revolution-era cartoon showing an electrical generator labeled
"Declaration of the Rights of Man" generating "Republican electricity" that promises to "give
despots a commotion that will overturn their thrones" (at right are the collapsing despots).
Bibliothèque nationale de France.

cautioned *Poor Richard's Almanack* for 1746.[136] By the time Physiocrats and
other reformers argued, during the Revolution, that the economic and political balance should be modeled upon the flow of electricity (see fig. 4.3), not
the springs of a clock, France was well acquainted with the science as well as
the morals of Poor Richard.

136. *Poor Richard, 1746. An Almanack for the Year of Christ 1746.* By Richard Sanders, Philom. Philadelphia: Printed and sold by B. Franklin, in *BFP* (1960), 2:350.

Figure 5.1. A lightning conductor that, according to its inventor, could be either grounded or insulated at will, so that one could choose either to be protected from lightning or to perform experiments with it. From the frontispiece to Beyer, *Aux amateurs de physique* (1809).

Chapter Five

THE LAWYER AND THE
LIGHTNING ROD

He seized the lightning from heaven and the scepter from tyrants.
—Turgot's epigram of Benjamin Franklin

In the old French province of Artois,[1] sandwiched between Flanders and the Strait of Dover, within the little town of Saint-Omer, halfway from Calais to Lille, there lived an elderly lawyer and amateur *physicien*[2] named Charles Dominique de Vissery de Bois-Valé. One day in May 1780, thrusting skyward from the tallest chimney of Vissery's house in the rue Marché-aux-herbes, the gilt blade of a sword appeared.[3] At the blade's base was a weathervane, done in a useful and appropriate image: a globe, lightning-struck, spewing forth burning rays.

The blade and weathervane were screwed into a sixteen-foot iron bar that stood in a funnel of tin. The funnel had a very long tail, a fifty-seven-foot tin tube, which snaked over the roof and down the wall of the neighboring house. At two or three feet from the ground, it turned and made for a nearby well, whose curb it pierced before terminating in an iron rod. The rod ended in a ring; from the ring hung a chain; the chain plunged at last into water.[4]

The sword and darting rays of the weathervane prettily accomplished a directive from Barbier de Tinan's translation of the abbé Toaldo's memoir on the design of lightning rods,[5] Vissery's inspiration and guide. They were pointy

---

1. The pre-Revolutionary province of Artois corresponds roughly to the modern Department of Pas-de-Calais.

2. Here and throughout, I use the eighteenth-century French term *physicien* rather than "physicist," which denotes a more specialized and professionalized kind of thinker. *Physicien* meant something closer to "natural philosopher." *Physiciens* worked on, but did not restrict themselves to, subjects that would be included in the more restrictive category of "physics" that emerged in the next century, such as electricity.

3. Robespierre, "Premier plaidoyer" (1783), in *Oeuvres*, 1:29.

4. "Extrait du procès-verbal des notaires, du 7 août 1780," "Extrait des registres de l'Académie de Dijon du 18 août 1780," in Robespierre, *Oeuvres*, 1:103–7.

5. Barbier de Tinan, *Mémoires sur les conducteurs* (1779). See Antoine-Joseph Buissart, draft of memoir, Archives Générales de Pas-de-Calais, Collection Barbier [hereafter AGPC, Coll. Bar.], 4J/120/30, 2.

protrusions, whose utility for attracting and dispelling electrical fire was well established.[6] One or another of these points would, in principle, capture a thunderbolt whatever its path of approach. Such points, atop a metallic continuity stretching from ridge of house to wet earth, made a lightning rod. (For a slightly later depiction of a lightning conductor for amateurs, see fig. 5.1.)

Vissery was, he said, "animated by an enthusiasm for the public good." Several of his inventions, dedicated to that end, he reckoned "worthy of being presented even to the King": an unspecified device to preserve royal troops in battle; a second, related mechanism, especially for troops "exposed to the perils of Water"; a method to preserve water from taint for a year or more; and a device that allowed a diver to breathe, at the greatest depths, "a fresh and fortifying air." He claimed also to have introduced the electrical generator and the air pump to the "good patriots" of Saint-Omer. His purpose in establishing the conductor on his roof was, as always, to "serve humanity and the State."[7]

The affair of the lightning rod of M. de Vissery de Bois-Valé did influence the fortunes of the French state, and even arguably of all humanity. It began, however, as the most local of disputes, between Vissery and the neighbor down whose wall he trailed his lightning rod's tail. It became a three-year court battle, arriving finally, during the summer of 1783, at the provincial court of last appeal, the Conseil provincial et supérieur d'Artois,[8] located in the provincial capital, the city of Arras. By then Vissery's case had become a political lightning rod, the talk not only of Arras but of Paris. Its culmination launched the career of an unknown member of the Arras bar, so young he had still been in law school when the affair began,[9] Maximilien Robespierre (fig. 5.2).

Robespierre argued and won the case. He persuasively resolved the prob-

6. On the power of points, see Benjamin Franklin to Peter Collinson, 11 July 1747, in Franklin, *Franklin's Experiments* (1941), 171–78; Cohen, "Franklin's Hypothesis on Lightning Rods Confirmed in France," and "The Introduction of Lightning Rods in England," in Franklin, *Franklin's Experiments* (1941), chaps. 2, 3; and Heilbron, *History of Electricity* (1979), 327–28, 352–53, 379–82.

7. Vissery to Buissart, 7 September 1780, AGPC, Coll. Bar., 4J/120/2; n.d., ibid., 4J/120/4; 7 September 1780, ibid., 4J/120/2.

8. Artois was one of four provinces with "sovereign courts" (*conseils souverains*) instead of parlements. The Conseil d'Artois in Arras dated from 1677 and was the court of last appeal in criminal matters, in affairs relating to taxes and subsidies, and in civil cases of up to two thousand livres. With more at stake, one could appeal the Conseil's decision to the Parlement of Paris. See Marion, *Dictionnaire des Institutions* (1923), 137–38. The Conseil d'Artois consisted of one or two *présidents* and two *chevaliers d'honneur*; twenty *conseillers*; two *gens du roi* (an *avocat général* and a *procureur général*), who represented the Crown; five substitutes; and a *greffier en chef* (town clerk). Of these, the important figure for our purposes is the *avocat général*, Foacier de Ruzé, who argued opposite Robespierre in the 1783 trial. See Robespierre, *Oeuvres*, I:xxi–xxii n. 2.

9. Robespierre received his law degree in May 1781, and the following August, at the age of twenty-three, was received as *avocat* at the Parlement of Paris. But he decided to practice law in his native city, Arras, and was admitted to the Arras bar the following November. See Robespierre, *Oeuvres*, I:xiii–xiv.

Figure 5.2. Pierre Roch Vigneron's portrait of Robespierre, after Adélaïde Labille-Guiard's
1790 painting, shows Robespierre at thirty-two, seven years after his triumph in the Vissery case.
Copyright Réunion des musées nationaux/Art Resource, NY.

lem at its heart, namely, the proper relations of scientific to legal authority.
This problem acutely interested the principals in the drama, who were all
professional lawyers and amateur experimental *physiciens*—Robespierre
himself; Vissery; and the Arras barrister Antoine-Joseph Buissart, who was
Robespierre's senior colleague and Vissery's original lawyer in the appeal.

Buissart had argued that judges must defer to *physiciens*. Robespierre framed a more satisfactory solution. He drew upon the familiar ideals of sentimental empiricism.

PROLOGUE: EMPIRICISM IN NATURAL SCIENCE AND LAW

Law and physics in 1780s France shared a common, contradictory dogma: General truths are founded in particular facts not because of the facts' places in general theories, but, on the contrary, because of their irreducible particularity. The genesis of this understanding of facts—as bits of knowledge, or of observation, or of the world itself made factual by their rugged independence of any theory—along with related notions of objectivity, evidence, testimony, and proof, has recently been the focus of much attention on the part of historians and social theorists.[10] Their studies have shown overwhelmingly that although practitioners of natural science were central players in the promulgation of this distinctive view of facts, they did not act alone. Other disciplines, too, notably political economy and law, embraced and helped to develop the ideal of the disinterested empiricist, uncommitted to any system, whether political, philosophical, economic, or governmental, and of the disaffiliated facts to which such an empiricist clung. The stubbornly autonomous fact was not the preserve of the natural sciences, then, but was instead the product of an early modern interaction among natural, moral, and social science.[11] In this interaction, jurisprudence played a pivotal role, exporting legal standards of persuasion and importing scientific ones.[12] The conviction

10. See especially the work of Lorraine Daston, including "Baconian Facts" (1991), "Marvelous Facts" (1991), "Objectivity" (1992); and Daston and Peter Galison, "The Image of Objectivity" (1992). Daston has proposed that scientific objectivity, during its early history as an ideal of natural science, was an attribute of the most obstinate particulars, that is, monsters, miracles, and all those things that resisted reconciliation with a general theory. Daston, "Marvelous Facts" (1991). For an overview of recent work in this area, see Chandler et al., eds., *Questions of Evidence* (1993), which collects essays on evidence, objectivity, and proof by historians, social scientists, literary theorists, and legal scholars.

11. In her recent *History of the Modern Fact*, Mary Poovey tells the "story of how description came to seem separate from interpretation or theoretical analysis" in the "sciences of wealth and society" of the early seventeenth through the early nineteenth centuries. Poovey, *History of the Modern Fact* (1998), xii, 4. On empiricism in early modern jurisprudence, see Shapiro, *Probability and Certainty* (1983), chap. 5; and Daston, *Classical Probability* (1988), chap. 1.

12. Ian Hacking proposes that law provided the model for Leibniz's understanding of probability, certainty, and proof, in Hacking, *Emergence of Probability* (1975), 88. Barbara Shapiro argues that the reorientation of the natural sciences around "matters of fact" during the sixteenth and seventeenth centuries brought them into active engagement with jurisprudence. For example, "Bacon's contributions to legal thought were closely connected with his scientific views. His approach toward both law and nature was inductive, for he argued that one should keep close to the

that facts were facts by virtue of their resistance to theories belonged, in other words, as much to lawyers as to natural philosophers.

Thus, the lawyer and royal historiographer Jacob-Nicolas Moreau wrote in 1780, "Our public law had been, like physics, abandoned to systems; we have returned to experience, we have established facts." [13] Precisely by being local, by resisting preconceived theory and entrenched tradition, a fact had universal implications. The legal reform movement of the 1770s and '80s was founded on the conviction that the local facts of legal cases written up in *mémoires judiciaires* indicated a new, universal jurisprudence, deriving social from natural law. Barristers plundered the facts of their cases to make this argument. "I resolved," wrote Pierre-Louis Lacretelle, ". . . to lift these cases, as much as I could, to their true dignity." [14] Sarah Maza, in her study of the Old Regime's closing causes célèbres, therefore shows how the "'intimate' or 'particular' stories" of private life told in the briefs were central to the emerging public political culture, undermining the recently influential distinction between a private and a public sphere. [15] Facts of physics, like facts of family life, could be made to carry greater political morals simply by being facts. The most particular of natural facts were the true basis of an enlightened jurisprudence, according to the Bordeaux parlementarian, philosophe, and academician Charles Secondat de Montesquieu.

Not only did Montesquieu advocate empiricism in jurisprudence, but his was an empiricism of a distinctly sentimental variety, inflected, like the

---

particulars of each." Shapiro, *Probability and Certainty* (1983), 168–69. Daston found that "more than any other single factor, legal doctrines molded the conceptual and practical orientation of the classical theory of probability"—mathematicians' and philosophers' attempt to formalize rational belief—in part by supplying the conception of probability as a "degree of certainty." Daston, *Classical Probability* (1988), 6, 14. For the more recent history of the epistemological engagement between science, technology, and law, see Golan and Gissis, eds., *Science and Law* (1999); Alder, "To Tell the Truth" (1998); Golan, "Authority of Shadows" (1998); and Mnookin, "Image of Truth" (1998). On the history of legal epistemology and conceptions of evidence, see Nye, ed., *Evidence and the Law* (1998).

13. Jacob-Nicolas Moreau, in Xavier Charmes, ed., *Le Comité des travaux historiques et scientifiques (Histoire et documents)*, 3 vols. (Paris, 1886), 1:159; quoted in translation in Baker, *Inventing the French Revolution* (1990), 77.

14. Pierre-Louis Lacretelle, "Un Barreau Extérieur à la fin du XVIIIᵉ siècle"; quoted in Bell, *Lawyers and Citizens* (1994), 164. Of the legal reform movement, Bell writes, "The idea was . . . to join '*philosophie*,' in the eighteenth-century sense of the word, to jurisprudence"; legal reformers "began to draw on the full range of ideas, motifs, and vocabulary popularized by the *philosophes*"; "they bolstered their arguments with citations drawn . . . from natural law theorists, John Locke and Algernon Sydney." See Bell, *Lawyers and Citizens* (1994), 164, 202, 204. On barristers' participation in the Enlightenment, see also Maza, *Private Lives* (1993), 234–35, 212–62; and Berlanstein, *Barristers of Toulouse* (1975), chap. 4. On the participation of parlementarians in the Enlightenment, see Stone, *Parlement of Paris* (1981); Stone, *French Parlements* (1986); and Doyle, *Parlement of Bordeaux* (1974).

15. Maza, *Private Lives* (1993), 14. See also Bell, *Lawyers and Citizens* (1994), 207.

empiricist principles of his contemporaries in natural science, by the ideals of sensibility. Like Diderot and Buffon, Montesquieu worried about the solipsism of rational system-building. The Stoics with their "speculative" philosophy, for example, had been too interested in the problem of "knowing oneself." [16] This self-absorption was common to speculative philosophers and to despots. A despotic political system, like a speculative philosophical one, was "self-sufficient; all around it is empty." Despots, like system-builders generally, had a taste for reductive simplicity, thus "when a man makes himself more absolute, he thinks first to simplify the laws." In a monarchy the multiplication of exceptions and particular cases reflected the sensitive administration of justice.[17]

Montesquieu was well schooled in the ideals and concerns of sentimental empiricism due to his active involvement in the natural sciences as director of the Academy of Bordeaux. Reporting on the annual prize competitions, he wrote short discourses on an eclectic array of natural phenomena including echoes, transparency, weight in solid bodies, and the functioning of the kidneys. He also summarized memoirs submitted to the Academy treating topics in natural history, medicine, and experimental physics, and regularly communicated his own observations in natural history to friends and correspondents. Montesquieu was particularly interested in anatomy and medicine and drew heavily upon these fields in framing his legal philosophy. His interest and admiration were reciprocated. D'Alembert, Charles Bonnet, and Maupertuis all bestowed the Enlightenment's highest honor upon Montesquieu by comparing him to Newton. And Buffon included Montesquieu in a list of history's greatest men alongside Bacon, Newton, Leibniz, and Buffon himself.[18]

Throughout *The Spirit of the Laws*, Montesquieu emphasized the empiricist particularity of his principles. He had studied the "infinite diversity of laws and mores," being careful never to "regard as similar cases with real differences, nor to miss the differences in those that appear similar." Uniformity, he found, made a strong impression only on the weak-minded. "Greatness of genius" consisted in recognizing that difference was more important than sameness. As long as "the citizens follow the laws, what does it matter whether

16. Montesquieu, "Eloge de la sincérité" (1717), in *Oeuvres*, 43.

17. Montesquieu, *Esprit des lois* (1748), vol. I, bk. VI, chap. 1, 145–46; chap. II, 147; see also Keohane, *Philosophy and the State* (1980), 398.

18. Montesquieu, *Oeuvres*, 45–59. Shklar writes that Montesquieu's involvement in the natural sciences gave "his mind its particular cast," and supported "the very core of his political theory." Shklar, *Montesquieu* (1987), 12. She suggests that the Academy of Bordeaux with its focus on natural philosophy, and not the Parlement of Bordeaux, was "the real centre of [Montesquieu's] social and intellectual life" (10, 5). On the importance of Montesquieu's involvement in natural philosophy, see also Keohane, *Philosophy and the State* (1980); Grimsley, "The Idea of Nature in Montesquieu's Lettres Persanes" (1974); and Kiernan, *Science and Enlightenment* (1968), chap. 5.

these laws are the same?" Montesquieu also hated abstraction: the "laws must not be separated from the circumstances in which they were made."[19]

A long central section of the work argues that laws must reflect the "degrees of sensibility" of the people they governed, which in turn depended upon climate and terrain. "The character of the spirit and the passions of the heart are extremely different in the various climates," Montesquieu wrote, and "the laws should be relative" to these differences. Cold air contracted the body's fibers, increasing the spring with which they sent blood back to the heart, while warm air had the opposite effect. Montesquieu cited the visible contraction of iron bars and human extremities in cold weather. This explained the difference in character between the vigorous and courageous dwellers of the chilly north and the lazy, timid, and vengeful inhabitants of the sultry south. These principles were drawn not "from my prejudices but from the nature of things." Their independence was assured by their basis in empirical observation. Microscopic scrutiny of a sheep's tongue, for example, supported Montesquieu's theory of the temperamental differences between northerners and southerners. When the tongue was frozen, its papillae, which Montesquieu took to be the "principal organ of taste," were much diminished; when thawed, its papillae reappeared. This "observation confirm[ed]" that in northern countries "the tufts of nerves are less open . . . [and] sensations are therefore less vivid."[20]

Diversity was the rule of nature and the basis of political freedom. According to the traditional parliamentary doctrine codified in *The Spirit of the Laws*, a doctrine known as the *thèse nobiliaire*, liberty resided in a noble estate whose function was to intercede between the king and the people. The "intermediate powers" must preserve their autonomy through the proliferation of traditional privileges and exceptions. Thus, the robe nobility justified its opposition to taxes such as the *dixième* that would be levied equally against the three estates, and to all encroachments on regional customs and privileges. In the same way, it justified upholding the venality of offices and the resulting autonomy of the seigneurial class. The robe nobility's deeply traditional politics rested upon philosophical particularism. Montesquieu wrote that uniformity must mean either despotism or anarchy: "Abolish the prerogatives of the lords, clergy, nobility and towns in a monarchy; you will soon have a popular state or else a despotic state."[21]

---

19. Montesquieu, *Esprit des lois* (1748), vol. I, preface, ii; vol. IV, bk. XXIX, chap. 18, 145; chap. XIV, 137. "Laws must not be separated from the purposes for which they are made" (vol. IV, bk. XXIX, chap. 13, 135).

20. Ibid., vol. II, bk. XIV, chap. 2, 192; chap. 1, 189; vol. I, preface, XII; vol. II, bk. XIV, chap. 2, 191–92.

21. Ibid., vol. I, bk. II, chap. 4, 45.

Historians have offered radically divergent characterizations of Montesquieu's politics. To some he was "the hero of the liberal tradition"; to others, "the rationalizer of reaction." He opposed religious intolerance, slavery, and the old penal code; yet his proposed reforms were not inconsistent with feudal tradition. Franklin Ford has suggested that one must distinguish between the "urbane and likeable man of letters and the political theorist whose views were destined to serve the most reactionary groups in France." Yet Ford's own study of Montesquieu and his colleagues during Louis XV's reign indicates that one can understand its political history only by recognizing that the enlightened philosophe and the feudal *parlementaire* were one and the same. Combined with their traditional privileges, parlementarians had the advantages of a modern education and enlightened interests. They were historians, literary scholars, geographers, philosophes. In robe writings and robe libraries under Louis XV, there began to appear works on all the favorite topics of the day: the arts, travel, and natural science.[22]

Montesquieu personified the intellectual and institutional association of political and natural philosophy, and he was not alone. Just over half the "ordinary" members and more than a quarter of the "associated" members of the provincial academies were robe nobility. In the Academy of Bordeaux in 1740, eleven of the twelve permanent members were parlementarians. The Academy of Dijon was established in 1736 by the will of Hector Pouffier, dean of councillors in the Parlement of Burgundy. All five trustees of Pouffier's will, the first directors of the new Academy, were councillors in the Parlement of Burgundy, and their successors in the Parlement were to succeed them also at the Academy. Parlementarians were thus members and founders of the provincial academies whose proliferation represented "the most important single element" in the transformation of aristocratic doctrine during the eighteenth century. The Enlightenment provided the parlementarians with the tools to effect the intellectual development of the *thèse nobiliaire* and to identify and exploit the vulnerabilities of the opposing royalist doctrine. Their modern philosophy served their traditionalist politics while their politics inspired their philosophical explorations, shaping the programs of the provincial acade-

---

22. Ford, *Robe and Sword* (1965), 243–44. Resistance, defensiveness, and the "nois[y] reenuncia[tion]" (20) of old claims based upon medieval tradition comprised the stance, according to Ford, from which the high robe systematically obstructed the Crown and so, ironically, provoked the political crises and transformations of the late ancien régime. The resuscitation of an old tradition was in fact what allowed the Parlement to exercise its political philosophy in the years following Louis XIV's death. Its members seized the opportunity offered by a five-year-old monarch and a vulnerable regent to reclaim an ancient privilege, the right to remonstrate. Ford describes *The Spirit of the Laws* as the culmination of a reactionary aristocratic defense of feudal privileges against the Crown's attempts at fiscal and administrative rationalization. See ibid., 84–87 and chap. 12.

mies.[23] The moral and social implications of sentimental empiricism were familiar in courts and scientific academies throughout France by the 1780s.

Exploiting these implications, Robespierre beguiled his audience in both the Conseil d'Artois and the court of public opinion. He invoked a set of ideals and corresponding evils that were familiar in the overlapping arenas of legal and natural philosophy: on the good side, sensitivity, empiricism, and particularism; and on the evil, abstract speculation, theory, and system-building. Speaking a language that was by then common to physics and jurisprudence, the language of sentimental empiricism, Robespierre argued that judges need not defer to scientific experts.[24] Theory was a matter of expertise, but fact was not. Let academicians argue their theoretical differences among themselves; these could have no bearing on the important questions to be decided in a court of law. Leave aside the theories of both physics and jurisprudence, Robespierre proposed, and the two sciences would meet in the truth: in sensible facts.

Thus bringing physics and jurisprudence together, Robespierre reinforced a quirk that sensibilist empiricism had engendered in each. Where theory might connect particular facts to general knowledge, its elimination left a notable gap. When lawyers and *physiciens* rejected mediating influences between local experience and universal truth as pernicious, they produced arguments with a curious lacuna between the narration of facts and the derivation of morals. The personal stories told in legal briefs of the 1770s and '80s "seemed to have little relevance to the 'big issues' faced by the French nation," but they were routinely attached to "broader public implications."[25] A similar gulf between the particular and the general lay at the crux of a political philosophy that was then rapidly growing in popularity, that of Jean-Jacques Rousseau. At the start of the Revolution, Robespierre would publicly dedicate himself to Rousseauism,[26] and elements of Rousseau's theory were already implicitly

23. Roche, *Le Siècle des lumières* (1978), 2:413–15; Ford, *Robe and Sword* (1965), 216–17, 221, 235, 238 (quote on 235); Bell, "Lawyers into Demagogues" (1991), 117–18; Baker, *Inventing the French Revolution* (1990), 119–20; and Maza, "Tribunal" (1987), 80–82. Roche has written that "enlightened provincial academicism could never choose between two types of society," one traditional, corporatist, and aristocratic; the other innovative, egalitarian, and rationalist. "In its origins, in its ideological choices, its hesitations, its ambiguities and its dreams, in its very practices, it participated in both the one and the other." Roche, *Le Siècle des lumières* (1978), 1:394.

24. The critical problem with expert testimony—the disagreement among experts and their partiality, caused by their commitment to the side that hired them—had become sufficiently evident by 1780 that Morveau wrote a treatise on the subject. Morveau recommended that experts be appointed by an agreement of the opposing parties or, failing that, by the judge. Guyton de Morveau, "Mémoire sur les rapports d'Experts" (1780), in *Discours publiques et éloges* (1775–82): 281–335. See also Golan, "History of Scientific Expert Testimony" (1999).

25. Maza, *Private Lives* (1993), 14, 10.

26. Robespierre, "Dedicace aux mânes de Jean-Jacques Rousseau" (1789), in *Oeuvres*, 1:1. In the "Dedicace," Robespierre claimed to have met Rousseau shortly before the latter's death in

present in Robespierre's plea for Vissery. Rousseau had repudiated expertise.[27] He had also, in contrast with Montesquieu and other proponents of the *thèse nobiliaire*, mobilized philosophical particularism *against* a system of political representation. He had argued that the corporate particularity of any individual or class of individuals rendered illegitimate their attempts to straddle the distance between individual wills and the general will. The alienation of sovereignty must be direct and absolute, the merging of all individual wills into the general will, with no mediating authority.[28]

Robespierre exploited the doctrine of sensibility to authorize such breaches between the particular and the general. The very disjunction between local fact and universal truth—the absence of any interloping theory—was what bound them together, just as, in Rousseau's theory, an individual will and the general will were united by the absence of any intermediary. Robespierre offered Rousseauian repudiations of both expertise and false political power, resting them on a new empiricist foundation. Experts and local officials alike falsely pretended to a general knowledge they could not achieve. No interpretive theory could reach from fact to universal truth, from individual experience to the general will. Judges, like *physiciens*, must therefore cling humbly to the facts. To be sure, as will be central to the story of the trial, Robespierre made this argument by invoking a most specialized variety of "fact."

In its first days, the lightning rod affair was about concrete particulars: a rod, a funnel, a wall, a gable, a well, three houses on the rue Marché-aux-herbes in Saint-Omer. In a flash it became instead a trial about Enlightenment and superstition, the right to property and the abuse of power. Ironically, because no interloping theory was to mediate between fact and implication, the empiricist mode of legal argument was characterized by a curious detachment from what one might have taken to be the facts of the case. In his plea, Robespierre spoke much about receptiveness to the testimony of the senses, and very little about the actual lightning rod of M. de Vissery de Bois-Valé.

ACT I: THE SENTENCE

Here is how, according to Vissery himself, the troubles began.[29] Mme. Renard-Debussy, whose wall Vissery had borrowed for the sake of humanity, was an

---

1778. On the historical possibilities regarding this encounter, see Nathalie Barbara Robisco, "Le Mythe de la rencontre avec Rousseau dans la formation du jeune Robespierre," in Jessenne et al., eds., *Robespierre* (1993), 36–43. On the influence of Rousseau on Robespierre, see Roger Barny, "Robespierre et les lumières," in ibid., 44–59.

27. Rousseau, "Discours sur les sciences et les arts" (1750).

28. Rousseau, *Du Contrat Social* (1762), bk. III, chap. XV, 307. On Rousseau's theory of the relations between individual wills and the general will, see Furet, *Penser la Révolution* (1978), 50–51; and Starobinski, *Jean-Jacques Rousseau* (1971), 256–57.

29. The following chronology is taken from Vissery to Buissart, 7 September 1780, AGPC,

"old quibbler"[30] with whom he had had litigious relations. She was moreover afraid of lightning, as was Mme. Cafieri, to whose gable the lightning rod was partly affixed. Mme. Renard-Debussy set about "augmenting" Mme. Cafieri's fear and spreading it among the other neighbors, declaring "that she would leave her house whenever it thundered etc.," until at last a "feminine council" decided to present a petition to the aldermen[31] of Saint-Omer to have Vissery's "dangerous invention" removed.

This petition, "dictated by ignorance and written by the very hand of the quibbler," predicted that Vissery's "electrized and magnetized machine would attract thunder from all over the city" to fall upon the roofs of the rue Marché-aux-herbes. (Whether the rod was itself "electrized and magnetized," in order to attract lightning, or more generally, whether and how it attracted lightning would later become a central point of scientific contention.)[32] M. Cafieri, the "small and tender husband of the fearful lady," was concerned particularly on account of the firewood in his attic. During "moments that these lady-advocates of the night know how to exploit," M. Cafieri was persuaded to carry the petition throughout the neighborhood, going as far abroad as the Dominican monastery, where he secured the signature of the Father Superior. Some, however, refused to sign. So it was that a "cabal" of "seven or eight timid and fearful persons" signed a petition to the aldermen to dismantle Vissery's monument to modern science.

When the town bailiff, Sr. Jacques Valour,[33] arrived at Vissery's door with the neighbors' approved petition, Vissery tried to dissuade him with explanations of the lightning rod, to which Valour responded "that he understood not a word" and that he "was only discharging his commission." Vissery stood firm,

---

Coll. Bar., 4J/120/1, 2; Buissart, draft of memoir, ibid., 4J/120/30, 2–10; see also Robespierre, "Premier plaidoyer" (1783), in *Oeuvres*, 1:28–34.

30. *"vielle chicaneuse"*; in Buissart to Cotte, 2 November 1780, AGPC, Coll. Bar., 4J/118/35.

31. *Echevins*, which I translate here as "aldermen," were municipal officers who, along with mayors, composed the government of a city or municipality. The term was used primarily in the provinces of Artois, Flanders, and Hainaut (east of French Flanders, now in the department of Nord). *Echevins* were usually named either by the king's intendant or the local lord (seigneur). A decree of the Conseil d'Artois of 20 September 1773 governed the composition and election of Artesian municipalities: for Saint-Omer it allocated a mayor and six aldermen. The nomination of the aldermen was carried out by the deputies to the provincial estates and two members of each body (*nobles, gradués, bourgeois, négociants*). In most cities, including Saint-Omer, the aldermen comprised the entire system of justice, short of appeal to the provincial sovereign court. See Marion, *Dictionnaire des Institutions* (1923), 195.

32. See infra, §3, §4.

33. Bailiffs were "minor officials of the judiciary . . . responsible both for maintaining order during court proceedings and for executing the decisions of the court for which they worked." See Mousnier, *Institutions* (1979), 1:448. Valour seems also to have been Mme. Renard-Debussy's father: "Here are the names you asked for: Mme. Aldegonde Fagez third wife of Sr. Jacques Valour deceased bailiff from their union came Adelaide Valour their only daughter,

insisting that the lightning rod merited gratitude, not opposition.[34] So the bailiff brought the petition to the aldermen to pronounce an official judgment. One of these aldermen happened to be Mme. Cafieri's brother, who equaled her husband in pliability. He stirred his colleagues to hand down a sentence "worthy of the 17th century," in which they forced the public prosecutor[35] to concur. Without regard "for the cited authority of the greatest *physiciens*, Academies, republics, and entire Realms that have adopted this admirable invention," demonstrating their "hostility toward the *monde savant*," and in flagrant violation of the right to property, the aldermen ordered Vissery to take down his rod within twenty-four hours, in the presence of the bailiff, who would otherwise do it for him. They thereby "augmented the ferment among the people, too susceptible to the impressions of their superiors," and incited, in the "limited minds of St. Omer," a general "terror-panic."[36]

Vissery's tale is one of cosmopolitan science battling provincial prejudice. But the sides cannot be so clearly drawn. Maza has written of the domestic quarrels and scandals filling the Old Regime's closing causes célèbres that these stories reveal political attitudes far "more complex than can be suggested by any simple dichotomy between 'insiders' and 'outsiders.'"[37] The same can be said of the attitudes exhibited in the Old Regime's causes célèbres of natural science. Consider the reasons listed in the aldermen's original sentence of 14 June 1780. They are practical rather than superstitious and express no general hostility toward modern science, but instead a dubious attitude toward Vissery's engineering credentials.[38] The aldermen demonstrated their own competent grasp of Vissery's project, to "establish over his house an electric Conductor, *to attract the thunder*, in the hope that this Conductor, ending in the well of his house, the thunder will be able to descend there and *drown* itself." They did not contest the philosophical validity of this plan. Instead, they

---

married to Sr. Renard De Bussi." Fragment in Vissery's handwriting, n.d., AGPC, Coll. Bar., 4J/120/46.

34. The abbé Bertholon's version of Vissery's encounter with Valour, cited in Walter, *Robespierre* (1961), 1:35–36; Buissart, draft of memoir, AGPC, Coll. Bar., 4J/120/30, 3.

35. *Procureur du roi syndic*—the *procureurs du roi* acted as public prosecutors, representing the Crown and acting as intermediaries between the king and the courts. See Mousnier, *Institutions* (1979), 1:761; Marion, *Dictionnaire des Institutions* (1923), 460. Vissery reported that the *procureur syndic* for Saint-Omer, "le Sieur Jacques," was "better educated" than the aldermen and had tried to restrain them in their haste to condemn the lightning rod. Vissery to Buissart, 7 September 1780, AGPC, Coll. Bar., 4J/120/1.

36. Vissery to Buissart, 7 September 1780, AGPC, Coll. Bar., 4J/120/1, 4J/120/2.

37. Maza, *Private Lives* (1993), 6.

38. Buissart himself later acknowledged that the aldermen "[did] not seem to want absolutely to condemn the invention in itself, but throw doubts upon the capacity of the author and on the construction of the machine." Buissart, draft of memoir, AGPC, Coll. Bar., 4J/120/32. See also Vissery to Buissart, n.d., ibid., 4J/120/4.

supposed that "le sieur de Vissery, being perhaps not a great *Physicien*, could also be mistaken in the dimensions of his machine." Since the rod had "cast alarm throughout the neighborhood," it was now a matter of "police and public safety," and the order to remove it must be fulfilled even in the event of appeal.[39]

Vissery stalled. Two days after the aldermen rendered their judgment, he appealed their decision and accompanied his appeal with a brief on the design and function of his lightning rod. This resulted in a summons to an audience with the public prosecutor on Wednesday, 21 June. That audience led to a second one, on the same day, with the aldermen. The "whole city" was by now talking of the affair, and a great crowd flooded the Saint-Omer town hall to hear Vissery's plea and the aldermen's response.[40] The aldermen issued a second sentence, affirming their first one, dismissing Vissery's appeal, and fining him almost five livres.

The aldermen expanded upon their concerns. Granted that a lightning rod attracts lightning from the clouds, how can one be certain it will not deposit that lightning on buildings and on people? They correctly pointed out that "*physiciens* are not themselves in agreement concerning the proportions" of a properly built rod.[41] In his plea Robespierre would cite a passage from Benjamin Franklin suggesting that a mere "wire of a quarter of an inch diameter" would be sufficient. The continuation of the passage, which Robespierre did not cite, confirms the aldermen's suspicion of reigning uncertainty: "However, as the quantity of lightning discharged in one stroke, cannot well be measured . . . and as iron . . . is cheap, it may be well enough to provide a larger canal . . . than we imagine necessary."[42] The aldermen worried that the "disproportion between the point and the body of the machine" in Vissery's design might cause some fluid to leak out and set fire to the house. They cited examples: a building with a lightning rod in Kent had been incinerated; sparks had flown from a lightning rod in Siena, knocking a man to the ground; and the tip of a lightning rod in Mannheim had melted.[43] (For a prominent lightning stroke earlier in the century, see fig. 5.3.)

---

39. "Jugement des échevins" (14 June 1780), in Buissart, *Mémoire* (1782), 5. See also De Pas, *A Travers* (1914), 165–67.

40. Buissart, draft of memoir, AGPC, Coll. Bar., 4J/120/30, 6.

41. "Extrait de la sentence des échevins de Saint-Omer, du 21 juin 1780," in Robespierre, *Oeuvres*, 1:102–3.

42. See Robespierre, "Second plaidoyer" (1783), in *Oeuvres*, 1:89–90; Franklin to Collinson, September 1753, in Franklin, *Franklin's Experiments* (1941), 277.

43. "Objections des échevins de St Omer" (1780?), AGPC, Coll. Bar., 4J/120/34. The Kentish example had taken place in 1774 and been cited by Jean-Paul Marat in his *Recherches physiques* (1782), 414. The second example refers to a stroke of lightning that hit a rod on the tower of the cathedral in Siena on 18 April 1777; the event was reported in an article entitled "Extrait d'une let-

Figure 5.3. Lightning strikes the Campanile of San Marco, Venice, 23 April 1745. From Figuier, *Les Merveilles de la science* (1870), 1:573.

If metal conducts electrical fire, what about the metal crampons attaching the rod to the roof? Might they not divert the lightning into the house? And was it wise to conduct fire down a rod placed against the chimney, which enclosed a column of air? Might the electrical fire not interact dangerously with fires made on the hearth? The aldermen wondered too whether a conducting rod might not cause volcanoes, earthquakes, and waterspouts by forcing volatile electrical fluid into the ground.[44] Vissery himself believed that electricity was "without doubt the veritable cause of the formation of waterspouts and terrestrial thunder," as did his barrister.[45] The Italian Franklinist Giambattista Beccaria had made the same proposal.[46]

tre de M. Pistoi, Professeur de Mathématiques à Sienne, du 25 avril dernier," in the *Journal de Physique* of November 1777 (379–81). The third example was a stroke that hit a rod in Mannheim, elevated by the abbé Hemmer, demonstrator at the *cabinet de physique* of the Palatine Elector. This stroke was reported in a letter from Hemmer, dated 4 October 1779, and printed in the *Journal de Physique* of February 1780 (116–17).

44. "Objections des échevins de St Omer" (1780?), AGPC, Coll. Bar., 4J/120/34.

45. Vissery to Buissart, 7 September 1780, AGPC, Coll. Bar., 4J/120/2; Bertholon to Buissart, 2 May 1781, ibid., 4J/119/3.

46. Beccaria, *Dell'elettricismo* (1753), bk. II. See also Heilbron, *History of Electricity* (1979), 365 n. 89.

The aldermen also introduced a statistical argument, using the contemporary theory of probabilities that considered the value as well as the likelihood of an outcome.[47] Lightning rods, they said, were much more dangerous than the small-pox inoculation, when measured against potential gains, for "out of 100,000 persons 99,000 are attacked by smallpox, while out of 100,000 edifices only a single one is struck by lightning."[48] Vissery, annoyed at the twenty-four-hour deadline given him to dismantle the rod, inadvertently conceded as much. The machine, he said, could not "be taken off like a shirt," and, anyway, "it has not thundered in this city all year, and thus we have no more to fear from thunder, than from the plague that reigns in Constantinople."[49]

In short, the aldermen gave, by contemporary standards, solid scientific arguments against Vissery's lightning rod. Their electrical queries, their empirical examples, and their statistical reckoning were perfectly in keeping with the most reputable of 1780s natural science—and indistinguishable, in terms of their scientific respectability, from Vissery's own arguments.[50]

Still Vissery stalled. Angry mobs arrived on the rue Marché-aux-herbes, threatening to break his windows, to rip off his lightning rod, to torch his house (see fig. 5.4). Vissery went to the commandant,[51] bringing an "English gentleman" and a recent issue of the *Journal de Physique.* The gentleman attested to (and the journal reported) the popularity and utility of lightning rods in England and across Europe. The commandant was sympathetic. He recommended that Vissery take down the sword blade while waiting for the sentence to be overturned. On 23 June Vissery removed the sword blade from the top of his lightning rod[52] "to satisfy the Judgment rendered by the Magistrate . . . and in order not to expose himself and his house to the popular insults with which he has been menaced . . . and finally to calm the uneducated public, and above all the sissies [*femmelettes*] of his neighborhood."[53] From the

47. See Daston, *Classical Probability* (1988), 17–18, 39, 243.

48. "Objections des échevins de St Omer" (1780?), AGPC, Coll. Bar., 4J/120/34.

49. Vissery to Buissart, 7 September 1780, AGPC, Coll. Bar., 4J/120/1; "Objections des échevins de St Omer" (1780?), ibid., 4J/120/34. The aldermen also implied that one should act one's age: "M. de Vissery, being almost 80, has no need to post a lightning rod on his house."

50. Vissery's representation of the aldermen's stance as antiscientific does, however, have a basis in their sentence. They write that "*physiciens* are enthusiasts" and that "the sciences, balancing their advantages and ill-effects, have produced no good for Society." "Objections des échevins de St Omer" (1780?), AGPC, Coll. Bar., 4J/120/34.

51. Commandants were officers of the *gouvernements*, military and political divisions of which there were thirty-nine in the late eighteenth century, including the *gouvernement* of Artois. The highest authorities of the *gouvernements* were initially the *gouverneurs*, but their authority gradually shifted to their underlings, lieutenants and commandants. See Marion, *Dictionnaire des Institutions* (1923), 113, 259–60.

52. Buissart, draft of memoir, AGPC, Coll. Bar., 4J/120/30, 8–9.

53. "Extrait du procès-verbal des notaires, du 7 août 1780," in Robespierre, *Oeuvres*, 1:103–4.

barbed weathervane and sixteen-foot iron rod down, he left the rest of the lightning rod illegally in place. In fact, "I consoled myself a little by substituting another shorter point, which . . . forms me a lightning rod, *voilà* how one deals with the ignorant multitude."[54]

ACT II: THE APPEAL

Vissery now began to court enlightened attention for his "disagreeable affair."[55] He wrote to "several great *physicien*s and to several Academies," those of Arras and Dijon. Hugues Maret, doctor and perpetual secretary of the Academy of Dijon, responded, recommending that Vissery produce an official statement on the construction of his lightning rod in the presence of Valour, the bailiff. Valour never appeared, but Vissery, undeterred, proceeded in the presence of two notaries and sent Maret the statement[56] and a detailed drawing. These were examined by a committee composed of Maret and Louis-Bernard Guyton de Morveau, who was a fellow member of the Dijon Academy, a lawyer at the Parlement of Dijon, and the author of the article "Thunder" in the 1777 supplement to the *Encyclopédie*.[57] Based on their report, the Academy sent Vissery an "ample and honorable certificate . . . proving that my lightning rod is made according to all the rules of the art."[58]

In their report Morveau and Maret noted the irregularity of partly substituting a tin tube for an iron rod. They emphasized, however, that any metal of a "considerable enough volume" would work and cited a case of a lightning stroke in Dijon, which Morveau had reported in "Thunder," to demonstrate the excellent conductivity of tin: lightning had struck a house and been conducted along its tin gutters, only "exploding" at their two ends. The lightning stroke had divided into two currents, and each had caused "a great hole" at the

54. Vissery to Buissart, 25 October 1782, AGPC, Coll. Bar., 4J/120/8.

55. Vissery to Buissart, 7 September 1780, ibid., 4J/120/1. Vissery was accustomed to rallying important people to his personal causes. Early in his correspondence with Buissart, he wrote, "I need a powerful protector to have access to the Throne, which I lack at present, my Parisian correspondent, despite his beautiful promises, no longer responds to me, not knowing whether he is dead or alive, I need to find another means: in reflecting upon this matter, it seems to me that the R.P. Cotte, having the good fortune and advantage of often speaking to the King, could mention to him something of my inventions to prick his curiosity." Buissart transmitted the message, to which the père Cotte replied to please tell Vissery "I myself do not present my own observations to the King, I have never had the honor of speaking to the King nor even of seeing him." Vissery to Buissart, 7 July 1780, ibid., 4J/120/2; Cotte to Buissart, 15 August 1780, ibid., 4J/118/34.

56. "Procès-verbal des notaires, du 7 août 1780," in Buissart, *Mémoire* (1782), 60–61.

57. Louis-Bernard Guyton de Morveau, "Tonnerre" (1777), in Diderot, ed., *Nouveau dictionnaire*, 4:948–52.

58. Vissery to Buissart, 7 September 1780, AGPC, Coll. Bar., 4J/120/1.

Figure 5.4. Rioting in Saint-Omer over Vissery's lightning rod. From Figuier,
Les Merveilles de la science (1870), 1:565.

end of the gutter, damaging the wall on either side "over a very large area."[59]
This example, though intended to show the safety of Vissery's conductor, confirmed two of the aldermen's suspicions: that a lightning stroke could divide
into separate currents; and that these could follow other metals in a house's

59. Guyton de Morveau, "Tonnerre" (1777), in Diderot, ed., *Nouveau dictionnaire*, 4:949.

construction. Morveau and Maret also invoked one of the aldermen's own examples to opposite effect: in the Siena lightning stroke of April 1777, they
pointed out, the cathedral itself had suffered no damage.[60]

Vissery's appeal to the Academy of Arras, meanwhile, brought his plight
to the attention of another barrister/*physicien*, Buissart. A member of the Arras and Dijon Academies and the Paris Society of Medicine, and a regular
contributor to the *Journal de Physique*, Buissart was accorded the dubious
honor of being the "greatest *physicien* of the Arras bar."[61] Like Vissery, Buissart was an inventor. He had devised a "hygrometer" for measuring water va-
por in the atmosphere, an instrument admired enough that its authorship was
fiercely and lengthily contested.[62] Buissart had also developed a new use of
lightning rods, to protect the fields from hail, on the theory that hail was
caused by electricity in the clouds, which could be continuously discharged
by the action of many rods at once.[63]

During the summer of 1780, Buissart followed Vissery's case with interest.
In September the case had gained enough prominence to be written up in the
Mercure de France,[64] and Vissery appealed to Buissart for help. Maret had advised Vissery that he had a case against the signers of the petition. "My design,"
Vissery wrote to Buissart, "is not to dwell much on the utility of lightning
rods...but to revolt against [*fronder*] the too-rigorous judgment of the judges."
Vissery proposed to argue on procedural, not scientific grounds, that the aldermen had violated their duty to consult scientific experts before making
their decision: "When prudent judges wish not to compromise themselves...
in matters they are not obliged to fathom they refer to so-called Experts—as
is done in surgery, Architecture, etc.—to pronounce in knowledge of the
facts: but this is precisely what the Echevinage of St. Omer did not do."[65]

The fields of medicine and architecture supplied Vissery's, and later
Buissart's,[66] examples of the importance of expert scientific advice in court. If
a "grumpy and ill-intentioned neighbor said this gable, this chimney is not
plumb, it was ordered torn down without a visit of Experts . . . what opinion

---

60. "Extrait des registres de l'Académie de Dijon du 18 août 1780," in Robespierre, *Oeuvres*,
1:104−7.

61. Walter, *Robespierre* (1961), 1:35−37.

62. See, for example, Bertholon to Buissart, 2 May 1781, AGPC, Coll. Bar., 4J/119/3; Walter,
Robespierre (1961), 1:35.

63. Maret, told of this idea, responded that it had already been proposed, but that the public
was "not yet advanced enough" for its execution. Maret to Buissart, 6 February 1781, AGPC, Coll.
Bar., 4J/120/11. Bertholon worried that "nasty jokers" would make "ill-placed jeers" at the prodigious number of rods in Buissart's plan. Bertholon to Buissart, 2 May 1781, ibid., 4J/119/3.

64. The *Mercure de France* write-up is mentioned in Cotte to Buissart, 15 September 1780,
ibid., 4J/118/31.

65. Vissery to Buissart, 7 September 1780, ibid., 4J/120/1.

66. See Buissart, draft of memoir, ibid., 4J/120/30, 66−67.

could we have of such justice?" Similarly, if a "machine or apparatus had the virtue of curing or dissipating the plague . . . would one need the permission of the police to erect it on one's house?" Here was no justice, but "barbaric despotism."[67] Because it had acted wrongly in failing to consult experts, the public ministry should bear the costs of Vissery's appeal. His lightning rod had not caused a "terror-panic"; the rash and inexpert judgment of the aldermen had been the cause.[68]

Buissart agreed to defend Vissery's case on appeal to the Conseil d'Artois. In preparing his appeal, Buissart corresponded, on the scientific side, with Maret; the abbé Pierre Bertholon de St. Lazare, an experimental *physicien* with "an apostolic zeal for electricity";[69] and with three affiliates of the Royal Academy of Sciences in Paris: Louis Cotte, an Oratorian,[70] meteorologist, and corresponding member[71] of the Academy; Jean-Baptiste Le Roy, an academician

67. Vissery to Buissart, 8 June 1783, ibid., 4J/120/9.

68. Vissery to Buissart, 7 September 1780, ibid., 4J/120/1. Vissery believed this last point to be the salient one and would later insist that Buissart emphasize it. At Vissery's urging, Buissart wrote to Cotte that he had not previously made it sufficiently "clear" that "it was not the lightning rod that caused the uproar and alarm, but the judgment of the magistracy of St. Omer." Buissart to Cotte, 8 December 1780, ibid., 4J/118/36. Vissery's subsidiary claims were as follows. First, the citizens of Saint-Omer, frightened by their aldermen, had since been calmed by a journal article about the case written by a friend of Vissery's (perhaps the *Mercure de France* notice mentioned above, n. 64), and it was therefore no longer a police matter. Second, the bailiff must serve him the neighbors' petition so that he could bring suit against his accusers. A note in Buissart's papers identifies a precedent stating that the public ministry was obliged to identify an appellant's denunciators so that the appellant could bring suit against them to recover the costs. "Journal historique le 7 fevrier 1772," ibid., 4J/120/17. However, the bailiff never served the petition to Vissery, a fact that became central to Vissery's claim against him. See Buissart, draft of memoir, ibid., 4J/120/30, 5, 74; "Consultation du 15 septembre, 1782," in Robespierre, *Oeuvres*, 1:114. Third, Vissery must not be held responsible for the rent of the neighboring house, which had been devalued by the scandal and threatened with abandonment. Vissery to Buissart, 7 September 1780, AGPC, Coll. Bar., 4J/120/1. Finally, the public ministry should bear the costs of his appeal, though it was not generally liable for the costs of appeals against it, because it had "lent its hand to childish and frivolous fears." Buissart to Cotte, 2 November 1780, ibid., 4J/119/35. In April 1782 the Conseil d'Artois held a Saint-Omer prosecutor (*procureur du roi de la maîtrise des eaux et forêts de St. Omer*) liable for the expenses incurred by an appeal that was granted against a sentence he had passed. The case involved hunting rights in a royal forest. See Buissart to Cotte, 16 April 1782, ibid., 4J/118/62; Buissart, fragmentary notes, n.d., ibid., 4J/120/50. As an afterthought, Vissery added another demand: "If an accident occurred to the neighboring house . . . I am not responsible." Vissery to Buissart, n.d., ibid., 4J/120/4.

69. Buissart to Cotte, 8 December 1780, ibid., 4J/118/36; 20 April 1781, ibid., 4J/118/43.

70. The Congregation of Priests of the Oratory of France was a teaching order founded in 1611 by the future Cardinal Bérulle and having close ties with the Academy of Sciences. On the importance of the Oratory in seventeenth- and eighteenth-century natural sciences, see Pierre Costabel, "L'Oratoire de France et ses Collèges," in Taton, *Enseignement et diffusion* (1964), part 1, chap. 3.

71. "Corresponding members," or *correspondants*, formed a class of nonresident members. See Hahn, *Anatomy of a Scientific Institution* (1971), 77.

and the leading French exponent of Benjamin Franklin's theory of electricity; and the marquis de Condorcet, perpetual secretary of the Academy.

These consultants were unanimously incredulous that "fifty leagues from the capital and at the end of the eighteenth century," as Le Roy remarked,[72] such a backward decision as the Saint-Omer aldermen's was possible. Bertholon declared it an "ignominious pretension and without any foundation to claim one cannot erect a lightning rod without police authorization." Cotte marveled that the aldermen seemed to possess not a "smattering of physics."[73] Did they not read the journals? And even if they "read only the gazette," they should know better than to mistrust Vissery's lightning rod.[74] Maret, upon learning that a fellow doctor was "one of the magistrate-destroyers of the lightning rod," wrote, "This gives me a very bad idea of him, and I hope he won't be long . . . in being punished for his indiscretion."[75]

The philosophes' advice to Buissart was authoritarian for an enlightened crowd. Maret supplied Buissart with a list of lightning rods across Europe, approving of Buissart's intention to cite as many examples as possible, for "the common man can be but an imitator, and there are so few who think for themselves that to crush them under the weight of authority is to perform a service for them."[76] Buffon counseled him not to "worry on account of those who are contradictors only because they are ignorant." Condorcet reportedly recommended "pouring ridicule over the whole affair."[77] This Bertholon did, in a performance at the Montpellier Academy of Sciences' regular public session before the Languedoc Estates General. He described the scene to Buissart:

> These Estates are always brilliant, as there are 22 Bishops, as many Barons, deputies of the Clergy, those of the nobility, the Commandant of the province, and the second commandants, the intendants, etc., a great number of people of all the estates, even distinguished ladies etc., the room is always full, and one of the grenadiers . . . was obliged to quit his post, because he was suffocating from the hot air.

In this overheated environment, Bertholon read a memoir on the efficacy of lightning rods.

72. Le Roy to Buissart, 9 November 1782, AGPC, Coll. Bar., 4J/120/21.
73. Bertholon to Buissart, 25 April 1783, ibid., 4J/119/23; Cotte to Buissart, 23 August 1780, ibid., 4J/118/29.
74. "Procès du paratonnerre," n.d., ibid., 4J/120/16.
75. Maret to Buissart, 16 February 1782, ibid., 4J/120/13.
76. Maret to Buissart, 6 February 1781, ibid., 4J/120/11. Buissart ingenuously quoted this advice verbatim in his memoir, along with most of the rest of his correspondence. Buissart, draft of memoir, ibid., 4J/120/30, 17.
77. Bertholon to Buissart, 2 May 1781, ibid., 4J/119/3; 29 January 1782, ibid., 4J/119/8.

I spoke of the matter of St. Omer, I heaped ridicule over the foolish judges of that place... the whole was heard with pleasure; and I will even tell you that the reading of my memoir was frequently interrupted by applause.[78]

Bertholon also recommended, "to succeed in your trial, you have to create a rumor that there is a *sçavant*, and a *littérateur* of Paris who proposes after the judgment in the trial to lampoon the judges or adversaries if there are any; and that Beaumarchais will add his touch: this is worth more than all the good reasons in the world, for those who are not made to hear them."[79] Bertholon himself promised a "bloody tirade against the ignorance of the alleged magistrates of St. Omer; those of Arras will have their part, if they participate in the sovereign omerian imbecility."[80]

Besides sympathetic outrage, Buissart's scientific consultants promised facts. Bertholon should have had many at his disposal, as a leading advocate and designer of lightning rods. He had recently invented a new variety, "ascending rods," which were meant to work in reverse, discharging strokes of lightning that went upward from the ground into the clouds.[81] Like Buissart, Bertholon had also proposed some bold new applications of electrical conductors, the "para-earthquake" and the "para-volcano," on the theory that electricity was a cause of these natural disasters as well.[82] (For other creative applications of the lightning-rod principle, see figs. *5.5* and *5.6*.) Buissart began his research by posing Bertholon four questions: in which regions of France had lightning rods been established, and when, and had the gazettes mentioned them, and what were the most striking examples of their effectiveness? But Bertholon cautioned that the "enumeration" Buissart requested "should not be made, because it will not be considerable enough to strike the ignorant. In France there are very few lightning rods." He calculated a total of eleven, all of

78. Bertholon to Buissart, 29 January 1782, ibid., 4J/119/8.

79. Bertholon to Buissart, 29 December 1782, ibid., 4J/119/16. The playwright Pierre-Augustin Caron de Beaumarchais had ridiculed the Parlement in a series of much-celebrated and imitated *mémoires* during the Maupeou Revolution of 1771–74, a power struggle between the Crown and the Parlement in which Louis XV and his chancellor, René-Nicolas-Charles-Augustin de Maupeou, exiled the recalcitrant Parlement and replaced it with a new, more pliant one. On the Maupeou Revolution, see Bell, *Lawyers and Citizens* (1994), 138–63; and Swann, *Politics and the Parlement of Paris* (1995), chap. 12. On Beaumarchais's attacks on the Parlement, see Lipatti, *Mémoires de Beaumarchais* (1974); and Maza, *Private Lives* (1993), 130–40.

80. Bertholon to Buissart, 3 April 1781, AGPC, Coll. Bar., 4J/119/1–2.

81. Bertholon to Buissart, 3 April 1781, ibid., 4J/119/1–2; 2 May 1781, ibid., 4J/119/3. "Ascending thunder" arose from Franklin's revised theory of lightning, that the ground was positive and the clouds negative, so it was the earth that struck into the sky rather than vice versa. See Franklin to Collinson, September 1753, in Franklin, *Franklin's Experiments* (1941), 268–72.

82. Bertholon to Buissart, 2 May 1781, AGPC, Coll. Bar., 4J/119/3.

them *en province*, which he listed to Buissart: one in Valence in the Dauphiné; three in and around Dijon, and one in Bourg-en-Bresse, in Burgundy; one on a country house in Anjou; Voltaire had had one at Ferney; and finally, there were four in Lyon, established by Bertholon himself the previous year.[83]

Nor had Cotte much to offer in the way of instances of working lightning rods. He wrote that he had a "Conductor 15 *toises* long that is not even arranged

83. Bertholon to Buissart, 3 April 1781, ibid., 4J/119/1–2.

Figures 5.5 and 5.6. Lightning rod hat (1778) and lightning rod umbrella designed by Barbeu-Dubourg, from Figuier, *Les Merveilles de la science* (1870), 1:569, 597.

in a manner to preserve our Church . . . as my unique object is to obtain Electricity during stormy weather," and so his rod was insulated, not grounded. It was arranged to connect with a grounded conductor nearby, but Cotte had seldom established this communication. (For an illustration of such an arrangement, see the frontispiece to this chapter, fig. 5.1.) He had only seen lightning strike once in Montmorency, where he lived, and "not even on my Conductor." It had fallen "in a body of Water, which has much more to fear than a conductor for attracting thunder." This thought inspired Cotte: Vissery might countersue Renard-Debussy to oblige him to dry out his garden pond, if he had one. "The Sentence that condemns him will be assuredly better founded than that which requires the demolition of lightning rods."[84]

Vissery's supporters looked expectantly toward the capital for salvation, but it proved difficult to secure official Parisian sanction for Vissery's case, perhaps because of the scarcity of experience with lightning rods. When Cotte consulted "one of the cleverest councillors of Châtelet," the lawyer told Cotte that the aldermen had not judged wrongly. Another tribunal might grant Vissery on appeal the right to reconstruct his machine, but would never inculpate the aldermen for having taken "necessary measures to stop the tumult and clamor of the people."[85] The public ministry would therefore not be responsible for the expenses of Vissery's appeal. Buissart was undeterred. Vissery had been wronged by the aldermen of Saint-Omer, who had violated their duty to consult experts. They should bear the costs of their error. Vissery's cause was "that of all *physiciens.* That is why I propose to defend it vigorously." In December Buissart wrote to Cotte, asking him to secure the approbation of the Paris Academy of Sciences for Vissery's lightning rod.[86]

Cotte obliged by consulting Condorcet. But Condorcet responded that since the Academy of Dijon had already pronounced on Vissery's rod, the Paris Academy of Sciences would decline to reexamine it. They would wait until the Conseil d'Artois asked for their advice. Condorcet therefore offered only some recommendations about how to maneuver the court into approaching the Academy. Buissart reported some months later that he had taken measures to ensure, and had been "led to believe," that the Conseil d'Artois would consult the Paris Academy of Sciences before rendering a judgment in the case.[87]

84. Cotte to Buissart, 23 August 1780, ibid., 4J/118/29.

85. Cotte to Buissart, 20 September 1780, ibid., 4J/118/32. The Châtelet of Paris was the most important and prestigious presidial court of France. Its jurisdiction included the city of Paris and surrounding *faubourgs,* but its influence extended throughout the kingdom. On the functions of the Châtelet, see Marion, *Dictionnaire des Institutions* (1923), 88–90; Mousnier, *Institutions* (1984), 2:350–54; and Andrews, *Law* (1994), 56–75.

86. Buissart to Cotte, 19 December 1780, AGPC, Coll. Bar., 4J/119/38; 8 December 1780, ibid., 4J/119/36. See also Buissart to Cotte, 19 December 1780, ibid., 4J/119/38.

87. Cotte to Buissart, 10 January 1781, ibid., 4J/118/41; Buissart to Cotte, 7 July 1781, ibid., 4J/118/46.

As for a lawyers' consultation, which Buissart had also requested, Condorcet thought it better to have a lawyer, "guided by *physiciens*," write a brief.[88] The lawyer whom Vissery and Buissart charged with this task was slow and unresponsive.[89] Vissery grew restive. In the spring of 1781, after waiting six months for the Paris brief, he wrote to Buissart that "for several days frequent and frightful flashes of lightning have been followed by dreadful noises." His "adversaries" would be to blame "if thunder falls and crushes my house." Vissery had grown tired of "the *demi-savans*" who teased that he may well be benefiting humanity but "they would not want to be my neighbor." Being of an "advanced age," he had begun to fear he would never see the end of the affair. "Does it not seem," he complained to Buissart, "that an evil demon is presiding over all these long delays?" Buissart reported Vissery's mood to Cotte. They would do without the Parisian lawyer's brief and would return to their original idea of a consultation. This would accompany Buissart's own brief, which they would publish instead.[90]

ACT III: THE BRIEF

Responding to his client's growing impatience, Buissart dashed off the eighty-six-page brief in eight days.[91] He took his central line of argument from Vissery: the subordination of legal to scientific authority. There were, Buissart considered, two questions to be decided: whether lightning rods were dangerous or advantageous, and whether Vissery's lightning rod was well or badly constructed. The answer to each rested upon expert knowledge.[92] By failing to consult experts on the first question, concerning lightning rods in general, the aldermen had been guilty of incorrect physics and rendered a judgment "irregular as to content." By failing to consult experts on the second question,

88. Cotte to Buissart, 10 January 1781, ibid., 4J/118/41. Buissart had requested the "reasoned opinion of one or two good lawyers of Paris" concerning the public ministry's liability for the costs of the trial. Buissart to Cotte, 8 December 1780, ibid., 4J/119/36; see also Buissart to Cotte, 19 December 1780, ibid., 4J/119/38. For the brief that Condorcet proposed, Vissery would pay only the printing costs, and the lawyer would be compensated by the sales of the brief. Trial briefs were published with increasing frequency over the course of the eighteenth century. On the importance of the *mémoire judiciaire* in French political culture, see Maza, *Private Lives* (1993), 35–38; and Bell, *Lawyers and Citizens* (1994), 87–89.

89. The lawyer in question is identified in several letters as "Frenais": Buissart to Cotte, 20 June 1781, AGPC, Coll. Bar., 4J/118/45; Vissery to Buissart, 17 May 1781, ibid., 4J/120/3; and Buissart to Cotte, 20 April 1781, ibid., 4J/118/43. The last also implies that Frenais was an assistant to the Cardinal Louis de Rohan.

90. Vissery to Buissart, 21 May 1781, ibid., 4J/120/3; date illegible, ibid., 4J/120/6; Buissart to Cotte, 20 June 1781, ibid., 4J/118/45; 19 July 1781, ibid., 4J/118/47.

91. Buissart to Cotte, 19 November 1781, ibid., 4J/118/54.

92. Buissart, draft of memoir, ibid., 4J/120/30, 67.

concerning Vissery's lightning rod in particular, the aldermen had been guilty of incorrect procedure and had rendered a judgment "irregular as to form."[93]

However, the second half of the brief, ostensibly devoted to the design of Vissery's lightning rod, in fact contains no discussion of the rod's design except an assertion that it is "invulnerable to all criticism."[94] Instead, Buissart here considers the death of Georg Wilhelm Richmann, who had been killed during an experiment with an insulated conductor in St. Petersburg in 1753 (fig. 5.7).[95] Buissart uses this accident as a context in which to emphasize the difference between what he calls an "electrometer" and a lightning rod. An electrometer, he explains, is an insulated conducting rod used to collect electricity and thereby "to indicate the greater or lesser amount . . . spread through the atmospheric air." The lightning rod, in contrast, is grounded; it "communicates with the damp earth or water," and so rather than collecting, it diffuses electricity into the ground. People "badly confuse these two instruments, which are very different." Vissery's instrument was a lightning rod, not an electrometer, and therefore entirely dissimilar to the dangerous instrument involved in Richmann's accident.[96]

Meanwhile, the first half of the brief, devoted to showing the efficacy of lightning rods in general, is a largely miscellaneous treatise on the science of electricity, including its history over the past century, its amusement value, and its medical applications (accelerating perspiration and the circulation of the blood).[97] There is only a brief passage on the more salient analogy between electricity and lightning. Franklin had had this analogy in mind, according to Buissart, during his experiments with the Leyden jar. The commotion when he discharged the condenser had reminded him of a stroke of lightning in miniature. Franklin had then "recognized" that "all electrized bodies have an atmosphere," that this atmosphere "extends farther at the angles of bodies than anywhere else," and that, therefore, "points attract the electrical material" and, by implication, lightning, "from farther and more efficiently than all other bodies." The argument was simple: lightning is electricity. (Though Buissart also lists as "secondary causes" of lightning "inflammable air" and phosphorus, to which he attributes the sound of thunder.) Points attract electricity. Ergo the

93. "Le Jugement du magistrat de St. Omer est irrégulier quant au fond"; Buissart, draft of memoir, ibid., 4J/120/30, 10, 48.

94. Here Buissart cites the approval of the Dijon Academy and the later concurrence of the père Cotte. Buissart, draft of memoir, AGPC, Coll. Bar., 4J/120/30, 55–64.

95. On Richmann's death, see Heilbron, *History of Electricity* (1979), 352.

96. Buissart, draft of memoir, AGPC, Coll. Bar., 4J/120/30, 49–55.

97. Buissart, draft of memoir, ibid., 4J/120/30, 11–14. Jean-Antoine Nollet had experimented on the effects of electricity on perspiration and transpiration; see Nollet, "Troisième Mémoire," in *Mémoires de l'Académie des Sciences* (1747), 230–42; and Nollet, "Concerning Electricity," in *Philosophical Transactions of the Royal Society of London* 45 (1748), 187–94.

Figure 5.7. The death of Richmann, from Figuier, *Les Merveilles de la science* (1870), 1:529.

lightning rod (see fig. 5.8).[98] This information, conveyed in two sentences, comprises Buissart's entire explanation of the electrical phenomenon most relevant to the design of lightning rods, the so-called "power of points." But this vagueness was not Buissart's fault. Franklin's theory of electrical action, including the power of points to attract and dispel electrical fluid, was itself

98. The quotes are taken from Buissart, draft of memoir, AGPC, Coll. Bar., 4J/120/30, 25, 13. "Inflammable air" was a contemporary name for hydrogen. The analogy between lightning and

riddled with causal gaps and inconsistencies. These had inspired much opposition, both to the theory and to its leading application, the lightning rod.[99]

Advocates of lightning rods called this opposition prejudice. Franklin himself, for example, saw prejudice in Nollet's refusal to accept the lightning rod. Franklin pointed out that Nollet must believe in the conduction of lightning since he warned against the customary ringing of church bells during storms, because the bell-pull could conduct electricity down to kill the ringer. Franklin remarked, "How long even philosophers, men of extensive science and great ingenuity, can hold out against the evidence of new knowledge that does not square with their preconceptions."[100]

Yet Nollet's objection was philosophically consistent. He pleaded "too great a disproportion between the effect and the cause"; erecting a lightning rod was like putting a narrow tube into a rushing torrent in the hope of preventing a flood.[101] Nevertheless, the Franklinist view of doubts like Nollet's was influential. Vissery and his allies adopted it, and Robespierre popularized it. Many since have accused early opponents of lightning rods of "prejudice" and called their arguments "pseudo-scientific."[102] Perhaps they were, but if so, they were no worse than the arguments in favor of lightning rods, whose proponents were altogether unable to explain how their devices were meant to work.

Franklin initially suggested that the rods drew electrical charge from the clouds, gradually and continually restoring equilibrium and so preventing

---

electricity was common among electrical experimenters by the 1740s. Nollet, for example, proposed it several years before Franklin. See supra, chap. 3, 101–2; Torlais, *Un Physicien* (1954), 112–13; and Nollet to Jean Jallabert, 2 June 1752, in Nollet, *Correspondance* (n.d.), 216. Thus, the analogy between lightning and electricity was independent of Franklin's account of electrical action. See Heilbron, *History of Electricity* (1979), 339–41. As for the role of the analogy in the development of Franklin's theory, he introduced the concepts of electrical atmospheres and the power of points in a letter to Peter Collinson dated 11 July 1747, which included descriptions of his first Leyden jar experiments. In that letter Franklin describes an electrical spark as being "like a flash of lightning." Franklin, *Franklin's Experiments* (1941), 177. But he did not expand upon the analogy until 1749. See "Letter V" (1749), in ibid., 201–11; "Opinions and Conjectures" (1749), in ibid., 213–36; and Cohen, "Franklin's Work in Electricity," in ibid., 111. After the first successful conducting rod experiment at Marly-la-Ville in May 1752, an experiment that Franklin had proposed, he and lightning rods were universally associated. But Jacques de Romas, a Nolletist electrician in Bordeaux, received certificates of priority for the electrical kite experiment from the Bordeaux Academy of Sciences and the Paris Academy. Romas even likely designed a prototype lightning rod and submitted the design to a member of the Bordeaux Academy, the baron de Montesquieu, the year before the Marly experiment. See Romas, *Oeuvres* (1911), 289, 73, 183–85. See also Heilbron, *History of Electricity* (1979), 351.

99. On the inconsistencies in, and opposition to, Franklinist electrical theory, see Heilbron, *History of Electricity* (1979), 344–402.

100. Franklin to Winthrop, 2 July 1768, in Franklin, *BFP*, 15:166–72.

101. Nollet, *Lettres* (1753), 1:162–63.

102. The quotations are from recent examples: Cohen, *Franklin's Science* (1990), 119; and Cohen, *Franklin and Newton* (1956), 511.

Figure 5.8. The first test of the power of points as applied to lightning, at Marly-la-Ville
under the auspices of Franklin's French translator, Thomas François Dalibard, on 10 May 1752,
from Figuier, *Les Merveilles de la science* (1870), 1:521.

a bolt. Then he decided that rods could also function even if they failed to
prevent a stroke, by channeling it into the ground and diffusing it. Later yet,
Franklin decided that when lightning struck, the ground was usually electrified positively and the clouds negatively, thus it was the ground that struck into
the sky rather than vice versa. In that case the relevant property would be the

power of points not to attract, but to dispel electrical fire, presenting yet a third way in which rods might function.[103] Franklinist advocates of lightning rods were unable to explain how points both attracted and dispelled electrical fire, and to decide whether one or the other (or both) of these powers was instrumental in protecting civilization from lightning.[104]

Buissart's few, equivocal comments on the power of points reflect the reigning theoretical uncertainty. For example, having initially claimed that pointed conductors attract electric material, he later seemed to contradict this earlier statement when he wrote: "[The idea] that electric conductors attract thunder from the clouds . . . is but a phantom easy to destroy." Conductors do not attract lightning; instead, they silently discharge the clouds' surplus electricity by channeling it into the ground. But Buissart's elaboration on this statement includes another about-face: "If an explosion does occur, the blade of fire escaping from the clouds is attracted by the conductor, provided it passes within the extension of its sphere of activity." [105] Buissart also mentions the theory of ascending thunder and Bertholon's ascending rods, which were meant to work by the power of points to dispel, not attract, electrical material.

Poor Buissart! The claim that lightning rods attract lightning to houses that would not otherwise have been struck was at the heart of his opponents' case. In response, he could only cite the leading authorities on lightning rods, who offered a resounding yes, no, and maybe. But never mind—he cut his electrical explanations short with a rhetorical query: "To what end shall we devote ourselves to the reasons of Science that establish the theory of thunder and consequently of Lightning Rods? The efficacy of this machine must be shown less by solid reasons, than by conclusive facts." Hence, it "matters

103. See Cohen, *Franklin's Science* (1990), 119, 126–27, 141.

104. "Thus the pointed rod either prevents a stroke from the cloud, or, if a stroke is made, conducts it to the earth." Franklin, "Of Lightning, and the Method (now used in America) of securing Buildings and Persons from its mischievous Effects" (September 1767), in Franklin, *Franklin's Experiments* (1941), 391. Such theoretical uncertainties about the functioning of lightning rods fueled an intense controversy in Britain during the 1770s concerning the relative advantages of pointed versus blunt conductors, the "comical battle of the knobs and points." Heilbron, *History of Electricity* (1979), 380. George III put an end to the dispute in 1777 by coming down on the side of Franklin's chief rival, Benjamin Wilson, in favor of blunt conductors. This decision may have reflected the king's view of Franklin's Revolutionary political activities as much as his view of electrical action, although, to be sure, an apparent failure of a pointed rod on the powder magazines at Purfleet and a convincing demonstration of Wilson's method in the great hall of the London Pantheon also worked in Wilson's favor. See Heilbron, *History of Electricity* (1979), 380–82; and Mitchell, "Politics of Experiment" (1998).

105. Buissart, draft of memoir, AGPC, Coll. Bar., 4J/120/30, 46. The proceedings of the Academy of Sciences for 1784 note that "unfortunately, electrical experiments have not yet taught us anything that could lead us to know the sphere of activity of the point of a conductor." Procès-verbaux, 103:91, AS.

little if we know the nature of the electric fluid; [or] its manner of acting."[106]
Here was a new and potentially powerful tactic: the irrelevance of explanations in the face of "facts." By misrepresenting his opponents as having categorically rejected lightning rods out of an indifference to facts, Buissart was
able to assume the mantle of sensitive empiricism.[107] He listed every instance
of a lightning rod's existence or functioning that he had been able to gather.

These "facts," however, were also equivocal. For example, Buissart excerpted the account of the Siena lightning stroke—the same event that both
the aldermen and the Dijon Academy had already claimed in support of their
opposing positions—from the *Journal de Physique*.[108] The passage in Buissart's
memoir included "a purple globe of fire" that descended along the lightning
rod and, before disappearing into the ground, "threw off great sparks"; a man
standing in the door of his shop across the street, knocked to the ground; and
large quantities of sulfurous smoke that poured through the streets afterward.[109] Buissart then excerpted the account of the Mannheim lightning stroke
from the same journal.[110] This, too, involved unsettling facts, such as the fact
that—as the aldermen had mentioned—the point of the Mannheim rod was
afterward observed to be damaged.[111]

Perhaps because of the ambiguity of the facts, Buissart did not fully develop the empiricist strategy. Instead, he rested his case on the subordination
of legal to scientific authority: "Judges are not supposed to be *Physiciens*.
Physics and Jurisprudence are two very different Sciences." Even judges well
informed in natural philosophy could not possibly keep up with the latest developments. They must therefore "address themselves to professionals."
Buissart adopted Vissery's own architectural analogy. An overhanging chimney is said to be solid by its owner, but a neighbor claims it is in imminent
danger of falling, crushing people beneath and part of his house. Would a
judge "immediately have the chimney demolished? no, without doubt, but he
would order by a preparatory judgment . . . that it be visited by architects. . . .

---

106. Buissart, *Mémoire* (1782), 29.

107. Buissart wrote that the lightning rod was "the happiest discovery made this Century, far
from being dangerous in itself as the aldermen of St. Omer announce." Buissart, draft of memoir,
AGPC, Coll. Bar., 4J/120/30, 45. In their first sentence, the aldermen had written: "This physics
experiment that the sieur de Vyssery wants to do is dangerous in itself, and casts alarm throughout the neighborhood." The context suggests that "dangerous in itself" was intended to mean not
that all lightning rods are necessarily dangerous, but that Vissery's lightning rod presented a direct danger as well as an indirect one through its effect on the residents of the city. See "Sentence
des échevins de Saint-Omer, du 14 juin 1780," in Robespierre, *Oeuvres*, 1:101.

108. See supra, n. 43.

109. Buissart, draft of memoir, AGPC, Coll. Bar., 4J/120/30, 31–33.

110. See supra, n. 43.

111. Ibid., 4J/120/30, 35–36.

[T]he judge would only decide based on the report of experts and would pronounce nothing by himself. The same goes for physics as for architecture."[112]

However, as with the crucial scientific question of whether lightning rods attract thunder, Buissart was equivocal concerning the crucial jurisprudential question of how much physics a judge could be expected to know. A judge's function, to "maintain order and harmony in the society," required that he "have a certain knowledge of all that is useful or harmful to the public." Judges therefore had a duty to "apprehend, so to speak, each day, the discoveries that have been made in this genre. The books, journals and political papers that are disseminated with such profusion have no other purpose."[113]

On the other hand, Buissart claimed that judges were required to consult experts by Title XXI of the Ordinance of 1667,[114] the civil half of Jean-Baptiste Colbert's recodification of civil and criminal procedure. In fact, Title XXI stated that "judges cannot make visits to scenes when all that is needed is a simple report of experts."[115] It did not constrain judges' decisions, but only their travels. The intent had patently been to economize and to simplify the judge's job, not to limit his authority.[116] Nevertheless, "*Voilà*," Buissart concluded, "the ordinary working of Justice. The Judge decides only on the basis of the Experts' report, and pronounces nothing on his own."[117]

Based upon scientific equivocation, jurisprudential ambivalence, and some dubious citations of fact and statute, Buissart concluded that the Arras judges must draw no conclusion: they must defer to the Academy of Dijon or seek the advice of the Paris Academy of Sciences.[118]

ACT IV: CONSULTATIONS AND INTRIGUES

In July 1781 Buissart sent his brief to Cotte. He need not have hurried. In September Cotte reported that it was at the Maison des Avocats, awaiting the addi-

---

112. Ibid., 4J/120/30, 66–67.

113. Ibid., 4J/120/30, 69.

114. Ibid., 4J/120/30, 61; Buissart, *Mémoire* (1782), 68–71.

115. *Ordonnance civile touchant la réformation de la justice,* Titre XXI, "Des descentes sur les lieux, taxe des officiers qui iront en commission, nomination et rapports d'experts," Art. 1: "Les juges, même ceux de nos cours, ne pourront faire descente sur les lieux dans les matières où il n'écheoit qu'un simple rapport d'experts" (April 1667). In Isambert and Jourdan, eds., *Recueil Général des Anciennes Lois* (1822–33), 18:103, 140–45. On the passage of the *Ordonnance*, see Andrews, *Law* (1994), 417–18.

116. Characterizing the principles of jurisprudence underlying the Colbert reforms, Richard Andrews cites Daniel Jousse: "Judges should be entirely free in their judicial opinions, free from any constraints or other pressures that could prevent them from acting according to their knowledge, wisdom and convictions." Andrews, *Law* (1994), 496.

117. Buissart, *Mémoire* (1782), 68–71.

118. Buissart, draft of memoir, AGPC, Coll. Bar., 4J/120/30, 77.

tion of a signed consultation.[119] In December Cotte encouragingly mentioned the names of two eminent jurisconsults: Lacretelle and Guy-Jean-Baptiste Target.[120] The following spring, "these Messieurs" still kept "the most profound of silences."[121] Buissart dispatched Bertholon in person to demand the memoir from Lacretelle; Lacretelle told Bertholon he had given it to Condorcet; Condorcet said he had sent it to Cotte; Cotte was out of town and, when Bertholon tracked him down, said he had already sent it back to Buissart. Finally in May, after almost a year, Lacretelle and Target signed a consultation and sent it, with the brief, to Buissart.[122]

The consultation affirmed Vissery and Buissart's argument that judges must defer to experts in matters of science. The jurisconsults wrote that it was "not up to the people, nor even to Judges to pronounce" on matters like Vissery's, but was the business of "*Savants.*" The alderman had broken "one of the principal rules of the judicial order. The Law forbids a judge to decide by himself all questions having to do with the Arts and Sciences. It orders him in this case to invoke the knowledge of Artists and *Savants*, and to judge according to their report." The alderman had also violated "the fundamental right to property" and had shown themselves "to oppose the progress of the Sciences." The barristers, moreover, agreed with Vissery and Buissart that the aldermen's failure to consult experts had provoked fear: "One could say that if the terror of the people was born of their ignorance, it grew by the imprudence of the Judgments."[123]

Indeed, more fully than Vissery and Buissart had done, the barristers developed a theory of the importance of expertise. This theory rested upon an assumed social hierarchy. The people must defer to the judges, while the judges deferred to the experts. By failing to carry out their part, the aldermen had dropped from their position in the ladder of authority and "become people themselves." The jurisconsults' social and legal theory of expertise was about authority, not knowledge. The role of experts was not so much to inform a judgment, as to legitimate it: "Even if the Judge is well enough versed in a

---

119. Buissart to Cotte, 19 July 1781, ibid., 4J/118/47; Cotte to Buissart, 3 September 1781, ibid., 4J/118/49. Cotte said the memoir had been brought to the Maison des Avocats personally by Condorcet, into whose hands Cotte had delivered it himself. Cotte to Buissart, 8 October 1781, ibid., 4J/118/52.

120. Cotte to Buissart, 22 December 1781, ibid., 4J/118/56. Cotte mentions three other possible collaborators: "Gerbier" (Pierre-Jean-Baptiste Gerbier), "Beaumont" (Elie de Beaumont), and "Le Gouvée" (Jean-Baptiste Le Gouvé). On Le Gouvé, Target, Gerbier, and Beaumont as jurisconsults, barristers, and men of letters, see Bell, *Lawyers and Citizens* (1994), 121, 132–34.

121. Buissart to Cotte, 16 May 1782, AGPC, Coll. Bar., 4J/118/63.

122. Bertholon to Buissart, n.d., ibid., 4J/119/14. The consultation was also signed by "Polverel" (the barrister Etienne de Polverel) and "Henry."

123. "Consultation du 3 mai 1782," in Robespierre, *Oeuvres*, 1:112.

Science or Art, to resolve the question according to his own lights, he should nonetheless consult people of the Art, because he has a mission as a Judge, and he has none as a *Savant*." However, the consultants also found that the aldermen would have been correct in responding to the popular alarm by provisionally banning Vissery's machine if only they had also sought "to reconcile the Inhabitants of their City" to the lightning rod. Like the Saint-Omer aldermen and the Châtelet lawyer Cotte had earlier consulted, Lacretelle and Target emphasized the importance of quelling a panic. Vissery's affair, they wrote, affirmed the natural tendency of the people to fear novel scientific discoveries. Their terror was an escalating contagion, quickly becoming "a Fanaticism." The consultation recommended that the superior court show prudence in reestablishing Vissery's lightning rod. In order to calm the people, the court should order an examination of the machine by the Paris Academy of Sciences. The Academy would name local "*physiciens*" to present it with an official description, upon which it would base its decision. This decision would then be published and distributed in Saint-Omer before the rod was replaced.[124]

Buissart received this consultation with mitigated gratitude. He would have liked an even stronger inculpation of the aldermen, one that would have required them to pay the costs of Vissery's appeal.[125] Vissery was even less pleased. The consultation was too deferential to the "alarms and chimerical terrors" of the people. Also, a capricious Vissery now decided he disliked the recommendation that the superior court order that his lightning rod be examined by experts at the Paris Academy of Sciences. True, acquiring this recommendation had been his and Buissart's original purpose. But Vissery had grown too impatient to tolerate such an inspection and was, moreover, reluctant to pay for it. He could almost believe that there had been "a connivance between [his adversaries] and the Lawyers of Paris."[126]

Nevertheless, the brief, with the accompanying consultation, was published at the end of 1782. Vissery had Buissart forward copies to Bertholon, Maret, Cotte, and Franklin.[127] These scientific consultants now began to

124. Ibid. I use *savant* rather than "scientist" for the same reasons that I use *physicien* rather than "physicist." The eighteenth-century category of *savant* differed from the twentieth-century category of "scientist" in being broader, more fluid, and not professional. See supra, n. 2.

125. Buissart to Cotte, 1 June 1782, AGPC, Coll. Bar., 4J/118/63.

126. Vissery demanded a second consultation with Arras lawyers, who proved more obliging, finding the aldermen personally liable for Vissery's legal bills. See Vissery to Buissart, 10 June 1782, ibid., 4J/120/7; "Consultation du 15 septembre, 1782," in Robespierre, *Oeuvres*, 1:112–14; and "Consultation," in Buissart, *Mémoire* (1782), 68–71. On Vissery's dissatisfaction with the first consultation, see also Vissery to Franklin, 10 December 1782, reproduced in Vellay, "Mélanges et documents" (1914), 135–37.

127. Vissery to Buissart, n.d., AGPC, Coll. Bar., 4J/120/4.

quibble with the brief's many facts, most of which they themselves had supplied. "Do not at all cite in your memoir," admonished Bertholon, "this supposed project of elevating lightning rods on the gallery of the Louvre, because
this, not being true, will hurt the cause of lightning rods."[128] The news of the
"supposed project" had come from Le Roy, who in turn pointed out some factual lapses. In one, Buissart sarcastically berated the aldermen for claiming,
in their original sentence, that the "famous Bernoulli, who died in his bed,"
had been killed in the accident that had in fact claimed the life of Richmann.
But Buissart then located Richmann's fatal accident in Moscow, as, later, did
Robespierre.[129] Both ignored Le Roy's correction, stating that Richmann died
in the St. Petersburg Academy of Sciences, where he lived. Le Roy, apologiz-
ing for introducing such a trivial concern, thought consistency important,
since "you justly reproach the magistrates for having killed the celebrated
Daniel Bernoulli."[130]

More to the point, Le Roy emphasized that there was no lightning rod on
the king's palace at La Muette, as Buissart's memoir and Morveau and Maret's
report[131] claimed, nor even, as Robespierre later maintained,[132] on the *cabinet
de physique* at La Muette. In the first place, the *cabinet de physique* was not at La
Muette, but nearby at a house in Passy. Secondly, there was no lightning rod
at the *cabinet de physique*, but only an "electroscope" (the instrument Buissart
had called an "electrometer"[133]), an insulated conductor that served to collect, rather than dissipate, electrical fire, and that was frequently confused
with the lightning rod. Le Roy himself had installed the electroscope in the
garden of the *cabinet de physique* in Passy.

The most, therefore, that could be said of the king and queen's devotion to
the cause of the lightning rod was that although a similar instrument could

128. Bertholon to Buissart, 25 April 1783, ibid., 4J/119/23. Le Roy had written to Buissart that
he planned to oversee the establishment of lightning rods along the Grand Galerie after renova-
tions, then in progress, were completed. Le Roy to Buissart, ibid., 4J/120/22. Bertholon's objection
arose partly from his competitive dislike of Le Roy, who, he insisted, "understands absolutely
nothing about the construction of lightning rods." Bertholon to Buissart, 25 April 1783, ibid.,
4J/119/23.

129. "Sentence des échevins de Saint-Omer, du 14 juin 1780," in Robespierre, *Oeuvres*, 1:101;
Buissart, draft of memoir, AGPC, Coll. Bar., 4J/120/30, 53; Robespierre, "Premier plaidoyer"
(1783), in *Oeuvres*, 1:40 (Robespierre also spelled the name "Rikman").

130. Le Roy to Buissart, 9 November 1782, AGPC, Coll. Bar., 4J/120/21.

131. See "Extrait des registres de l'Académie de Dijon, du 18 Août 1780," in Robespierre, *Oeuvres*, 1:107.

132. Robespierre, "Premier plaidoyer" (1783), in *Oeuvres*, 1:59–60. Robespierre ambiguously
calls the instrument an "electric rod."

133. Le Roy wrote that the instrument should properly be called an "electroscope" and not,
as Buissart called it in his memoir, an "electrometer," because it did not measure electricity, but
only made its presence known.

plainly be seen from a spot on the grounds of La Muette where the queen frequently lunched, neither she nor the king had ever complained of it.[134] In fact,
despite Vissery, Buissart, and Robespierre's insistence that only a provincial
backwater would hesitate to install a lightning rod, there were none in the
capital before December 1783, when Bertholon wrote to Buissart that he had
just erected the first two.[135]

The publication of the memoir inaugurated a new stage of the debate, one
that took place in print. The aldermen began to seek published corroborations
of their arguments against lightning rods;[136] and Buissart became concerned
to refute these and find his own corroborators. When, for example, he discovered that Grandidier's history of the Strasbourg cathedral[137] contained a
"bad idea of conductors" and that Marat, in his treatise on electricity,[138] "also
speaks disadvantageously of lightning rods," he urged his collaborators to respond through the journals.[139] Bertholon declined with emphasis: Grandidier
was "ignorant" and Marat, "lunatic."[140] Another author, a potential supporter,
was "a man with no judgment, who does not know how to write, and who is
singularly ridiculous, and very old."[141] None merited a response or reference.
So Buissart himself, under an assumed name, wrote the refutations, which he
later deemed to have "worked marvelously."[142] He sent two letters to the
Affiches de Flandres, signed "Nostradamus." In the first, Nostradamus explained

134. Le Roy to Buissart, 9 November 1782, AGPC, Coll. Bar., 4J/120/21. Bertholon added spite-
fully that "nothing is worse made than this instrument," and that "15 days later the wind overturned the upper apparatus." See Bertholon to Buissart, 22 January 1783, ibid., 4J/119/17; and Bertholon to Buissart, 29 December 1782,ibid., 4J/119/16.

135. Bertholon to Buissart, 14 December 1783, ibid., 4J/119/15 bis. Bertholon's establishment
of the first two lightning rods in Paris, on the *hôtel* of the duchesse d'Ancenis and the convent of
the Religieuses Augustines Angloises, is announced in the *Mercure de France* (28 December 1783),
188–89.

136. For example, the Saint-Omer aldermen publicized their discovery of two articles in the
Journal de Luxembourg making a case against lightning rods. Buissart responded by sending a letter to the *Affiches de Flandres* and wrote to Cotte that it was important "to combat [these two ar-
ticles] but this task cannot be mine, since I am the lawyer for M. de Vissery, it must be that of
M. Le Roy or yours." Buissart to Cotte, 11 December 1782, AGPC, Coll. Bar., 4J/118/68.

137. Grandidier, *Histoire de l'église* (1776–78).

138. Marat, *Recherches physiques* (1782).

139. Buissart to Cotte, 27 March 1783, ibid., 4J/118/70.

140. Bertholon to Buissart, 29 December 1782, ibid., 4J/119/16; 15 December 1782, ibid.,
4J/119/15 bis.

141. Bertholon to Buissart, 22 January 1782, ibid., 4J/119/17. The potential supporter was
Charles Rabiqueau, who wrote *Spectacle du feu élémentaire, ou Cours d'électricité expérimentale* (Paris,
1753). Buissart was unimpressed by Bertholon's admonishment: he quoted Marat in his memoir
and Rabiqueau at great length in Nostradamus's first letter. Buissart, draft of memoir, AGPC,
Coll. Bar., 4J/120/30, 26; "Lettre de Nostradamus au Redacteur des affiches de flandres," n.d., ibid.,
4J/120/27.

142. Buissart to Cotte, 17 July 1783, ibid., 4J/118/73.

the difference between a "lightning rod" and an "electrometer," to show that a lightning rod can never be positively charged, and so can never positively "electrize" the house it protects. The second letter presented a distinction without a difference: lightning rods did not attract electricity, but, on the contrary, electricity sought out lightning rods.

Nostradamus invoked the purposefulness of natural phenomena in the manner of sentimental empiricists. He argued that electricity's desires and preferences would lead it to compensate for all manner of structural problems in a lightning rod. Let the bar be too thin, "the thunder will follow it even while melting it." Let it be broken, "the thunder, due to its affection for metallic materials, will jump to carry itself from one Bar to the other." Let the gap be wide, "the thunder will furrow the wall a bit to arrive" at its beloved bar. Finally, Nostradamus explained that a lightning rod acted only within its "sphere of activity," a funnel-shaped area of electrical "void" extending from the point into the clouds above. The faultiest lightning rods, ungrounded or with rusted points, had no sphere of activity, so no effect, making lightning rods innocuous at worst.[143] These were the facts, unsullied by explanations and fully certified by experts.

Buissart's memoir made a strong impression. At home in Saint-Omer, Vissery told Buissart, he had discussed it with the lieutenant general,[144] who before reading it had shared "the general opinion that the blade of the Sword was Electrised and magnetized to attract the Thunder." The officer was sufficiently converted by Buissart's text to ask Vissery if he could come see the apparatus, "to which I gladly consented." Meanwhile in Paris, Bertholon reported having talked of the trial with Jean d'Alembert, at whose house "all the best assemble each evening."[145]

There were, however, some lingering areas of dissatisfaction. Vissery, who now disliked the idea of an inspection by members of the Paris Academy, as

143. Buissart, "Lettre de Nostradamus au Redacteur des affiches de flandres," n.d., ibid., 4J/120/27.

144. The Crown named "lieutenant generals" to the provinces to keep the provincial governors in check. In practice, however, the authority over the provincial government rested more with the commandant than with either the governor or the lieutenant general. See supra, n. 51; and Marion, *Dictionnaire des Institutions* (1923), 336.

145. Vissery to Buissart, n.d., AGPC, Coll. Bar., 4J/120/4; Bertholon to Buissart, 15 December 1782, ibid., 4J/119/15 bis. Bertholon promoted the brief tirelessly: "I have spoken to many people about your excellent memoir in favor of M. de Vysseri, it is much enjoyed and much applauded, and does you great honor. I will continue to procure for it as many sales as possible." Bertholon to Buissart, 25 March 1783, ibid., 4J/119/20. He offered to get it announced in the *Journal de Paris* and the *Mercure de France*, where it received a very favorable mention. Bertholon to Buissart, 29 December 1782, ibid., 4J/119/16; see infra, n. 182. Gérard Walter suggests that the *Mercure de France* announcement was, however, inserted by Lacretelle, who collaborated in editing the journal at the time. See Walter, *Robespierre* (1961), 1:38. A review in the *Feuilles de Flandres* is mentioned in Vissery to Buissart, 8 June 1783, AGPC, Coll. Bar., 4J/120/9.

recommended in the Parisian barristers' consultation, also found Buissart's brief, which he had read upon its return from Paris, "a little voluminous" with "too great a display of Erudition." Vissery was certainly thinking of the costs of printing a ninety-page document. But his argument was stylistic. Vissery invoked Morveau's "Of the Style of the French Bar," which celebrated the simplicity of legal writing.[146] Most readers, Vissery insisted, would understand nothing "of abstract matters . . . one must therefore limit oneself to the facts." After a testy response from Buissart, Vissery apologetically retracted his editorial suggestions.[147] But Buissart himself had gestured, in his brief, toward forgoing "reasons" in favor of "facts." [148] This gesture, together with Vissery's admiration for a "fact"-based legal style and his new disenchantment with the idea of an examination by experts, prepared the ground for Robespierre's change of strategy.

ACT V: THE TRIAL

Maximilien Robespierre, while still a schoolboy, had demonstrated such a "pronounced taste for the exact sciences" that all Arras had taken to calling him by the nickname "Barometer." [149] He was familiar not only with the principles and demonstrations of contemporary natural science, but with its central epistemological dogma: knowledge of nature resides in sensibility, not theory. Early acquaintance with this dogma prepared Robespierre to develop upon, and eventually to depart from, his senior colleague's strategy for beguiling the judges of the Conseil d'Artois.

In the fall of 1782, having finished his brief, Buissart handed the lesser work of oral argument to his junior colleague.[150] The case came before the Conseil d'Artois the following May.[151] Robespierre presented his plea, which lasted an hour and a quarter, on 17 May.[152] He did not argue the importance of con-

146. Guyton de Morveau, "Du style du barreau français," *Discours publics* 3 (1775–82): 137–99.

147. Vissery to Buissart, 10 June 1782, AGPC, Coll. Bar., 4J/120/7; Vissery, "Observations sur le Mémoire," ibid., 4J/120/32; Buissart to Cotte, 15 July 1782, ibid., 4J/118/64; Vissery to Buissart, 25 October 1782, ibid., 4J/120/8.

148. See supra, §3.

149. Jacob, "Un Ami de Robespierre" (1934), 278.

150. Robespierre's appointment is mentioned in Vissery to Buissart, 25 October 1782, AGPC, Coll. Bar., 4J/120/8; see also Walter, *Robespierre* (1961), 1:35–37.

151. Robespierre, *Oeuvres*, 1:23 n. 2; see also Paris, *La Jeunesse de Robespierre* (1870), 56. Valour, the Saint-Omer bailiff, had died in the interim, so Vissery made his claims for recovering the costs of the appeal against Valour's widow and against the Renard-Debussy family. In claiming that the bailiff's heirs should be liable for the costs of Vissery's appeal, Buissart argued that the bailiff had made himself liable by withholding the names of Vissery's accusers. See Buissart, draft of memoir, AGPC, Coll. Bar., 4J/120/30, 75.

152. "Lettre d'un Ancien professeur de physique au Redacteur des feuilles de flandres, le 25 mai 1783," in AGPC, Coll. Bar., 4J/120/23.

sulting experts in scientific cases. He did not emphasize the difference between physics and jurisprudence. Instead, he began with a flattering reassurance: "Do not fear, Sirs, that I will engage in an infinite discussion of a theory alien to the Bar (if there are any completely alien to it)." He promised to speak a common language: "I will dwell above all on facts and experience." [153]

Although the historical and scientific material and the many examples in Robespierre's plea were drawn from Buissart's brief, Robespierre departed sharply from Buissart in his legal strategy. He abandoned Buissart's claim that the aldermen had violated Title XXI of the Ordinance of 1667 by failing to call for an expert consultation. Nor did Robespierre claim, as Buissart had done, that experts must decide the appeal and that the Arras judges should defer to academicians. He instead made the opposite argument. The assessment of Vissery's lightning rod, he said, was fully within the competence of the judges of the Conseil d'Artois.[154]

The "Public," knowing that the lightning rod "was due to Physics" and each realizing his own ignorance in that science, might imagine "that this affair was placed outside the sphere of the Bar, and that without being a *Physicien* by profession, it was impossible to decide whether electric conductors were harmful or advantageous." However, Robespierre assured the judges that "far from requiring a specialized study of Physics," the knowledge of lightning rods rested entirely upon "daily experiences . . . our most familiar amusements; [and] phenomena that offer themselves to our eyes in each storm." [155] In short, the lightning rod was a matter of sensible fact, and "magistrates can pronounce boldly on this point." *Savants* might be required to settle questions of theory, but not questions of fact. The virtues of the lightning rod were fully proven by "experience." It sufficed "to have common sense and eyes to recognize them." [156] A magistrate's duty to the people—which the Saint-Omer

---

153. Robespierre, "Premier plaidoyer" (1783), in *Oeuvres*, 1:36–40.

154. The "infinite relations of the . . . sciences with the power and prosperity of States" made distinguishing "a dangerous citizen" from a "useful *savant*" central to the process of governing. Ibid., 1:36–38. There were also practical reasons for this change of tactics. In addition to responding to Vissery's growing impatience, Robespierre himself was probably eager to win the case quickly and definitively, rather than merely winning it provisionally, subject to an expert examination. See Robespierre, "Second plaidoyer" (1783), in *Oeuvres*, 1:97.

155. Robespierre, "Premier plaidoyer" (1783), in *Oeuvres*, 1:62–63; "Second plaidoyer" (1783), in ibid., 1:88–89.

156. Robespierre drew a further distinction between new controversial techniques of science and older established ones. If Vissery's lightning rod "were the first instrument of this sort . . . if on one side they praised the utility of this kind of machine, while on the other they represented it as a pernicious invention, and if these two opposed systems rested on principles of Physics," then the judge would have no choice but to turn to the academician. But the lightning rod was as established in physics as inoculation in medicine, which "enjoys, since a considerable time, the confidence of the public," and no longer required an official sanction by the Society of Medicine. Robespierre, "Second plaidoyer" (1783), in ibid., 1:85–87.

aldermen had failed to discharge—was to consult not the experts, but their own sensations.

In arguing against a consultation with experts, Robespierre combined the principles of sentimental empiricism with Rousseauian political theory. In his 1750 "Discours sur les sciences et les arts," Rousseau had railed against the sterile abstractions of academic science, "the ratios in which objects attract one another in a void; . . . what curves have conjugate points." Out of "idleness," experts generated such "futile" knowledge and brought about a general dissolution of morals and corruption of taste. This was the fault not of science, but of expertise: "The prince of eloquence was Consul of Rome, and perhaps the greatest of philosophers Lord Chancellor of England." If the "first had only occupied a chair at some university, and the other had achieved only a modest academic pension, [would] their works . . . not have suffered from their situation?" The sciences needed not leisured specialists, lost in the depths of their own speculative fancies, but politically engaged practitioners, open to the world and active within it. Such civic-minded natural philosophers would combine knowledge with power, legitimating both.[157]

In the same way that no discrete group of experts ought to practice the arts and sciences, Rousseau argued that no such body could legitimately govern. The *Social Contract* (1762) accordingly rests upon the rejection of a form of political expertise, that is, representation. Once again, the experts' corporate particularity was their ruin. No class of deputies could represent the true sovereign. This sovereign, the general will, was a "collective being" formed of the individual wills of all the citizens. No subset of the citizenry could stand for this collective being, because "the particular will tends, by its very nature, to partiality."[158]

No expert could speak for the facts; no deputy could speak for the people. Rousseau repudiated both sorts of extrapolation from particular experience to general knowledge. These repudiations resonated with the empiricist teachings of contemporary natural science. The *Social Contract* was a "crash-

157. Rousseau, "Discours sur les sciences et les arts" (1750), 18, 29. The references in the quote are to Cicero and Francis Bacon.

158. A particular will might accidentally *coincide* with the general will, but its inability to *represent* the general will was definitional, the general will being defined by its generality. Rousseau, *Du Contrat Social* (1762), bk. III, chap. XV, 307. Rousseau developed his rejection of political representation in response to the crucial role that the authors of traditional liberal political theory (notably Montesquieu) assigned to administrative elites. In liberal theory, it was the constitutional role of traditional elites to discern and defend the interests of the people. See Baker, "Calculus of Consent," in *Condorcet* (1975), 225–44. On the role of the robe nobility in Montesquieu's political philosophy, see Ford, *Robe and Sword* (1965). On the constitutional function of elites in traditional liberalism, see Manent, *Intellectual History of Liberalism* (1994). On the development of Rousseau's political philosophy in response to those of his predecessors and contemporaries, see Dérathé, *Jean-Jacques Rousseau* (1970).

ing failure," Robert Darnton has observed, Rousseau's "least popular book before the Revolution." But, Darnton has persuasively argued, Rousseauian political theory was pulled from obscurity and transmuted into a popular force
by amateur natural science, the "greatest fashion of the decade before 1787."
Scientific amateurs developed a "vulgar kind of Rousseauism" in which the
contractual origins of society shrank to the vanishing point, while the rejection of experts and their sterile "rationalism" loomed large. Rousseau's reference to a precontractual state of nature fueled a "mystical" notion in popular
science of sensitive intimacy with the "primitive" natural world. A romantic
empiricism thus informed the genesis of popular Rousseauism.[159] Robespierre's brief for Vissery was a milestone in this development.

For here, Robespierre accomplished a potent connection between natural
and social philosophy. He founded Rousseauian repudiations of both expertise and false political power upon the empiricist principles of contemporary natural science and legal theory.[160] The further development of this
sentimental-empiricist rendering of Rousseau would be central to Robespierre's later career. In his Revolutionary speeches, he would maintain that
Rousseau's "doctrine" was "drawn from nature." The sentiments of civic re-
sponsibility were, Robespierre insisted before the National Assembly, "more
natural than you think."[161] He would thus attach the general will to purported
facts of the individual psyche, producing an empiricist, anthropological political theory.[162]

Legislators must accordingly renounce the eternal disputes of the
"metaphysicians." They must operate inductively, always "composing [their]
general principles from particular observations." Particularity was itself a

159. Darnton studies these developments in one particular fad of popular natural science,
mesmerism. See Darnton, *Mesmerism* (1968), 3, 161, 124, 115–16.

160. Robespierre's empiricist version of Rousseau's political philosophy is one moment in a
long history of interactions between epistemology and political philosophy. In this history, epistemologies and styles of political philosophy have aligned differently at different times. For example, Steven Shapin and Simon Schaffer have studied an earlier moment, during the English civil
wars of the mid-seventeenth century, and a very different alignment. In their reading of Robert
Boyle's and Thomas Hobbes's dispute about the social value of philosophical empiricism, Boyle
joined empiricist principles to a liberal theory of government, and Hobbes, rationalist principles
to an absolutist theory of government. See Shapin and Schaffer, *Leviathan and the Air-Pump* (1985).

161. Robespierre, "Sur les rapports des idées religieuses et morales avec les principes républicains et sur les fêtes nationales. Rapport présenté au nom du Comité de Salut public" (18 floréal
an II [7 May 1794]), in Robespierre, *Ecrits* (1989), 321; "Sur la rééligibilité des députés de l'Assemblée nationale (18 May 1791), in ibid., 117.

162. Barny writes that Robespierre "adheres to the anthropology of Rousseau," and that "the
expression 'the nature of things'... that returns incessantly to Robespierre's lips testifies to his desire to treat the objects of society in the most objective and scientific manner possible, following
the lesson of Montesquieu." See Barny, "Robespierre et les lumières," in Jessenne et al., eds.,
Robespierre (1993), 48, 50.

guarantee of generality. Robespierre's favorite of Rousseau's writings had been dedicated to this point. The author of the *Confessions* had announced himself unique and universal, each in virtue of the other. "Myself alone. I feel my heart and I know men." Nature had broken the mold in which it cast him and, by revealing that mold's most singular idiosyncrasies, Rousseau would "show my fellows a man in all the truth of nature; and this man will be myself." Echoing both Rousseau and Montesquieu, Robespierre demanded, "Is it not the case that the more general a thing is, the more it is subject to exceptions?"[163]

Empiricist governing, like empiricist natural philosophy, meant the sensitive acceptance of contradiction, mystery, and the weakness of reason, since the "moral world, even more than the physical world, seems full of . . . enigmas."[164] No abstract theory of representation must intervene between the enigmatic passions of the individual and the actions of the state. Instead, representatives must identify absolutely with the represented.[165] The state must accept as givens the beliefs of the individual and inductively generalize these beliefs into national institutions. In particular, Robespierre would claim that the popularity of religious belief was an incontrovertible fact. The *"aristo-cratic"* idea of atheism was an artificial product of "philosophism," while the "entirely popular" idea of an "incomprehensible power" arose naturally. In obeisance to this fact of human nature, the state must institute a civic cult for the promotion of virtue and unity. Moreover, Robespierre would argue, his own empiricist respect for the fact of popular belief in a supreme being allowed him to "speak neither as an individual, nor as a systematic philosopher, but as a representative of the people."[166]

Robespierre's notion of an empiricist basis for political authority was first developed in his plea for Vissery. There he rehearsed its several corollaries: the condemnation of rational theory, expertise, and false political power; the embrace of enigma as a rebuke against rationalism; and the insistence that particular experiences are related to general truths not by theoretical explanation, but by acceptance and inductive generalization. Added together, these summed to the claim that a sound argument must rest upon acceptance and generalization of enigmatic particulars, and that the maker of such an argument could lay claim to a dual authority: he spoke for nature and for the people.

163. Rousseau, *Confessions* (1770), 43; Robespierre, "Sur la fuite du Roi" (14 July 1791), in *Oeuvres*, 7:571–72.

164. Robespierre, "Sur les rapports des idées religieuses" (18 floréal an II [7 May 1794]), in Robespierre, *Ecrits* (1989), 317, 306.

165. Here, Robespierre was developing upon Rousseau's criticism of political representation. Robespierre, "Sur la réélection des membres de l'Assemblée nationale (16 May 1791), in ibid., 115.

166. Robespierre, "Contre le philosophisme et pour la liberté des cultes (Aux Jacobins, le 1er frimaire an II [21 November 1793])," in ibid., 284; "Sur les rapports des idées religieuses" (18 floréal an II [7 May 1794]), in ibid., 317.

Ingeniously, in his plea for Vissery, Robespierre professed to let the facts speak for themselves, while at the same time hardly touching upon most of what one would assume were the relevant facts. He said nothing about the design of Vissery's lightning rod. As we have seen, details about any particular lightning rod or stroke did not speak for themselves. Quite the contrary, they spoke with the greatest promiscuity and ambiguity for several opposing theories of the lightning rod at once. Instead of offering particular details as facts, Robespierre adopted a more persuasive strategy. He presented the judges with a series of inductive generalizations about electrical behavior. According to Robespierre's implicit, negative definition, these generalizations were "facts" because they were "not-theories."

They derived their status as not-theories from the absence of any interpretive system mediating between them and the particular observations in which they were founded. These generalizations were facts, in other words, thanks to their stubborn refusal to participate in theoretical explanations.[167] Quite the opposite of explaining, Robespierre's facts marked explanatory limits, the points at which understanding must give way to feeling. One did not explain a fact: it was itself the last word in the matter; beyond it was out of bounds to the intellect. A fact was the unexplained explainer, the absolute sovereign. Facts, defined as particulars generalized and as the unsurpassable ends of explanation, were checks on power and reason, on the particular will of the expert, the system-builder, the usurper of the people's sovereignty.

Thus, according to Robespierre, "experience has shown that the electric material tends toward metals and aqueous fluids," and "observation convinced us that metallic points have the virtue of drawing off electric material." Certain materials were "suited by their nature to receive" electricity, while others were not. The electrical material would "necessarily seek" a metal bar; electricity had a "predilection" for metal; it was thus "physically impossible" for electrical matter to jump from a metal bar to a wooden house. These were facts. Electrical action was "no longer a mystery" but was now "an elementary principle"[168]—the difference being the renunciation, not the achievement, of understanding.

Foacier de Ruzé, the Arras prosecutor (*avocat général du roi*), who inherited the case in its appeal, spoke the next week, on 24 May.[169] Robespierre had embraced empiricism in principle, but made only principled arguments in its name. Ruzé said nothing about empiricism, but his arguments were purely practical: each one concerned the design of Vissery's lightning rod. Ruzé

167. See supra, p. 142 n. 10.

168. Robespierre, "Premier plaidoyer" (1783), in *Oeuvres*, 1:36–40, 52.

169. "Lettre d'un Ancien professeur de physique au Redacteur des feuilles de flandres, le 25 mai 1783," AGPC, Coll. Bar., 4J/120/23.

supposed the bar to be a half inch in diameter, the width "recommended by the *physiciens*," and suggested that in some cases its capacity might be too limited to accommodate the electrical matter. Might not the thunder then "divide itself, and direct one of its divisions into the house"?[170] In fact, Le Roy had raised a similar problem in an offhand remark in a letter to Buissart: "I forgot to tell you that all is well with the conductor of M. de Vissery, except the channel of tin, . . . [which] being very thin, can easily be pierced, that is what happened with a tin pipe in the lighting stroke at Brest."[171]

Ruzé showed, furthermore, that he, too, could deploy empirical evidence. "Experiences and observations have taught," he wrote, that "the necessary precautions in this regard are not yet perfectly fulfilled." He also cited the Viennese example, the sparks the rod had been seen to throw, a clear indication of its insufficient capacity. Ruzé concluded, not with hostility toward the sciences but, on the contrary, with an implicit assumption of their continued progress. Lightning rods, he said, must be regarded as dangerous "until *physiciens* have found the means of guaranteeing them with absolute certainty." Finally, Ruzé, like Buissart, argued that "magistrates are not at all *Savants*; their mission is not to decide questions that concern the sciences." Like Vissery's own barrister and Parisian consultants, the prosecutor recommended that the court consult a scientific academy before pronouncing on the design of Vissery's lightning rod.[172]

Buissart, now in the guise of a "former professor of physics," reported the first stages of the trial to the *Feuilles de Flandres*. He either did not notice or did not object to Robespierre's change of strategy. Indeed, Buissart said nothing of Robespierre's plea except that it was "a masterpiece of erudition and eloquence." On the other hand, though he called Ruzé's defense "weak," he faithfully reproduced its "anti-conductorist" arguments:

> Lightning rods attract thunder to the armed building, the crampons that attach the machine to the wall, the unknown thickness of the Rod, the accidental disintegration, the luminous spray that shines at the point in stormy weather, the sphere of activity, the armed Buildings on which thunder has fallen, the points of conductors that have melted, &c., &c. presented successive dangers; the crampons and the disintegrating Rod, *lateral explosions*; the thickness of the rod and the melted points, *an uncertainty*; the luminous spray at the point, *an engorgement*; the sphere of activity, *a void in the clouds*; houses armed and incinerated, *a proof of attraction*, &c. . . . [I]t was even advanced, that the lightning rod, perform-

170. "Objections nouvelles de M. l'Avocat général" (1782?), ibid., 4J/120/36.

171. Le Roy to Buissart, 9 November 1782, ibid., 4J/120/21.

172. "Objections nouvelles de M. l'Avocat général" (1782?), ibid., 4J/120/36; Robespierre, "Second plaidoyer" (1783), in *Oeuvres*, 1:68.

ing its function, condensed the electric material in the bosom of the earth, which could produce volcanoes, earthquakes, waterspouts, &c.[173]

Robespierre was granted a response on the thirty-first of May.[174] He began by characterizing the prosecutor's arguments as "reasonings of theory." His message was simple: Forget reasoning and theory. "Experience is on my side," and "against experience . . . what good are all the reasons?" Robespierre proposed neither to support theory with facts, nor to derive theory from facts, but to set facts against theory. The "natural tendency" of electrical matter toward metal bars in preference to tile roofs was a fact. It represented all the explanation that physics required and superseded any question about its mechanism. A "Poet or Orator" might call it a miracle. A philosopher would recognize it as no prodigy, but a "law of nature" and an "ordinary phenomenon."[175]

Taking on Ruzé's arguments in succession, but never refuting them, Robespierre instead dismissed each one as theoretical, and therefore invalid. His opponent reckoned that there was a comparatively low probability of being struck by lightning, making the potential risks of a lightning rod less worthwhile. Robespierre dismissed the calculation—"I have not verified this calculus of probabilities"—and replaced it with a fact guaranteed by his own sensitivity: "What I know with certainty, is that the victims of thunder are unhappily too numerous." Ruzé wondered how to know the dimensions of a lightning rod's sphere of influence. Robespierre replied casually that "all the facts" showed it was "very considerable." But, he demanded, "what does it matter to us to measure it with a geometrical exactitude?"[176] Calculations and geometry were theory, and theory was irrelevant.

To every theoretical query, Robespierre had a "factual" answer, that is, an undefended assertion. What if the rod attracted lightning that would not otherwise have struck? Robespierre responded in the casuistic style of "Nostradamus": to say the rod "attracts" electricity was only a figure of speech. In fact, the rod was "purely passive," and it was the lightning that sought the rod. What if the rod were too small to accommodate all the electricity and released some onto the roof? Robespierre responded that "such an engorging . . . could not take place," since the lightning would only seek the rod for as long as it offered easy passage. When the rod became full, the lightning would look elsewhere.[177] Ruzé also posed Nollet's question, How to conceive that such a vast quantity of volatile material could enter such a small point? Robespierre chastised, "How to conceive this phenomenon? . . . [I]t matters little how to

173. "Lettre d'un Ancien professeur . . . , le 25 mai 1783," AGPC, Coll. Bar., 4J/120/23.
174. Ibid.; Robespierre, *Oeuvres*, 1:22–23.
175. Robespierre, "Second plaidoyer" (1783), in *Oeuvres*, 1:68–69, 79.
176. Ibid., 1:70–72.
177. Ibid., 1:72–74.

conceive it, if experience attests that it exists. If it were inexplicable, it would have that in common with most other effects that nature presents to us." Sensibility meant acceptance of the inexplicable. Lightning would escape from a conductor when rivers climbed mountains, when iron fled the magnet, when dropped rocks forgot to fall. Theirs was not to reason why.[178]

After Robespierre's response to Ruzé,[179] the court rendered its judgment: Vissery was permitted to reestablish his lightning rod, but he must desist in all claims against his accusers and must bear the costs of the reestablishment and the appeal himself.[180] Vissery wrote to Buissart, "You have given me . . . victory." But he added, "I would have wished . . . it to be more complete."[181]

EPILOGUE: THE POLITICAL TRIUMPH OF SENTIMENTAL EMPIRICISM

Vissery's triumph was promptly written up in the *Mercure de France.* The notice praises Robespierre as "a young lawyer of rare merit" and advertises Buissart's brief as "an estimable Memoir that can be regarded as a treatise of physics."[182] In the wake of their shared triumph, Vissery, Buissart, and Robespierre formed a mutually beneficial alliance. "The three of us share the glory," Vissery wrote to Buissart after the judgment, "you, Monsieur, for your well-Written Memoir, Monsieur the Orator for his Eloquent Plea, and I by the winning of a cause that is no longer controversial."[183]

Robespierre seized the moment, applying to Vissery for funds to publish his pleas.[184] The plea was to be printed in Arras, announced in the *Mercure de France,* the *Journal Encyclopédique,* and *L'Année littéraire,* and sold in Arras and

178. Ibid., 1:74–75, 77.

179. According to an undated document, Vissery's legal team demanded in conclusion that he be permitted to reestablish his lightning rod; that the current bailiff and the heirs of Valour be held liable for the cost of restoring the lightning rod and for Vissery's legal bills; that two hundred copies of the decision be posted and distributed in Saint-Omer and Arras; and finally, that if the court could not pronounce definitively on the basis of the expert opinions it had been presented, it should forward the description and plan of the lightning rod, together with the Academy of Dijon's report, to the Paris Academy of Sciences to solicit its pronouncement. "Consultation des avocats de M. de Vissery," in Robespierre, *Oeuvres,* 1:114–16.

180. "Lettre d'un Ancien professeur . . . , le 25 mai 1783," AGPC, Coll. Bar., 4J/120/23; Robespierre, *Oeuvres,* 1:22–23. The court's decision is quoted in Robespierre, *Oeuvres,* 1:23.

181. Vissery to Buissart, 8 June 1783, AGPC, Coll. Bar., 4J/120/9. "M. de Vissery has won his trial." Buissart to Cotte, 17 July 1783, ibid., 4J/118/73.

182. *Mercure de France (Journal politique de Bruxelles)* (21 June 1783), 135–37; reproduced in Robespierre, *Oeuvres,* 1:116–18.

183. Vissery to Buissart, 8 June 1783, AGPC, Coll. Bar., 4J/120/9. Another letter includes plans for Buissart and Robespierre to pay Vissery an extended visit. Vissery to Buissart, 3 July 1783, ibid., 4J/120/10.

184. Walter, *Robespierre* (1961), 1:37–39.

Paris, at bookstores at the Palais Royal, the quai de Gesvres, the Jardin du Luxembourg, and the Tuilleries. With some reluctance, Vissery handed over four louis toward the production and placement of five hundred copies.[185] When the pamphlet appeared the following fall, Robespierre and Buissart marketed it aggressively. Robespierre sent a copy to Franklin. Buissart forwarded one to Cotte, writing that it did "much honor here to the young lawyer who wrote it," and later asking eagerly, "What do you think of it? [A]re you as happy with it as I?" Cotte responded that he had read the pamphlet with pleasure, whereupon Buissart supplied him with another copy for distribution.[186] Buissart also dispatched a copy to Bertholon, who deemed the pleas "superiorly and masterfully done," and who responded with a copy of his own memoir on lightning rods for Buissart to give to Robespierre.[187]

The *Mercure de France* gave a brief review of the pamphlet: it did great honor to its author, who was "barely emerged from adolescence."[188] However, Des Essarts's *Causes célèbres* reprinted long excerpts, calling the pleas "among the most interesting productions of the bar."[189] And the month after his triumph in Vissery's case, the callow careerist succeeded in gaining admission to the Academy of Arras, thus solidifying his newly won philosophical credentials.[190] It is considered to have been a first step on his journey toward "the Estates General, and into history."[191]

Robespierre's legal triumph was, however, less definitive than his personal triumph. He himself was cautious, provoking Vissery's indignation by warning him "not to be in too much of a hurry to reestablish [the] lightning rod."[192] Vissery could restrain himself no longer than two months, and on 31 July 1783, he reinstalled the sword blade at the pinnacle.[193] People came to admire it; but children also came to throw rocks and sing satirical songs. The aldermen began to insist that Vissery "would not be in the clear without weathering a

---

185. Vissery objected that he wanted to read the pleas first, as he had read Buissart's memoir before having it printed. Vissery to Buissart, 8 June 1783, AGPC, Coll. Bar., 4J/120/9. Buissart sent an emissary to persuade him and to arrange for printing. For the production and distribution plans, see C.-J.-B. d'Agneaux Devienne to Buissart, n.d., ibid., 4J/120/28; reproduced and annotated in Robespierre, *Oeuvres*, 1:118–19.

186. Buissart to Cotte, 7 September 1783, 20 October 1783, AGPC, Coll. Bar., 4J/118/ 73; Cotte to Buissart, 8 November 1783, ibid., 4J/118/74; Buissart to Cotte, 3 February 1784, ibid., 4J/118/76; see also ibid., 4J/118/76, /80.

187. Bertholon, *Nouvelles preuves* (1783); Bertholon to Buissart, 14 September 1783, AGPC, Coll. Bar., 4J/119/28.

188. *Mercure de France* (1 May 1784), cited in Walter, *Robespierre* (1961), 1:39.

189. Anon., "Suite," in *Causes célèbres* 117 (1784), 146.

190. Walter, *Robespierre* (1961), 1:42.

191. Counson, *Franklin et Robespierre* (1930), 8.

192. Vissery to Buissart, 8 June 1783, AGPC, Coll. Bar., 4J/120/9.

193. See Robespierre, *Oeuvres*, 1:99.

visit by Experts." Vissery resisted. He invoked "Nostradamus"; a lightning rod, "well or badly made," he insisted, was "not at all dangerous in itself: if well-made, it results in good effects, *e contra*, it is as though null." He added inconsequently that "if there is a danger, it is only for me."[194]

In the fall of 1783, a third appeal, against the judgment of the Conseil d'Artois, was signed. The appeal was in the name of one Bobo, a merchant in the rue Marché-aux-herbes who sold salad from a double bag he wore over his shoulders. Buissart suspected Bobo of being a convenient new front for the old opposition as, being entirely without funds, he was immune to Vissery's counterclaims against him.[195] Bobo's appeal reinvigorated interest in the Vissery affair. Briefly, there was talk of its being heard in the Parlement of Paris.[196] An editorialist in Des Essarts's *Causes célèbres* wrote sarcastically that Bobo showed a surprising discernment

> of the merit of the academy of sciences: you even assess the levels of learned companies; you fix the degree of confidence that each should obtain; you make a subtle distinction between lightning rods of the city and lightning rods of the countryside; you want the *physiciens* to explain themselves with precision on each of these objects. So much erudition seems to me suspect in a salad merchant.

The illustrious Franklin and the immortal Buffon, the editorialist predicted, would surely race each other to Saint-Omer, eager to take on such an important function. "How I love to imagine them crossing the market," he scoffed. The Conseil d'Artois ultimately vacated Bobo's appeal.[197]

Death, when it came to Vissery the summer after his legal triumph, did not suffice to make him relinquish his philosophical mission, and he left instructions in his will that the future inhabitants of his house must preserve the lightning rod; furthermore, they must pay an annual rent of twelve livres for its maintenance. An alderman bought the house, cheaply in view of the unusual charge, and promptly called in experts to inspect the contraption. These experts concluded unanimously that Vissery's rod was "erected contrary to the rules of the art and could not remain in its present state." So the aldermen got their revenge: they tore down the lightning rod of M. de Vissery de Bois-Valé, in the name of public safety and modern science.[198]

194. Vissery to Buissart, 3 July 1783, AGPC, Coll. Bar., 4J/120/10.

195. Buissart to Cotte, 21 November 1783, ibid., 4J/118/75.

196. "Mr. Brunel told me that the affair of the lightning rod has been brought to Paris." Delas to Buissart, 15 October 1784, ibid., 4J/120/26; "The affair of M. Vissery will not go, I believe, to the parlement; people are certainly cooling on this item; my trip to paris is not a settled thing." Buissart to Cotte, 28 May 1784, ibid., 4J/118/77.

197. Des Essarts, *Causes célèbres* 117 (1784), 171–72, 188.

198. Vissery died 9 July 1784. De Pas, *A Travers* (1914), 167.

Nevertheless, the notion that Robespierre triumphantly defended in the summer of 1783, that political decisions must be grounded in a pure sensitivity to nature's ways, unclouded by any intervening theory or explanation— uncorrupted by expertise—remained powerful in France throughout the tumult of the following decade. Lightning would continue to illuminate, at least in political rhetoric, the foolishness of the people and the experts alike. *L'homme du peuple* and *l'homme du monde* were equally bamboozled by lightning, explained a pamphlet published in 1789, entitled *Thunder Considered in Its Moral Effects on Men.* While the people retreated into superstition, the experts adopted "ingenious systems, but mistaken." The author of the treatise revealed his secret for transcending systems and superstitions to arrive at the truth about thunder: being neither peasant nor *physicien*, he was a "good observer." [199]

In the summer of 1793, ten years after Robespierre's plea for Vissery, the Jacobin-led National Convention, with Robespierre at its helm, abolished experts from French officialdom by the following decree: "All academies and literary societies established or endowed by the nation are eliminated." [200] Some months later Robespierre told the Convention: "Eh! what do they matter to you, legislators, the diverse hypotheses by which certain philosophes explain the phenomena of nature?" [201] Ten years earlier a compelling story about the relations between physics and jurisprudence had won a small victory for a young provincial lawyer and an old provincial tinkerer. That story now attained its institutional culmination as the official policy of the French Republic. From the particular, the universal was born.

199. Lanteires, *Essai sur le Tonnerre* (1789), 5–9, 14.

200. *Comité d'Instruction Publique,* 2:240, cited in translation in Hahn, *Anatomy* (1971), 238. On the closing of the Academy of Sciences, see ibid., chap. 8.

201. Robespierre, "Sur les rapports des idées religieuses et morales avec les principes républicains, et sur les fêtes nationales," 18 floréal an II, in *Oeuvres,* 10:452.

Le Magnétisme dévoilé

Figure 6.1. *Le Magnétisme dévoilé*: Franklin, co-chair of one of the royal commissions, is on the left, wielding the commission report to cast away the mesmerists, who flee in confusion. Bibliothèque nationale de France.

Chapter Six

THE MESMERISM INVESTIGATION AND THE CRISIS OF SENSIBILIST SCIENCE

Happy the systematic philosopher to whom nature has
given ... a strong imagination.
—D. Diderot (sarcastically), *De l'interprétation
de la nature* (1753)

When imagination speaks to the multitude, the
multitude no longer knows either dangers
or obstacles. ... Nations follow sovereigns,
and armies their Generals.
—J. S. Bailly, *Exposé des expériences qui
ont été faites pour l'examen du magnétisme animal* (1784)

Men united are no longer subject to their senses.
—B. Franklin et al., *Rapport des commissaires chargés par le Roi de
l'examen du magnétisme animal* (1784)

Like the spring of 1783, the spring of 1784 saw an official investigation of great notoriety, also concerning a man's alleged ability to channel a weightless fluid using a pointed rod. The new affair was that of Franz Anton Mesmer, who for five years had entertained, titillated, and ostensibly cured a growing segment of Parisian society, channeling their animal magnetic fluids by means of his wand, conductive bathtubs, magnetic eyes, and healing touch. His inquisitors were the members of two royal commissions appointed to investigate his practice (fig. 6.1).

Like Robespierre's defense of Vissery's lightning rod, mesmerism was an application of the credo of sentimental empiricism outside the ranks of established natural science (though once again, academic philosophers acted as participants and advisers). Both events, the Vissery affair and the mesmerism investigation, tested the social authority and cultural prominence of sensibilist science. In many ways, they offer complementary pictures of the relations among natural science, popular culture, and institutional power. The lightning rod case applied the authority of natural science in a legal setting; the mesmerism case exercised the dominion of the state over a scientific question. The lightning rod case created a political reputation out of a philosophical argument; the mesmerism case deployed philosophical reputations toward a political end. The lightning rod case involved a popular protest against a philosophical application; the mesmerism case, a philosophical protest against a popular fad. The Janus-faced creed of sensibility acted at the heart of each case. Being a subjectivist creed, it was particularly protean. If it supported an argument, it generally supported its opposite as well; so sensibility could uphold, willy-nilly, the competing claims of academic science, amateur science, and popular science, the individual, the state, and the rioting horde. Carrying the burden of defining the relation between the true and the good, the claims of sensibility traveled promiscuously across the volatile cultures of science and politics during the last years of the Old Regime.

It was Mesmer who made dramatically apparent the problematic promiscuity of a subjectivist epistemology, by appropriating to his own ends the central tenets of sentimental empiricism. Because he used sentimental-empiricist arguments to defend his practice, the commissioners charged with investigating him were forced to refute these arguments, explaining mesmeric effects by means of a new theory of their own that undermined sentimental empiricism. The commissioners argued that in certain cases, including the case of mesmeric patients, the most vivid of sensations did not, in fact, reflect the action of any external, physical agent. Instead, these sensations were the result of a mysterious, internal power, the power of imagination.

This chapter is centered upon the royal investigation of mesmerism in 1784. Here was a landmark event in several respects. It involved the first formal, psychological tests using what would come to be called a placebo sham. The investigators devised a method for, in their terms, isolating the action of Mesmer's hypothetical animal magnetic fluid from the action of the patient's imagination. In addition to being the first recorded instance of the use of a placebo and of, in modern terms, a method of blind assessment, the mesmerism investigation was the first known formal investigation of scientific fraud.[1] It was therefore a crucial episode in the history of psychology, medical testing, experimental practice, and state authority to police scientific conduct. In each domain the mesmerism investigation represented an institutional recognition of the deeply problematic nature of sensory evidence: sensations would

---

1. On the importance of the mesmerism investigation in the history of blind assessment and placebo controls, see Kaptchuk, "Intentional Ignorance" (1998), 393–99. On the emergence in early modern medicine of a professional and state role in policing the boundaries of legitimate science, see Bynum and Porter, eds., *Medical Fringe* (1987); Porter, *Health for Sale* (1989); and Lingo, "Empirics and Charlatans" (1986).

no longer be considered direct inscriptions of an outside world upon the mind of the observer.

My primary interest here, however, is not in the importance of the mesmerism investigation to subsequent developments in psychology, medicine, experimental practice, or the relations between science and the state, but instead in its relations to what came before. I will be locating it at the height of the Age of Sensibility. Mesmerism itself, its popularity, the appointment of the royal investigating commissions, and the content of their radical conclusions—all reflected the prior elevation of feeling as the basis of both natural knowledge and social union.

The mesmerism investigation has received surprisingly little attention from historians not because it has seemed unimportant, but because it has seemed unproblematic. The tendency has been to assume that the commissioners simply recognized a charlatan for what he was.[2] But here I am interested precisely in why the commissioners decided that Mesmer's application of sentimental empiricism was illegitimate. I understand the mesmerism investigation as a crisis, a seismic event along the fault lines in sensibilist natural science, triggered by its friction against popular culture on one side and political authority on the other.[3]

The argument I develop here has two parts. First, mesmerism posed a problem for established natural science not by departing from it, nor by violating its rules, but, on the contrary, by too literally applying its central credo, the credo of sensibility. Mesmerism was a kind of caricature of natural science in the sentimental-empiricist idiom. Like any caricature, it worked by exposing and magnifying the vulnerabilities of its subject—in this case, the sentimental-empiricist elevation of feeling as the ultimate arbiter of truth. An argument from feeling cannot be refuted except by undermining the principle on which it rests: if people felt a thing, either it existed or feeling was not the measure of truth. Writhing and groaning, Mesmer's patients dramatized the process of feeling. Moreover, what they felt, according to Mesmer, was

2. Robert Darnton, in his classic *Mesmerism and the End of the Enlightenment in France* (1968), devotes only two pages (62–64) to the official investigation of mesmerism; Charles Gillispie similarly passes over the event in *Science and Polity in France at the End of the Old Regime* (1980), as having successfully delineated pseudoscience from science (281).

3. In her study of mesmerism in Victorian Britain, Alison Winter characterizes mesmerism as a central instance of Victorians' preoccupation with "the influence they felt from each other" and the "sympathies that bound them." This preoccupation lay at the heart of Victorian worries about the relations between men and women, between doctors and patients, and among members of different social classes. Thus Winter uses mesmerism as a lens to study these areas of Victorian social life. Winter, *Mesmerized* (1998), 12. I propose that mesmerism was similarly situated in late Enlightenment France at a most fraught intersection: that of sentimental empiricism with popular and academic science and state control.

the etherial medium of sensation itself, which permeated the cosmos. They thereby demonstrated—in sentimental-empiricist terms—the real existence of this imponderable fluid of sensibility. To deny it was to undermine the authority of sensation.

Second, I suggest that the commissioners charged with investigating mesmerism, confronted by its wild popularity, composed an explanation of mesmeric effects that challenged the central axiom of sentimental empiricism, the axiom that feelings were responses to a world outside the mind and were therefore the bedrock of natural knowledge. Mesmerism drove the commissioners to develop a theory of how one could have feelings that were not responsive to the outside world. Philosophical consensus recognized a mental faculty responsible for *in*sensibility, detachment from the outside world and its proper action upon the five senses. This was the faculty of imagination.[4]

Rejecting mesmerism, the commissioners turned toward this faculty of the imagination and conjured it into a formidable power. In the mesmerism commission reports, imagination became sensibility's nemesis. Imagination could overwhelm the sensible body so literally as to throw it into fits of convulsions. And while sensibility was the basis not only of knowledge but of moral sentiment and sociability—as previous chapters have shown—imagination now became the root of social pathology. Imagination, the commissioners would ultimately warn, could release the audiences of popular science displays from the tenuous grip of their senses and turn them into a revolutionary mob.[5]

4. On "distrust of the imagination in science" see Daston, "Fear and Loathing" (1998) (quote from 87). On distrust of the imagination in the French Enlightenment, see Jan Goldstein, "Enthusiasm or Imagination?" (1998). Goldstein argues that the imagination was uniquely suspect, "believed to be the principal entryway for error and disorder, and potential site for the capture of the will and loss of self-control." Fear of the imagination did not originate in the eighteenth century, Goldstein writes, but it did importantly transform at the hands of sensationist epistemology: "The dangers of the imagination had already been underscored by the Cartesian psychology of the seventeenth century. But the sensationalist psychology of the eighteenth century reiterated those dangers and, more important, gave them a new inflection," by making environment crucial to the control, or loss of control, of the imagination (30). Goldstein develops this argument more fully in "Imagination and the Problematization of the Self at the End of the Old Regime," the first chapter of her forthcoming book from Harvard University Press, entitled *The Post-Revolutionary Self: Competing Psychologies in Nineteenth-Century France.* In what follows, I suggest that one way in which sensationism inflected French Enlightenment fear of the imagination was by setting the imagination up as the antithesis of sensibility. While sensibility focused the mind outward through the senses, imagination drew the mind inward, away from its sensory interface with the outside world.

5. Goldstein describes the way French Enlightenment writers viewed the faculty of imagination as socially dangerous unless held in check by "the consensual force of the community." Therefore solitary activities such as "reading novels, masturbating, pursuing one's trade outside the supervision of a guild, and speculating on the stock market" were particularly hazardous. At the same time, crowds also "harbored danger because they encouraged imagination's potential for contagion." In both cases—in solitude and in an unruly crowd—the individual imagination was loosed from communal control. Goldstein, "Enthusiasm or Imagination?" (1998), 30. Here,

I. MESMERISM AS A CARICATURE OF SENTIMENTAL EMPIRICISM

Imagine for a moment, wrote A. J. M. Servan, legal philosopher and mesmerist member of the Parlement of Bordeaux, that

> from the depths of America, an almost unknown land, a man even more unknown than his country, stood up to cry: "men listen to me! I have the power to draw thunder from the sky, and I can often force it to fall upon any point on earth it pleases me to choose": what mockery from one pole to the other!... Franklin... you would have been condemned to eat crow, and to abandon right then and there your physics and your genius.[6]

Franklin had been spared this fate, Servan suggested, despite the obvious implausibility of his claim, because its truth had been empirically confirmed. Consider, then, he urged, the no less plausible, and equally empirically supported, claims of Mesmer.

Historians debating whether Mesmer was a charlatan have sometimes defended him by arguing that he sincerely believed in his doctrine.[7] But it would be difficult to deny that Mesmer, whether or not deliberately deceitful, had certain characteristics of the charlatan. A charlatan's manipulation lies not merely in presenting a false theory as true, but in making the false theory plausible. A successful quack exploits the preoccupations and uncertainties of established science, turning its own foibles against it. Mesmer's theory is not so much a departure from credible philosophy as an exaggeration of it. Even Lavoisier, Mesmer's chief inquisitor, remarked upon the "skill and confidence with which animal magnetism is presented." Mesmer's admixture of "truths of fact and observation" with "pretended results of a purely hypothetical principle" comprised a body of doctrine that, Lavoisier conceded, "inspires awe, even among enlightened doctors."[8] In other words, Mesmer was a master, perhaps unconsciously, in the art of quackish parody.

To see mesmerism as a parody of contemporary empiricist science can explain an apparent contradiction revealed in Robert Darnton's 1968 study of the meaning of mesmerism for the political culture of the late Enlightenment.

---

drawing on Goldstein's observation, I take up the same worry from the opposite direction. While French sensationists warned that the absence of community could turn the imagination into a source of instability, as Goldstein points out, they simultaneously cautioned that the imagination undermined social cohesion. It did so by acting counter to that attribute that opened and connected the individual to the world around him or her and to its other inhabitants, sensibility.

6. Servan, *Doutes* (1784), 88.

7. See, for example, Zweig, *Mental Healers* (1932), 12–14; Saussure, "Le Caractère de Mesmer," in Vinchon, *Mesmer et son secret* (1971), 11; and Lopez, *Mon Cher Papa* (1966), 169.

8. Lavoisier, "Remarques" (1784), in *Oeuvres*, 3:508.

Darnton argues that mesmerism marked no clear departure from established science in the climate of the 1780s, a decade that he says had almost succeeded, with its succession of philosophical crazes, in erasing the line "dividing science from pseudoscience." At the same time, however, he presents mesmerism as a subversive affront to academic science.

Darnton attributes the "anti-establishment" cast of mesmerism, despite its continuities with established science, to two related factors. The first was its popularity: mesmerism was rejected by the academies but raised enormous popular interest. However, as his account reveals, the academies' rejections of mesmerism were ambivalent and internally disputed. The second subversive quality Darnton identifies in mesmerism was its style, which was "mystical" and "romantic," a reaction against the "cold rationalism of midcentury." Yet previous chapters have suggested—and Darnton's own discussion affirms— that established French natural science during the second half of the eighteenth century was itself shaped by a general reaction against what its practitioners considered the cold rationalism of an earlier generation.

In other words, mesmerism was both in keeping with *and* a threat to established natural science.[9] Ultimately, Darnton suggests, the alliance between mesmerists and political radicals was one of expedience. Chronicling the emergence of a partnership between academic science and government in late-eighteenth-century France, Charles Gillispie has characterized their relations as purely instrumental, an opportunistic exchange of "weapons, techniques, information" for funding, institutions, and authority, with no involvement of philosophy or principle.[10] Darnton tells a complementary story, describing a marriage of convenience between "radical" natural philosophy, in the form of mesmerism, and radical politics in the form of Revolutionary Rousseauism. Prominent Revolutionaries, including Brissot de Warville, Jean-Paul Marat, and Jean-Louis Carra, had been denied official sanction for their endeavors in natural science, leaving them bitter toward the academies, and embraced mesmerism as a "vehicle" for their political programs. Marat's "desire to avenge himself against the Academy of Sciences," Darnton claims, "provided the main thrust behind his . . . revolutionary career," as well as the occasion for his interest in mesmerism.[11]

While academic science and government bureaucrats established the pragmatic alliance that provided the moral to Gillispie's story, a "curious" allegiance between "scientific and political extremism" animates Darnton's.

9. Darnton, *Mesmerism* (1968), 11–16, 29, 37–38, 42–45,165, 60n. On proto-romanticism in midcentury natural philosophy and its appeal to those excluded from academic science, see also Hahn, *Anatomy* (1971), 139–40.

10. Gillispie, *Science and Polity* (1980), 549.

11. Darnton, *Mesmerism* (1968), vii, 3–5, 90–100, 161, 163–64, 110–11.

Both engagements are essentially institutional rather than philosophical, and perhaps this explains why neither Gillispie in his study of the scientific and political establishments, nor Darnton in his analysis of the scientific and political anti-establishments, devoted much attention to the royal investigation of mesmerism in the summer of 1784. The important thing in both discussions is the simple fact that the investigation took place, drawing a distinct line between science and pseudoscience. For Gillispie, there was "no point in recounting" the commissions' procedures, for they merely demonstrated the obvious truth that "susceptibility to magnetism was a function of suggestibility, poverty and ignorance." [12]

But how was the line between science and pseudoscience drawn? If Mesmer's theory seems largely continuous with those of the most established natural philosophers, what ultimately made it quackery to most (though not all) of the commissioners? And if Marat, Carra, and Brissot were attracted to mesmerism because of their bitterness toward the scientific establishment, why were half the members of the Parlement of Paris and a defecting contingent from the Faculty of Medicine similarly seduced? [13] What made mesmerism both absurd and plausible? How did it appeal to subversives and establishmentarians alike? From a contemporary rather than historical perspective, these were the very questions that motivated the royal commissions' 1784 investigation of mesmerism, and their proceedings contain an implicit answer: mesmerism was related to established natural science neither by being antithetical to it, nor quite by being continuous with it, but by taking its central tenets, the tenets of sentimental empiricism, so very literally as to produce what amounted to a caricature of them. [14]

Sensationists said that the natural world physically inscribed knowledge and sentiment upon the soul through the five senses. Mesmer offered them the medium by which this inscription took place: a universal, imponderable fluid of sensibility. As a leading mesmerist pointed out, invoking the chemists' phlogiston, there was nothing new about hypothesizing a universal fluid. Another remarked that people "incessantly say that all is linked in nature; . . . they never stop talking of the great chain of being." Here was a physical basis for the great chain, a single fluid uniting all the parts of nature. After all, what

12. Gillispie, *Science and Polity* (1980), 281.

13. La Harpe, *Correspondance littéraire*, 4:272–75; cited in Darnton, *Mesmerism* (1968), 87.

14. In her study of sensibility in literature and medicine, Anne Vila characterizes mesmerism as "more than just pseudoscience or charlatanism; rather, it was a cunning recasting of mainstream Enlightenment medical ideas." Vila, *Enlightenment and Pathology* (1998), 297. I suggest that mesmerism appropriated not only mainstream medical ideas, but, more generally, the central preoccupation of contemporary natural and moral science—the primacy of sensibility as the source of natural knowledge and social union—and that Mesmer's recasting of this principle took the form of carrying it to its logical extreme.

could this general unity mean without a material foundation? The conformity of Mesmer's theory with this idea "adopted by all the centuries and all the enlightened men" made it, his supporters argued, "extremely plausible." [15]

Franklin himself might easily have allowed the plausibility of Mesmer's claim that a "universal fluid exists in nature, a fluid which penetrates all animate or inanimate bodies." [16] In the midst of his summer spent investigating mesmerism, Franklin wrote out some "loose thoughts" on his own "universal fluid," a proper quantity of which, he said, constituted health in human bodies. Franklin credited his universal fluid with comprising light, heat, and the "greatest Part" of combustible bodies; with causing growth in animals and vegetables; with separating the particles of fluids and airs; and with maintaining smokes and vapors in their ethereal state.[17] Though he left unspecified the relation of his universal fluid to electrical and magnetic phenomena, he shared Mesmer's belief in the celestial origins of magnetism. Franklin speculated that the magnetic fluid existed "in all space; so that there is a magnetical North and South of the universe." He concluded fancifully that were it "possible for a man to fly from star to star, he might govern his course by the compass." [18]

Many academic philosophers, including Franklin, also found credible the clinical exploitation of imponderable fluids. From the earliest days of the Leyden jar, electricians, among them Nollet, had tested its efficacy in treating paralysis and reported some success (fig. 6.2). Nollet had ultimately judged dubious the evidence of good effects from electrical shocks but had suggested another medicinal application of electricity: using electrification to quicken circulation and to hasten the "evaporation" of disease from the body.[19] In the early 1750s, in response to European trials of the medicinal effects of electricity, Franklin employed two very large Leyden jars to dispense powerful shocks, three times daily, to the paralyzed limbs of several patients. He reported having observed beneficial effects, but only limited and never permanent. Nevertheless, he remained modestly optimistic that, "under the direction of a skilled physician," electrical treatments could effect full and permanent cures. Franklin's faith in the medicinal potential of electricity was not shaken by his findings against mesmerism. The year following the investigation, he recommended electrical shocks as a possible treatment for insanity.[20]

15. Deslon, *Observations* (1784), 2–3; Servan, *Doutes* (1784), 74–75, 81.

16. Mesmer, "Discourse" (1784), 33.

17. Franklin, "Loose Thoughts" (25 June 1784), in *Writings*, ed. Smyth, 9:227–30.

18. Franklin to Soulavie, 22 September 1782, in *Writings*, ed. Smyth, 8:599–600.

19. Torlais, *Physicien* (1954), 68; Heilbron, *History of Electricity* (1979), 353–54. On contemporary medical uses of electricity and magnetism, see Gillispie, *Science and Polity* (1980), 270–72.

20. Franklin to Pringle, 21 December 1757, in *Writings*, 3:425–27; Franklin to Ingenhousz, 29 April 1785, in ibid., 9:308–9; Franklin to Stiles, 9 March 1790, in ibid., 10:85; Franklin to Ingenhousz, 29 April 1785, in ibid., 9:309.

Figure 6.2. A paralytic being cured by an electrical generator, from the abbé Sans, *Guérison de la paralysie, par l'électricité, ou cette expérience physique employée avec succès dans le traitement de cette maladie regardée jusques à présent comme incurable ... Avec figures ...* (Paris, 1772).

Thus Mesmer did not depart from the standard wisdom by relating health to the regulation of imponderable fluids in the body; he merely stated a commonly held belief among natural philosophers. He was acquainted with such beliefs as a result of his training, having received his medical degree in 1766 from the University of Vienna for a (possibly plagiarized) thesis on the influence of the motions of the planets upon the human body. Mesmer then hesitated to take up a medical career.[21] For six years he depleted his fortune (or, rather, his wife's) by hosting sumptuous musical soirées at their luxurious estate and by passing his time restlessly reading in and among the several branches of natural science. From these explorations, Mesmer must have derived his intuitive grasp of contemporary natural philosophy's overarching preoccupation: the origin of knowledge in sensibility and of error in system-building. He adopted, for example, the fashionable position regarding the misleading artificiality of words.[22] In his autobiographical sketch, he reported having once, in order to eradicate linguistic contamination from his ideas, formed a "bizarre plan" that, thanks to a strenuous "effort" had proved successful: "I thought for three months without language."[23]

Mesmer began his memoir on animal magnetism with a series of avowals of sensationist orthodoxy: "Man is by nature an Observer"; the child's first task is to learn to employ his sensory organs; the "primary source of all human knowledge" is "experience." He also drew the commonplace connection between rationalism and arrogance, condemning the "ambition for knowledge" that led philosophers to replace observations with abstract systems.[24] Mesmer and his followers made much of his avoidance of medical systems. Doctors, one wrote, "believe more what they imagine than what they see: systems are always infinitely dearer to them than experiments."[25] Claiming to have discovered the power by which nature herself effected cures, Mesmer said his function was merely to help and hasten the natural reestablishment of harmony. His fondest wish was to "preserve my fellow-man ... [from] the incalculable hazards of drugs and their application."[26] This was a selling point even with Mesmer's critics, including Franklin, who perceived a hidden advantage in the mesmeric craze if those who were given to fancying

21. The following sketch of Mesmer's career draws upon accounts in Darnton, *Mesmerism* (1968), 47–81; Vinchon, *Mesmer* (1971), 21–63; Gillispie, *Science and Polity* (1980), 261–89; and Mesmer's own telling of the story in Mesmer, "Précis historique" (1781).

22. The sentimental-empiricist opposition between language and sensibility is the focus of chapter 7.

23. Mesmer, "Précis historique" (1781), 102.

24. Mesmer, "Dissertation" (1779), 44–45; "Dissertation" (1799), 96.

25. Servan, *Doutes* (1784), 89.

26. Mesmer, "Dissertation" (1799), 130; "Dissertation" (1779), 46–48. See also Servan, *Doutes* (1784), 96, 116; Bonnefoy, *Analyse* (1784), 39, 72, 86–87.

themselves ill might be persuaded "to forbear their drugs in expectation of being cured by only the physician's finger or an iron rod pointing at them."[27]

When Mesmer first turned to the practice of medicine, he adopted a technique of the ex-Jesuit astronomer Pater Hell (who moonlighted in medicine), applying magnets to the ailing parts of his patients. Mesmer soon elaborated this practice, adding a theory from his doctoral thesis, which hypothesized a fluid from the stars that flowed into a northern pole in the human head and out of a southern one at the feet, a "tide [that] takes place in the human body."[28] He also added more magnets, to channel the ebb and flow of the astral current, before dispensing with magnets altogether, leaving the doctor's bare hands and magnetic personality as the principle therapeutic instruments. Illness, Mesmer surmised, resulted from bodily obstructions of the universal fluid, which he claimed to remedy by touching his patients' bodies at their magnetic poles. The cures, paralleling the Leyden commotion during the restoration of equilibrium, involved violent "crises," or fits of writhing and fainting. These contributed to the notoriety of Mesmer's methods. In 1778, following a succession of public and bitter disputes over the authorship, the efficacy, and the moral rectitude of his science, Mesmer fled his native city. He left his last two patients in the care of his estranged wife. Having exhausted her tolerance—and Vienna's credulity—he headed for Paris.[29]

There he encountered a public already primed for his performances. Parisians, like Londoners and other cosmopolitan Europeans, had been attending popular science courses since the turn of the century, and in rapidly increasing numbers since the 1740s[30] (figs. 6.3 and 6.4). In these courses, they had witnessed the extraordinary powers of the physicists' imponderable fluids, particularly electricity. Nollet taught one of the most fashionable popular science courses in the capital and filled it with electrical displays (figs. 6.5 and 6.6). Audiences for philosophical amusements had also learned the dogma of sentimental empiricism. They had been taught that knowledge resided in sensory experience and responsive feeling—in sensibility. Public lecturers continually announced their purpose to educate their pupils by striking their senses. Nollet promised in his course never to "pass beyond a sensible physics"; and Priestley, who gave popular lectures in London, wrote that "the curiosity and surprize of young persons should be excited as soon as possible; nor should it

27. Franklin to La Condamine, 19 March 1784, in *Writings*, 9:182–83.

28. Mesmer, "Physical-Medical Treatise" (1766), 15.

29. Vinchon, *Mesmer* (1971), 26–27, 24, 46–47.

30. On the emergence of popular science courses in the late seventeenth and eighteenth centuries, see Sutton, *Science for a Polite Society* (1995); Stewart, *The Rise of Public Science* (1992); and Schaffer, "Natural Philosophy and Public Spectacle" (1983). On the relations between mesmerism and popular science in Paris in the 1770s and '80s, see Darnton, *Mesmerism* (1968), chap. 1.

Figures 6.3 and 6.4. Electrical amusements from the repertoire of William Watson. In the first plate, a lady generates electricity at the right, which passes through a boy suspended by silk cords and into the left hand of an insulated girl standing on a tub of dried pitch. She uses her other hand to attract and repel chaff in a dish. Above, electricity from an electrified gun barrel causes alternate attraction and repulsion of a clapper between two bells, and thereby rings the bells. In the second plate, an insulated gentleman ignites spirits of wine in a spoon held by a grounded lady, while above, chaff floats between two plates, one held by an insulated boy and the other by a grounded girl. From Watson, *Expériences et observations, pour servir à l'explication de la nature et des propriétés de l'électricité* (Paris, 1748). Courtesy of the Bancroft Library, University of California, Berkeley.

be much regarded whether they properly understand what they see, or not.... We are, at all ages, but too much in haste to *understand*... the appearances that present themselves to us." [31] By the 1770s the audience for popular science had learned to cultivate not so much a rational understanding of natural phenomena, as a sensitivity to them. They were ready for Mesmer.

Arriving in Paris in February 1778, Mesmer established a clinic that became an overnight success. Soon mesmeric salons had sprung up throughout the city. Inside, their atmosphere was murky and suggestive, with drawn curtains, thick carpets, and astrological wall decorations. Mesmer himself dressed impressively in a lilac taffeta gown. Patients gathered, joined by ropes, around *baquets*, tubs filled with miscellaneous bits of glass, metal, and water, from which flexible iron rods protruded (figs. 6.7 and 6.8). They pressed these rods to their left hypochondria (upper abdomens) and joined their thumbs to increase the communication of the magnetic fluid. Alternatively, they opposed their own magnetic poles to those of the magnetizer by placing their knees between his. He then pressed and prodded their bodies with a mesmeric wand or, more often, his fingers. By means of these titillating practices, he provoked the notorious mesmeric crises. For especially violent crises, mesmeric salons included separate rooms lined with mattresses.[32] Unable to attend to all the ailing Parisians who arrived in droves on his doorstep, Mesmer was forced to designate a surrogate: he "magnetized" a tree near the porte Saint-Martin to accommodate the overflow.

His quest for official sponsorship met with more mixed results. Mesmer applied in succession to all of the relevant learned bodies—first the Academy of Sciences, next the Society of Medicine, and, finally, the Faculty of Medicine. Like the ebb and flow of the astral tide, the philosophes were attracted and repelled by Mesmer's doctrine. Le Roy, the Franklinist electrician, then director of the Academy of Sciences, invited Mesmer to present his theory at a meeting of the Academy and hosted a demonstration of it in his own laboratory. This first display of Mesmer's science in Paris was met not just with skepticism but with outright laughter. Afterward, Le Roy would have nothing to do with Mesmer, despite Mesmer's repeated applications for his attention.[33] Félix Vicq-d'Azyr, then perpetual secretary of the Society of Medicine, rapidly developed the same attitude, as did the delegation of twelve members of the Faculty of Medicine who agreed to witness a series of Mesmer's treatments.[34]

31. Nollet, *Leçons* (1754–65), 1:237; Priestley, *Experiments and Observations* (1779–86), 1:x.
32. Darnton, *Mesmerism* (1968), 6–10.
33. Mesmer, "Précis historique" (1781), 106–7; Lopez, *Mon Cher Papa* (1966), 170; Vinchon, *Mesmer* (1971), 54–55.
34. Vinchon, *Mesmer* (1971), 49–60, 69–72, 80; Gillispie, *Science and Polity* (1980), 266–70.

Figures 6.5 and 6.6. Scenes from Nollet's public lectures at his own *cabinet* (the first plate), and at Versailles (the second), where an electrified suspended boy attracts and repels chaff with his hands, offers shocks with his nose, and causes a glow at the tip of a conductor held over his head. From Nollet, *Leçons de physique expérimentale*, 5th ed., vol. 1 (1759), and *Essai sur l'électricité des corps*, 2nd ed. (1750).

But the chemist Claude-Louis Berthollet joined the mesmeric Society of Harmony and persevered for a fortnight before storming out in midsession, proclaiming that he had been duped.[35] The mathematician, naturalist, and explorer Charles-Marie de La Condamine was intrigued by mesmerism and wrote to Franklin on the eve of the investigation for information, hoping to discover a new means to "comfort the poor inhabitants of the countryside."[36] And the botanist and doctor Antoine-Laurent de Jussieu, having served on the Royal Society of Medicine commission to investigate Mesmer, dissented from its negative final report. He judged the commissions' own explanation—attributing mesmeric crises to the power of imagination—insufficient to explain the dramatic effects. Jussieu sought a more material cause in the "principle of heat," permeating the air, constantly active, and "insinuating itself into bodies." He proposed the Franklinist-sounding hypothesis that this principle, inhabiting all bodies in their normal state, was induced by rubbing to form "atmospheres" around them. These, he said, were the "physical influence of man upon man."[37] Meanwhile, on the Faculty of Medicine, as men-

35. Berthollet, "Déclaration," 20 May 1784, in Lavoisier, *Oeuvres*, 3:505–6; Vinchon, *Mesmer* (1971), 125–26; Darnton, *Mesmerism* (1968), 52.

36. La Condamine to Franklin, 8 March 1784, FPA.

37. Jussieu, *Rapport* (1784), 188–89. See Vinchon, *Mesmer* (1971), 135; Darnton, *Mesmerism* (1968), 107; Zweig, *Mental Healers* (1932), 70–71; Duveen and Klickstein, "Joint Investigations" (1955), 297.

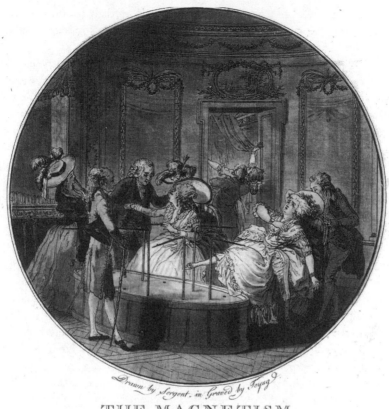

THE MAGNETISM.

Figures 6.7 and 6.8. Two scenes of mesmeric séances, in which one sees the "conductive" rods and ropes and, in each, a lady in the throes of a "crisis." Bibliothèque nationale de France.

tioned above, a significant contingent was converted to mesmerism, including Charles Deslon, physician to the comte d'Artois. "Never," the Academy mesmerism commission would ultimately tell their colleagues, "has a more extraordinary question divided the minds of an enlightened Nation." [38]

Deslon became for a time Mesmer's leading disciple. When he branched off to form his own mesmeric practice, two other patients/students took over the role. They were Nicolas Bergasse, a lawyer from Lyon, and Guillaume Kornmann, a banker from Strasbourg. Bergasse and Kornmann, with Mesmer, started the Society of Harmony, and within two years they had earned almost 350,000 livres and spawned three provincial societies. Since only those who

38. Bailly et al., *Exposé* (1784), 4.

could afford the fee of a hundred louis appeared on their roster, they were mostly from the ranks of the nobility or wealthy entrepreneurs, bankers, lawyers, and doctors.[39]

Mesmerists claimed that their doctrine was empirical, despite the fact that the animal magnetic fluid could not itself be directly sensed by the regular five senses. Some attempted unsuccessfully to render the mesmeric fluid perceptible, but for the most part, with Deslon, they allowed that it was "neither visible nor palpable" and could only be known by its effects.[40] They argued, not unreasonably, that other imponderables were similarly known only through their effects. One could see neither mineral magnetism, nor the imponderable cause of heat, nor the force of gravity.[41] The same was true for the action of the will upon the body. One mesmerist argued that since the commissioners admitted this influence without being able to see it other than by its effects, why should they deny the same courtesy to animal magnetism?[42]

Also, Mesmer defined animal magnetism as a "sixth sense" and, invoking a standard sentimental-empiricist axiom, cited its sensory nature to explain why he could neither describe nor define it. Senses were prior to ideas and could only be "experienced."[43] The marquis de Puységur, a loyal disciple who invented the technique of hypnotism and was therefore probably the most influential mesmerist with regard to the later history of psychology,[44] adopted this line of argument. As a sixth sense, animal magnetism had to be felt to be understood. "To *feel*," he wrote, "one needs neither *intellect* nor *science*, and [the science] of M. Mesmer is *felt* better than it can be expressed."[45] Both Mesmer and Puységur invoked the Enlightenment's favorite epistemological metaphor, suggesting that it was as useless to try to explain the magnetic "sense" as it would be "the theory of colors to someone blind from birth."[46] They thus effectively exploited a vulnerability of sentimental empiricism: to make feeling the ultimate test of truth was to render unanswerable all claims founded in feeling. If people felt animal magnetism, then according to sensibilist logic, it must be real.

Not only was animal magnetism a sixth sense, but it was the basis of the

39. Darnton, *Mesmerism* (1968), 51–52; Vinchon, *Mesmer* (1971), 108–20, 139–40.

40. Deslon, *Observations* (1784), 4.

41. Bonnefoy, *Analyse* (1784), 34–35; Deslon, *Observations* (1784), 3–5; Bergasse, *Observations* (1785), 22.

42. Anon., *Réflexions* (1784), 24.

43. Mesmer, "Précis historique" (1781), 103.

44. On the origins of modern psychology in mesmerism, see Crabtree, *From Mesmer to Freud* (1993), especially chaps. 5 and 17.

45. Puységur, *Mémoires* (1786), 73–74, 147.

46. Mesmer, "Précis historique" (1781), 135; Puységur, *Mémoires* (1786), 74–75.

other five and of sensibility itself. Puységur announced that it was "concerning our *sensations* that [Mesmer] has come to enlighten us," and, therefore, "his doctrine tends to lend support to all the truths that, until now, spoke only to the *mind*." Bergasse concurred; until Mesmer, the "theory of our sensations" had been "still unknown, despite the efforts of the Leibnitz's, the Lockes, and the Condillacs." Mesmer claimed that the animal magnetic fluid provided a material foundation for sensation, as "an agent acting on the inmost substances of the nerves of the animal body."[47]

Here was a direct extrapolation from contemporary sensory physiology, from the nervous ether common to post-Newtonian theories of sensation. As we saw in chapter 2, physiologists widely subscribed to an impression theory of sensation, according to which sensations were the motions of a nervous fluid propagated inward from the senses through the nerves. They arrived ultimately at the *sensorium commune*, a junction of nerves in the brain, where they met and combined, impressing themselves upon the brain's matter.[48] In keeping with these principles, Buffon called the brain itself an "internal, general and common sense,"[49] and Mesmer similarly hypothesized that his sixth sense was an "internal sense" located at the nexus where sensory impressions met after traveling inward along the nerves, a "union and interlacing" of the external senses.

Mesmer's inner sense, the basis of all sensation, revealed things that common sensory experience withheld. The internal sense was "related to the whole universe" and could perceive distant past and future events.[50] This idea, too, was arguably in keeping with contemporary physiological theory, which made sensation the interface between the motions of etherial fluids inside and outside of the brain. While any one sense conveyed only a specific sort of impression, the *sensorium commune* received whole impressions of the world outside. Its role in sensationist physiology and psychology was to embody the mind's openness to the world.[51] Accordingly, Mesmer's sixth sense was an organ that registered the movements of the universal fluid through which all events reverberated. Why should these reverberations not reflect the past, foretell the future, and even receive the imprint of human thoughts or "wills"?[52]

47. Puységur, *Mémoires* (1786), 147–48; Bergasse, *Observations* (1785), 50; Mesmer, "Dissertation" (1799), 89, 93, 127.

48. On the notions of a nervous ether and a *sensorium commune* in eighteenth-century sensory physiology, see chap. 2, pp. 25–27.

49. Buffon, "Discours" (1753), in *Oeuvres philosophiques*, 323.

50. Mesmer, "Dissertation" (1799), 111.

51. See chapter 2.

52. Ibid., 121.

Puységur claimed his hypnotized subjects, or "somnambulists," perceived hidden facts about their own and others' states of health by means of a "true *sensation*." And they were able furthermore to "*pre-sense*" their future sufferings and the dates of their cures. Puységur quoted one of his somnambulists as having said, "It is that I *feel* in advance . . . the ill that will befall me."[53] By means of the internal sense, Mesmer reasoned, people were "in *rapport*" with all of history and with the wills of others. The internal sense was the instinct by means of which people were "able to understand either the 'harmony' or the 'dissonance' which substances exert upon our structure." Like Diderot, Mesmer judged sensitive instinct to be truer than reason.[54]

The sixth sense, the basis of the other five—sensation itself in its purest form—was also capable of functioning independently. Mesmer proposed that for the internal sense to function at its peak, the other senses must in fact be silent. According to his theory of somnambulism, sleep was the interruption of the link between the *sensorium commune* and the external senses. During sleep the internal sense became the "sole organ of sensation; its impressions turn out to be independent of the external senses." The impressions of the internal sense were revealed to the sleeper, at last left undistracted by common sensation. Somnambulism resulted when sleepers attempted to express or act upon their internal sensations as though they were external. Mesmer also understood madness in terms of the internal sense as "gradations of imperfect sleep." Madness was a sensory disorder whose cure took the form of "perfecting the sensations."[55]

But madness and dreaming were both also forms of truth telling, since the internal sense reflected the most basic and universal truths. Mesmer and most of his disciples claimed that the magnetic fluid could only be sensed by a conscious subject in a state of disharmony or ill health. "The nature of our sensations," Mesmer explained, "is that they are nothing else but the perception of differences in proportions," thus the magnetic fluid—like Franklin's electrical fluid—must be in a state of imbalance to be felt. Bergasse thought that "sick beings are the only subjects suited" to reveal the reality of universal magnetism. And Puységur confirmed that once his somnambulists were cured, they "admit that they no longer *feel* anything." Illness and madness, weakening the grip of the external five sensations, magnified the internal sixth. Mesmer compared animal magnetism to the microscope, lending philosophers a view of the invisible, and madmen and sleepwalkers were also like scientific instru-

---

53. Mesmer, "Précis historique" (1781), 135; Puységur, *Mémoires* (1786), 74–75.
54. Mesmer, "Dissertation" (1799), 120–21. On Diderot's preference of sensitive instinct to reason, see chap. 3, p. 98.
55. Mesmer, "Dissertation" (1799), 122–25, 126, 127.

ments: "These subjects can, in a sense, be compared to a telescope," for they magnified the effects of the stars through their fluid medium.[56]

Unconsciousness was sensibility, and madness lucidity. These apparent contradictions were familiar from the writings of sentimental empiricists, as in *D'Alembert's Dream*, in which Diderot's fictional d'Alembert, talking in a feverish sleep, realizes a truth to which his waking self is blind, that sensibility permeates the material world.[57] His own sensibility, connecting him with the feeling world around him, speaks clearly only when his rational mind is silenced. Mesmer merely carried the logic of sentimental empiricism to its extreme with his animal magnetic fluid and the sixth sense upon which it acted: bypassing the familiar, weak, and imperfect five senses, they connected the individual mind directly to the universe as a whole. Through the reverberations of the animal magnetic fluid—the medium of sensibility—one could sense the past and future, the motions of the cosmos, the action of the wills of others. Mesmer also took the sentimental empiricists at their word by giving them a cosmic ether of sensibility that could be known only and utterly by feeling its effects. To refute mesmerism, established natural philosophy had to undermine two axioms of its own most cherished doctrine: the immediate and absolute sensory connection between mind and world; and the ultimate authority of feeling.

II. THE INVESTIGATION AND ITS RESULT: THE IMAGINATION

In March and April 1784, the baron de Breteuil, minister of the Department of Paris, appointed two commissions to investigate mesmerism. The first, from the Academy of Sciences and the Faculty of Medicine, included Le Roy, the astronomer Jean-Sylvain Bailly, the chemist Jean Darcet, and the doctor Joseph-Ignace Guillotin, and was chaired by Antoine Lavoisier and Benjamin Franklin; the second, from the Royal Society of Medicine, included Jussieu and was chaired by Pierre-Isaac Poissonier. Deslon was better placed than Mesmer to negotiate with government ministers and maneuvered himself into the position of primary subject of the examination. It was he who collaborated with the two commissions, despite Mesmer's and Bergasse's angry protests to Franklin.[58]

56. Mesmer, "Précis historique" (1781), 135–36; "Letter from M. Mesmer, Doctor of Medicine at Vienna, to A. M. Unzer, Doctor of Medicine, on the medicinal usage of the magnet" (1775), in *Mesmerism*, 25–29, quote on 29; "Dissertation" (1799), 124; Puységur, *Mémoires* (1786), 75; Deslon, *Observations* (1784), 17; Bergasse, *Considérations* (1784), 108–11.

57. Diderot, *Le Rêve de d'Alembert* (1767).

58. Vinchon, *Mesmer* (1971), 134; Duveen and Klickstein, "Joint Investigations" (1955), 291; Schaffer, "Self Evidence" (1992), 84.

From the beginning of their investigation, the commissioners operated on the hypothesis that mesmeric effects were due to no fluid of sensation, but instead to the faculty of imagination. This had been Berthollet's pronouncement when he angrily abandoned the Society of Harmony.[59] He had blamed a most controversial faculty. The imagination was the subject of profound ambivalence on the part of sentimental empiricists. Lorraine Daston and Jan Goldstein have each called attention to the eighteenth-century preoccupation with the imagination. Goldstein argues that the prevailing sensationalist psychology made the imagination, "situated at the point of intersection between mind and body," newly suspect. And Daston observes that the "errors that most terrified Enlightenment savants … were errors of construction, of a world not reflected in sensation but made up by the imagination."[60]

According to the scheme set forth in d'Alembert's "Discours préliminaire" to the *Encyclopédie*, imagination was one of three branches of the human mind, along with memory and reason, and was responsible for a corresponding third of human knowledge, the fine arts.[61] Diderot defined imagination as the "faculty of painting for oneself objects that are absent, as though they were present."[62] Closely allied to memory,[63] the imagination was the more sensitive, passionate faculty. While memory was cool and calm, recalling only the "signs" and "words" associated with sensible objects, imagination was warm and

---

59. Berthollet, "Déclaration" (20 May 1784), in Lavoisier, *Oeuvres*, 3:505–6.

60. Daston, "Fear and Loathing" (1998), 76; Goldstein, "Enthusiasm or Imagination?" (1998), 30. See supra, nn. 4, 5. Eighteenth-century ambivalence regarding the imagination can be seen in the fact that, as G. S. Rousseau has shown, "Enlightenment physiologists centered their attention on the diseased rather than the healthy imagination," producing "a preponderance of works concerned with madness and the malfunctioning imagination." Rousseau, "Science and the Discovery of the Imagination" (1969), 118, 123. On eighteenth-century suspicion of the imagination as the source of sexual fantasies, see Rousseau, ed., *Languages of Psyche* (1990), 41–42. On the feminine gendering of the imagination in Enlightenment natural philosophy, see Terrall, "Metaphysics, Mathematics" (1999), 258–59.

61. History was the province of memory, meanwhile, and philosophy of reason. D'Alembert, "Discours préliminaire" (1751), in Diderot and d'Alembert, eds., *Encyclopédie*, 1:xlvii–li.

62. Diderot, *Eléments de physiologie* (1784), 250.

63. "Imagination is the memory of forms and colors." Diderot, *Le Rêve de d'Alembert* (1784), 367. "Imagination is the faculty of recalling images." Diderot, *De la poésie dramatique* (1758), 61. On the relation between imagination and memory, see also Condillac, *Essai* (1746), part I, §2, chap. 2: "On Imagination, Contemplation and Memory." For an argument that Lockean sensationism reconfigured the relations between imagination and reason, see Tuveson, *Imagination* (1960), especially chap. 4, "The Imagination in the New Epistemology." Tuveson writes that "a certain passivity of the experiencing mind and an enhancement of the importance of extrarational mental activities" in sensationist epistemology transformed imagination from an active, creative faculty to one that merely received and preserved sensory impressions. This new imagination supplanted reason as the central mental faculty (88–91).

vivid;[64] it recalled the objects themselves, by "resuscitat[ing]" the very sensations originally provoked.[65] These features of the imaginative faculty made it crucial to the mind's impressionability and therefore to its functioning. La Mettrie reckoned that "all the parts of the soul can justly be reduced to imagination." Condillac credited the imagination with "all the fecundity . . . of which [the mind] is capable."[66] Diderot made imagination responsible for the mind's capacity to grasp the entirety of an object. Without its ability to make extrapolative leaps from parts to whole, he reckoned, one would see only a muddle of details.[67] He accordingly extolled the imagination as that "quality without which one is neither a poet, nor a philosopher, nor an *homme d'esprit*, nor a reasonable being, nor a man."[68]

On the other hand, the visceral images called forth by the imagination represented "as many occasions to go astray."[69] Differing from madness "only by more and less,"[70] the imagination was easily bamboozled. "There is a great difference," Diderot's protagonist in the *Entretien sur le fils naturel* (1757) observes, "between painting something in my imagination, and putting it in action before my eyes. One can make my imagination adopt any notion one likes. . . . It is not the same with my senses."[71] Here, indeed, was a central area of ambiguity, the relation of the imagination to the senses. On the one hand, imagination and sensation worked collaboratively. The imagination drew upon sensory experience, making use "of the eye in showing objects where they are not; of taste, of touch, of the ear." It was "the internal eye," and those lacking imagination were "hard" and "blind in the soul, as blind people are in the body."[72] On the other hand, the imagination and the senses were in perpetual conflict. The senses had always to be "on guard against our imagina-

64. Diderot, "Eloge de Richardson" (1761), 45; Condillac, *Essai* (1746), part I, 44–45. Imagination was "that vivid memory, that makes appear present what is absent." Condillac, *Traité des sensations* (1754), part II, 221.

65. Diderot, *Eléments de physiologie* (1784), 250. Condillac distinguished memory, which "recalls things only as past," from imagination, which "redraws them with such force, that they appear present." Condillac, *Traité des sensations* (1754), part I, 37. And Helvétius wrote that the imagination sought "to reclothe in sensible images the abstract ideas and principles of the sciences." Helvétius, *De l'esprit* (1758), 491.

66. La Mettrie, *L'Homme-machine* (1748), 112; Condillac, *Traité des systèmes* (1749), chap. 13, 238.

67. Diderot, *Eléments de physiologie* (1784), 226.

68. Diderot, *De la poésie dramatique* (1758), 218.

69. Diderot, *Eléments de physiologie* (1784), 250. "The imagination goes quickly when it goes astray, because nothing is as fertile as a false principle." Condillac, *Traité des systèmes* (1749), chap. 5, 53.

70. Condillac, *Essai* (1746), part I, 122.

71. Diderot, *Entretien sur le fils naturel* (1757), 157.

72. Diderot, *Eléments de physiologie* (1784), 252, 250, 254.

tion" and to "ceaselessly warn us of the absence of objects that we want to imagine."[73] Worse, the imagination was the more powerful element in this struggle, since it, unlike the senses, was "without limits."[74] Ultimately, the imagination was untrustworthy, "a faculty which exaggerates and misleads," and that "sees all that pleases it, and sees nothing more."[75] Imagining could therefore be most "dangerous," Condillac counseled; "if we do not make ourselves master of this operation, it will inevitably mislead us."[76]

Thus, sentimental empiricists were suspicious of the imagination for much the same reason that they distrusted the rational faculty. Both imagination and reason, sensibilists feared, could lead the mind deep into itself, away from the sensory channels that opened it to its surroundings. "The man of imagination strolls through his [own] head like a visitor to a palace," Diderot wrote. ". . . He goes here and there, he does not come out."[77] La Mettrie similarly warned that the imagination, if "too much abandoned to itself" and allowed merely to "look at itself in the mirror," would quickly lose its purchase on the world. "See that bird on the branch, it always seems ready to fly away," he cautioned, and "the imagination is the same."[78]

Along with solipsism, sentimental empiricists suspected the overly imaginative—like the excessively rational—of arrogance. Imagination allowed philosophers to believe that nothing lay "beyond the range of their mind," and that they could "know everything." It led them therefore to construct all-encompassing, "bold," and "extraordinary" systems.[79] Together, imagination and reason, seducing the mind into the intricacies of its own recesses, constituted the antithesis of sensibility: the spirit of system. Descartes, sentimental empiricists' favorite exemplar of the misguided system-builder, had a "vivid" and "fertile" imagination.[80] D'Alembert wrote that the taste for systems was "suited to flatter the imagination," and Diderot sarcastically proclaimed, "Happy the systematic philosopher, to whom nature has given . . . a strong imagination."[81] Condillac advised that those in whom "imagination dominates" were "accustomed to seeing badly" and ill-suited to philosophical re-

73. Condillac, *Traité des sensations* (1754), part I, 39–40.

74. Condillac, *Essai* (1746), part I, 127.

75. Diderot, *Eléments de physiologie* (1784), 255; Condillac, *Traité des systèmes* (1749), chap. 12, 222.

76. Condillac, *Essai* (1746), part I, 115, 118, 113.

77. Diderot, *Eléments de physiologie* (1784), 250. See also *Réfutation d'Helvétius* (1774), 309: enthusiasts of "gigantic imagination" could see only "the phantoms of their heads."

78. La Mettrie, *L'Homme-machine* (1748), 117.

79. Condillac, *Traité des systèmes* (1749), chap. 7, 67; chap. 2, 16.

80. Condillac, *Essai* (1746), part II, 280–81.

81. D'Alembert, "Discours préliminaire" (1751), in Pons, ed., *Encyclopédie*, 1:116; Diderot, *De l'interprétation* (1753), 192.

search, yet their imaginings "dazzle"[82] the mind. These "men of imagination," he lamented, "do not fail to get their systems adopted."[83]

The imagination's suspect relations to the operations of the five senses offered the mesmerism investigators an opportunity to resolve a problem that had been forcefully raised by Mesmer's results. Sentimental empiricists had often contrasted the outward focus of sensation, the body's response to the outside world, with the inward tendency of reason, which, Diderot wrote, "tends to live within itself."[84] But mesmerism challenged this easy contrast by making manifest the very internal nature of sensations: they take place, after all, utterly and only inside the person experiencing them. Only a mesmeric patient could say whether he or she felt the presence of the animal magnetic fluid. In an effort to recover the correlation between sensory impressions and the world outside the mind, the commissioners from the Society of Medicine argued that what Mesmer had revealed was not the basis of sensation, but the extent to which the mind—through its imaginative faculty—could hijack the senses, reversing them in their tracks. His "inner sense" was nothing but the turning of the senses inward. The commissioners therefore introduced their own distinction between "internal" and "external" sensations in opposition to Mesmer's. Accordingly, external senses remained the basis of all knowledge, while internal sensations were "equivocal, often illusory" impressions produced inside the mind by the imaginative faculty.[85]

The members of the Academy commission similarly sought to distinguish sensations produced from within, by the imagination, from sensations produced by external sources. When they began their investigation, by undergoing their own series of treatments at Deslon's clinic, they took a "necessary precaution" that would become highly controversial after the reports were published. They took care "not to be too attentive" to their own impressions. In response to Mesmer's argument for the reality of his animal magnetic fluid from the sensibilities of his patients, the commissioners adopted a policy of deliberate insensibility. They professed to avoid having too "fixed" an attention. Animal magnetism, if real, should forcibly "fix their attention" for them. To focus too intently upon one's own sensations was to risk producing imagined ones from within.[86] One historian has summarized the commissioners'

82. Condillac, *Traité des systèmes* (1749), chap. 13, 239, 237. "By excess or by lack of imagination, the intelligence is ... very imperfect" (chap. 13, 240).

83. Condillac, *Traité des systèmes* (1749), chap. 5, 138.

84. Diderot, *De l'interprétation* (1753), 185. On the contrast between inward-looking reason and outward-looking sensibility, see chapter 2.

85. Poissonier et al., *Rapport* (1784), 7–8.

86. Franklin et al., *Rapport* (1784), 18–21. A critic of the commissioners' methods argued that in empirical research one must "dispose the human machine in the most favorable manner to receive the impression." Servan, *Doutes* (1784), 36–38.

efforts to ignore their internal sensations as the strenuous avoidance of "self-absorption."[87] Not only reason but sensation, they now realized, could, if misdirected by the imagination, have an inward focus.

The results of this deliberately inattentive self-experimentation amounted to little: one commissioner on a single occasion felt a pain in his belly that lasted all day and was accompanied by fatigue and malaise. But he attributed these symptoms to the "powerful pressure" that had been exerted on his chronically delicate stomach by a zealous magnetizer. Otherwise none of them felt a thing even when, in the interests of thoroughness, they subjected themselves to a marathon course of three consecutive days.[88]

The commissioners now turned from self-experimentation to experimentation on others. Here their procedure reflected their sense that the solipsism and self-absorption of the overly imaginative—and Mesmer's manipulation of his patients' imaginations—were urgently social problems. Once again, the moral and the epistemological were inseparable. Sentimental empiricists made sensibility the basis not only of natural knowledge but of moral feeling and community (see chapters 2 and 3). Imagination, appropriating the senses and turning them inward, was a potential source of social instability and fragmentation.[89] To operate, as the commissioners believed Mesmer did, on the

87. "Self-absorption" is a term in Simon Schaffer's "Self Evidence" (1992), 80–85. Although Schaffer describes the commissioners as having worried about the pitfall of "self-absorption" in philosophical research, he interprets their primary concern to have been the opposite problem, the vulnerability of the researcher to the influence and manipulation of others. In the face of this vulnerability, Schaffer writes, the commissioners were concerned to preserve their self-possession. They understood mesmerists to have assumed control of their subjects' bodies as though they were puppets or automata. According to Schaffer, by demonstrating their immunity to mesmeric manipulation, the commissioners reaffirmed the sanctity of individual autonomy. However, because the commissioners maintained their immunity to mesmerism by scrupulously diverting their attention away from their internal sensations, it seems to me that avoidance of self-absorption, and not preservation of self-possession, was their dominant concern. Or, to put it differently, the commissioners feared losing their self-possession not simply to an outside manipulator, but to that manipulator's exploitation of their own tendency to become absorbed in themselves.

88. Franklin et al., *Rapport* (1784), 20–22; and Poissonier et al., *Rapport* (1784), 5–6, 23.

89. On the sensationalist correlation between the imagination and social fragmentation, see Goldstein, "Enthusiasm or Imagination?" (1998), 30; and supra, nn. 4, 5. In her discussion of the mesmerism debates, Lindsay Wilson has focused upon critics' perception of mesmerism as a social problem. She argues that mesmerism, used predominantly to treat the *maladies des femmes*, appeared as part of a more general phenomenon, an increased influence of women on late Enlightenment French culture. This cultural prominence of women, anti-mesmerists feared, undermined "traditional values of hierarchy, privilege and patriarchy." In making this argument, Wilson distinguishes the social from the epistemological dimension of the mesmerism debate, writing, for example, that the commissioners' strategy was "to avoid plunging into the problem of epistemological certainty at all cost. Instead [they] focused on public relations." Wilson, *Women and Medicine* (1993), chap. 5 (quotes from 123, 107). However, I mean to suggest that just as

imaginations of crowds of people presented a serious threat to the social order.

The investigators assumed that sensibility and imaginativeness varied by social class, and they divided their subjects accordingly.[90] They first brought seven sick members of the "*classe du peuple*" to Franklin's house at Passy. Four of these—an asthmatic widow, a woman with a tumor in her leg, and two children, one scrofulous and the other a convulsive—felt nothing. A couple showed some effects: a man with a tumor in his right eye had pain and watering in his left; and a woman who had been knocked over by a cow exhibited shoulder movements "similar to those of a person on whose face one sprinkles drops of cold water." Next, the commissioners invited to their private room at Deslon's establishment some "invalids from society" and also had Deslon magnetize Franklin and his entire household at Passy.[91] Only two of these "society" patients felt anything, both minor and otherwise explicable effects, a passing sensation of warmth in one case and alternate feelings of sleepiness and agitation in the other.

The results of these initial tests would seem inconclusive for both the popular and the patrician samples. Nevertheless, the commissioners decided that mesmeric treatments elicited a greater response from the popular classes than from enlightened society (or, for that matter, from the commissioners themselves, who were "armed," as they said, "with philosophical doubt"). The people, out of ignorance, a fervent desire to be cured, and an eagerness to please their betters, were rendered less able "to realize their sensations" than wealthier patients, and so more subject to their imaginations.[92]

Bailly's supplementary report, "to be placed before the eyes of the King and reserved for His Majesty alone," presented a similar argument concerning the effects of mesmerism upon women. This top-secret document treated a subject so delicate that Bailly, having sent the original to Franklin to look over before the official signing, requested him to burn the covering letter.[93] Noting that mesmeric patients were usually women and mesmerists always

sentimental empiricism was inseparable from a social theory that founded social unity in sensibility, so, reciprocally, anti-mesmerists' social worries about instability were inextricably tied to their epistemological worries about insensibility.

90. This taxonomic principle provoked little comment; one critic did object that the greater credence the commissioners gave to the testimony of subjects from the "*classe plus élevée*" than from the "*classe du peuple*" reflected "the most marked partiality." Bonnefoy, *Analyse* (1784), 48.

91. The Society of Harmony's lodges were necessarily attended almost exclusively by noble and wealthy bourgeois clients because of the admissions fee. See Darnton, *Mesmerism* (1968), 73.

92. Franklin et al., *Rapport* (1784), 23–32.

93. Bailly to Franklin, 8 September 1784, in FPA; Duveen and Klickstein, "Joint Investigations" (1955), 293.

men, Bailly described a certain "convulsive state that has been confused with the other crises." This state resulted from a cause not detailed in the public reports, a cause that was "hidden . . . but natural." It was the "dominion that nature gives to one sex over the other, to engage and arouse it." The intimacy of doctor and patient during mesmeric procedures, Bailly wrote, was especially hazardous when combined with the effects of the imagination, whose action rendered women unable to "realize what they are experiencing."[94]

That the imagination could have dramatic effects upon the body was by no means a new idea in 1784; it had been gaining currency over the course of the eighteenth century. Doctors and physiologists had catalogued many cases, for example, in which a pregnant woman's imagination influenced the shape of her fetus.[95] In an affair of the 1720s and '30s, of which mesmerism would later be reminiscent, the imagination was even charged with causing convulsions. The earlier episode involved a Jansenist cult that arose at the grave site of the deacon François de Pâris. Its adherents trembled and moaned, and claimed to be cured of their physical ills, in much the same manner Mesmer's patients later would. The so-called convulsionaries of Saint-Médard said they were the recipients of God's will and the beneficiaries of divine miracles. Doctors who disbelieved these claims, like Mesmer's investigators, found that the dramatic effects were caused instead by the subjects' imaginations.[96]

When the mesmerism commissioners set out to measure the extent to which "the imagination can influence our sensations," they themselves noted that the great power exerted by the imagination over animal functions had already been well appreciated: "It revives with hope, it chills with terror. In a single night, it whitens the hair."[97] Indeed, the notion that the imagination could have physical effects made good sense in the context of the predominant physiological understanding of the imagination as a bodily function of

94. Bailly, "Rapport secret" (1784), 43–44. On eighteenth-century physiological theories regarding sensibility and the passions in women, particularly in relation to the work of Pierre Roussel, see Williams, *The Physical and the Moral* (1994), 54–56. On the commissioners' theory of female susceptibility to mesmerism, see Wilson, *Women and Medicine* (1993), chap. 5.

95. See Rousseau, "Science and the Discovery of the Imagination" (1969), 120–23; "Pineapples, Pregnancy, Pica, and the *Peregrine Pickle*" (1971); Huet, *Monstrous Imagination* (1993); Goldstein, "Enthusiasm or Imagination?" (1998), 40 and n. 24.

96. On the Saint-Médard convulsionaries and the power of imagination, see Goldstein, "Enthusiasm or Imagination?" (1998), 40–48; and Wilson, *Women and Medicine* (1993), chap. 1. On the mind/body problem in eighteenth-century physiology and medicine, see Staum, *Cabanis* (1980), chaps. 2, 3.

97. Lavoisier, "Remarques" (1784), in *Oeuvres*, 3:510; Bailly et al., *Exposé* (1784), 15. The sexual effects of the imagination provided another example of its manifest influence upon the body. See Morris, "The Marquis de Sade and the Discourses of Pain," in Rousseau, ed., *Languages of Psyche* (1990), 291–330 (especially 318–20).

nervous fluids and fibers. But the commissioners' explanation of mesmeric crises differed importantly from previous assessments of the corporeal effects of imagination. They claimed not just that the imagination could imprint itself upon the body, but that it could hijack the senses, redirecting them inward. Undermining the trustworthiness of sensation, the commissioners' power of imagination presented an urgent problem for sentimental-empiricist epistemology.

The problem at hand, that is, as the commissioners understood it, was epistemological rather than physiological. Midway through their investigation, Bailly reported, the commissioners "ceased to be *Physiciens* becoming nothing more than *Philosophes*." They abandoned fluids and took up epistemology. But they retained the sober methods of the *physicien*. "We operated, as one does in Chemistry," by means of decompositions and recompositions.[98] The second phase of their investigation, therefore, combined an epistemological subject—the influence of the imagination upon bodily experience—with a scientific method, to create the first deliberately psychological tests. These tests involved the first instance of blind assessment using a placebo.[99]

The experimental protocol, drawn up by Lavoisier, centered upon two procedures. The first would isolate the fluid from the imagination, magnetizing subjects without their knowledge. The second would isolate the imagination from the fluid, persuading them that they were being magnetized when they were not. Lavoisier gave detailed directions for deceiving test subjects. He instructed the examiner, for example, to "appear from time to time to address M. Deslon in an undertone" in order to give a blindfolded subject the impression that Deslon was present. After thirty minutes of this pretense, a commissioner would announce that the trial was finished but would ask the patient, ostensibly no longer being magnetized, to retain the blindfold and report his or her sensations. Then Deslon would enter the room and begin a treatment in earnest.[100]

The primary instrument for separating the subject's imagination from the animal magnetic fluid was the blindfold. The commissioners included an exhaustive description of the device they invented:

> This bandage was composed of two rubber caps, whose concavity was filled with eiderdown; the whole was enclosed and sewn into two round pieces of fabric. These two pieces were attached to one another; they had strings that tied in back. Placed upon the eyes, they left in their

98. Bailly et al., *Exposé* (1784), 9–10.

99. See Kaptchuk, "Intentional Ignorance" (1998), 393–99.

100. Lavoisier, "Remarques" (1784), in *Oeuvres*, 3:509–10; "Plan d'expériences" (1784), in ibid., 3:511–13; "Résumé du rapport" (1784), in ibid., 3:519–20.

interval a space for the nose, and complete liberty for breathing, without one's being able to see anything, even the light of day, neither through, nor over, nor under the bandage.[101]

Acting in concert with the blindfold was a rule of silence that the commissioners imposed upon Deslon and all but one of themselves (who conducted the interrogations) to make it impossible for subjects to know who was in the room at any moment.[102]

In a letter to Franklin several years later, Guillotin recalled their earlier collaboration in the "very important though highly ridiculous affair of animal magnetism."[103] Consider the comedy they enacted before a woman from the "*classe du peuple*" with ailing eyes. Having arranged her in a room with the carefully constructed blindfold, they persuaded her that Deslon had arrived to magnetize her. In fact, Deslon was in another room attempting to magnetize the gouty and kidney-stone-ridden yet healthily skeptical Franklin. A commissioner played the role of the magnetizer, entering the room "affecting the stride of M. Deslon." His collaborators then "pretended to speak to M. Deslon, praying him to begin; but we did not magnetize the woman." Instead, they "sat calmly" and, by the power of suggestion alone, triggered violent convulsions that were loudly audible to their colleagues in the next room. There another young woman suffering from a "nervous affliction" had been told that the ubiquitous Deslon was magnetizing her from behind a closed door; she had responded similarly.

Sigaud de la Fond, no doubt having heard about these experiments from his colleagues on the Faculty of Medicine, joined in. In order to study the extent of the imaginative influence, he began to advertise himself as a mesmerist and soon acquired a reputation throughout the city. By means of his notoriety, he acted upon the imaginations of passing Parisians. Once he cured the migraine of a young artist he met on the Pont Royal. Another time a young lady in the rue Colombier asked him if he was on his way to Mesmer's hotel. "Yes," he responded, "and I can magnetize you [right here]." Pointing his finger at her, he later told the commissioners, he could have thrown her into convulsions had she not begged him to stop.[104]

While isolating the imagination from the hypothetical fluid, the commissioners complementarily isolated the fluid from the imagination. In one case they invited a laundress reputed to be sensitive to magnetism into a room in Franklin's house under the pretext of wanting some washing done. They ma-

101. Franklin et al., *Rapport* (1784), 37–38.
102. Lavoisier, "Remarques" (1784), in *Oeuvres*, 3:509–10.
103. Guillotin to Franklin, 18 June 1787, in FPA.
104. Franklin et al., *Rapport* (1784), 46–47, 39–41.

neuvered her into a chair in front of a doorway from which they had previously removed the door and replaced it with paper. While she was magnetized from behind the paper, she sat "conversing gaily" for half an hour about her health and other topics. Meanwhile, a commissioner sat at a desk taking surreptitious notes, pretending to write out a catalog of books.

The commissioners magnetized (and pretended to magnetize) their subjects indirectly as well as directly, through objects such as a basin of water and a set of china cups. Curiously, and almost without fail, the unmagnetized objects induced violent crises while the magnetized ones had no effect. Franklin's grandson, Benny Bache, an unimpressed fifteen-year-old, matter-of-factly recorded one such experiment in his diary: "My grandpapa . . . [and] the commissioners are assembled today with Mr. Delon [*sic*]," he wrote, and "they are gone into the garden to magnetize some trees." On this occasion the patient in question, a boy of twelve, collapsed at the farthest tree from the magnetized one. Deslon reportedly explained that trees were all naturally magnetic, to which the commissioners dryly responded that in that case "a person sensitive to Magnetism could never hazard to go into a garden without risking convulsions." [105]

But two could play that game. For example, the medical commission had suggested that mesmeric convulsions had perfectly ordinary causes, like the warmth, perspiration, and agitation of air occasioned by mesmeric massage. A mesmerist responded: "Eh! what will become of these unfortunates . . . unable to take a step, execute the slightest movement, suffer the opening of a door, tolerate the approach of any being, without falling into convulsions?" [106] And what about the imagination as the cause of mesmeric effects? A mesmerist responded, "Why do we not see crises multiplying at our tragic plays, at theaters?" [107] If the imagination had the power to effect such a response, Deslon wrote more seriously, then "there will no longer be anything certain, either in our ideas or in our sensations." [108]

Indeed, the commissioners' attribution of mesmeric effects to the imagination did call everything in doubt. According to the central axiom of sensationism, sensations were the material impressions upon the brain of an external

---

105. Lavoisier, "Résumé du rapport" (1784), in *Oeuvres*, 3:520–23; Bache, "Diary," 15 January 1784–15 January 1785, in FPA; Franklin et al., *Rapport* (1784), 44. Deslon's own account of his explanation differs: he claims to have argued that the boy was feeling the delayed effects of an earlier mesmeric session, in *Observations* (1784), 22.

106. Bonnefoy, *Analyse* (1784), 64; Deslon likewise protested that the touch employed in mesmeric massage was "always soft and light." Deslon, *Observations* (1784), 27.

107. Anon., *Réflexions* (1784), 17. See also Deslon, *Observations* (1784), 29.

108. Deslon, *Supplément aux deux rapports de mm. les commissaires* (1784), 54–55; cited in Schaffer, "Self Evidence" (1992), 87.

world. Yet Mesmer's investigators claimed that his patients' sensations did not originate outside themselves, but instead had an internal source. Furthermore, both mesmerism commissions were at a loss to explain the material basis of this internal source of sensations. Contemporary, sensationist physiology generally attributed imagination to something much like Mesmer's animal magnetic fluid: an etherial medium flowing through the fibers of the nerves. In sensation, this nervous fluid was set in motion by external ethers, in turn moving the nerve fibers through which it traveled; in the operations of imagination and memory, the same movements of nervous ether and nerve fibers were re-created by the brain itself.[109]

But the power of imagination that the mesmerism commissioners cited corresponded to no such material agent. The commissioners rejected Mesmer's claim that he could manipulate the fluid responsible for sensation; they therefore denied that mesmeric effects were due to the motions of a sensory ether. In addition to subverting the sensationist principle that sensations necessarily originated in the world outside the mind, the commissioners also undermined, in their reaction against Mesmer's blunt application of it, the materialism of sensationist psychology. They distinguished what they called "moral" from physical causes of physical feeling.[110] Insisting that mesmeric patients were responding to no material medium, but instead to a "moral" force, their own imaginations, the commissioners opened themselves to an obvious question: What precisely was this faculty of "imagination"? It seemed a more mysterious and troubling cause than Mesmer's magnetic fluid. How could one account for its undeniable, material influence upon the body?

The commissioners cast about unsuccessfully for plausible explanations. They speculated that the "intimate *rapport* of the intestine[,] the stomach and the uterus with the diaphragm" bore upon the question. "There exists a certain sympathy," they hazarded, "a communication, a correspondence between all the parts of the body, an action and reaction," whereby sensations received at the center would radiate outward. The "affections of the soul," perhaps, made an impression upon the central nerve center, which weighed, in turn, upon the stomach. In the end the commissioners capitulated, attributing

109. See, for example, Bonnet, *Essai de psychologie* (1754), chaps. IV, V, VI; *Contemplation de la Nature* (1764), part V, chap. VI; *La Palingénésie philosophique* (1769), 178–79. For secondary discussions of the physiology of the imagination in the eighteenth century, see Rousseau, "Science and the Discovery of the Imagination" (1969), 109–11; and Ilie, *Cognitive Discontinuities* (1995), 3, 209, 232–33, 301, 356. In an alternative to the etherial theories, La Mettrie proposed that the imagination was "a sort of medullary screen on which objects painted in the eye are projected as by a magic lantern." La Mettrie, *L'Homme-machine* (1748), 113.

110. Lavoisier, "Remarques" (1784), in *Oeuvres*, 3:510; Franklin et al., *Rapport* (1784), 23–32.

everything to the unspecified "power by which the imagination is able to act upon the organs and trouble their functioning."[111]

The problem of what the imagination might be, in material terms, so troubled the Society of Medicine commissioners that they hesitated to cite it as the essential cause of crises and ultimately assigned it only the secondary position of "accessory cause," giving priority to unproblematic, physical causes: the "extended application of hands, the heat produced by this application, the irritation excited by rubbing." In order to fit this conclusion to the facts, the medical commissioners were forced to consider only those facts that appeared "ordinary, constant." They deliberately set aside those that were "rare, unusual, marvelous"—the most striking effects of mesmerism—such as convulsions caused by "a finger or a conductor, [pointed] through the back of a well-padded seat . . . sensations felt while approaching a tree, a basin, a body or area that has previously been magnetized."[112] Mesmerist supporters understandably objected to this policy of selective empiricism. "You have thus seen facts, facts that surprised you, extraordinary facts...—[and] *You ignored them.*"[113]

Mesmerists were also quick to point out that the Academy commissioners' faculty of imagination begged the question: If mesmeric effects were due to imagination, to what was imagination due? One mesmerist demanded how the "*imagination,* without any other intermediary agent," could cause such dramatic results. Surely it was impossible, another critic wrote, "that a body exercises an action upon another without an intermediary being." Some "medium or milieu" must transmit motions and modifications among bodies. "What is *imagination?*" Deslon demanded to know; for "Messieurs the Commissioners put it a great deal into play without defining it." He claimed a proper physicist could only understand the imagination as "a fluid that flows through us." Otherwise, what might it mean to "*strike,*" or to "*disturb,*" the imagination?[114]

If the imagination were a fluid, it might participate in producing or hindering mesmeric effects. Or perhaps the magnetic fluid moved the imaginative fluid. Servan proposed that the imagination and magnetism might be merely two different names for the same phenomenon. Could the magnetic fluid not be the source of all "intellectual functions"? Imagination was one branch of a tree whose trunk was sensation; the unknown sap flowing through the tree, the "minister of sensation," was the fluid to which the mesmerists referred. In

111. Franklin et al., *Rapport* (1784), 61–63.

112. Poissonier et al., *Rapport* (1784), 17, 24–25.

113. Bonnefoy, *Analyse* (1784), 53.

114. De Saint Paul to Franklin, 10 December 1784, in FPA; Anon., *Réflexions* (1784), 15, 49; Bonnefoy, *Analyse* (1784), 22; Bergasse, *Observations* (1785), 41; Deslon, *Observations* (1784), 29, 31; and Servan, *Doutes* (1784), 65. On the imagination theory begging the question, see Miller, "Going Unconscious" (1995), 59.

that case the imagination would not be a "source from which to derive your proofs against Mesmer's fluid." On the contrary, the commission's experiments would have served to confirm the strength of Mesmer's force. The commissioners had done nothing more than "oppose Mesmer's agent to itself" and arrive at the same principle by a different route.[115]

While the commissioners treated the facts selectively and dispensed with material causes, critics of their reports responded like sober empiricists. They referred their detractors to solid "fact" and reminded them that "observations of the effects that nature operates . . . are not the exclusive property of the *Philosophes*; universal interest has made almost all individuals into so many Observers."[116] James Hutton had written to Franklin the year before the investigation that it seemed to be a "fact" that Mesmer had cured Court de Gebelin. "Now you are philosopher enough, if a Fact really is, not to dispute the Fact, though the *quo modo* has all the appearance of Quackery."[117] The facts did put the commissioners in an uncomfortable spot, for the observed effects of mesmerism, whatever their causes, were uncontested and the immaterial force of imagination seemed an implausible explanation. At the end of August, Franklin observed that the "report makes a great deal of talk. Everybody agrees that it is well written, but many wonder at the force of imagination described in it as occasioning convulsions."[118]

In a sense, then, it was the mesmerism investigators, and not Mesmer or his followers, who acted as the true radicals. Mesmerism itself was a faithful extrapolation from sensationist doctrine, a theory of sensibility according to which the human body was an instrument vibrating in a universal, material medium, its feelings directly responsive to the cosmos as a whole. It was the commissioners who extracted from mesmerism a new and radical force, the power of imagination: an immaterial force capable of causing physical sensations. It was they who derived from mesmerism a conclusion that seriously undermined their own sensationist methods and principles. For if the imagination were capable of such feats, how (as mesmerists pointed out) could the

---

115. Anon., *Réflexions* (1784), 31, 17, 19–20; Bonnefoy, *Analyse* (1784), 50, 65; and Servan, *Doutes* (1784), 31, 52–59, 60–68, 100–2. Another mesmerist pointed out that the commissioners themselves had said that moral causes exerted a powerful and disruptive influence upon the physical body. In that case, why had they introduced moral causes so liberally into their own experiments? By toying with their subjects' imaginations, had they not rendered their own investigation invalid? Rather than choosing the most suggestible subjects possible, they should have sought the least sensitive. For example, they should have chosen men rather than women—or better yet, plants and vegetables. Bergasse, *Considérations* (1784), 122–29.

116. "Dialogue," in Bergasse, *Receuil* (1784), 187; Mesmer, "Mémoire," in ibid., 8–9.

117. Hutton to Franklin, 2 May 1783, in FPA.

118. Franklin to William Temple Franklin, 25 August 1784; cited in Duveen and Klickstein, "Joint Investigations" (1955), 299.

senses ever be trusted? Moreover, the commissioners had nothing to offer in its place, no plausible explanation for their mysterious force of imagination.

They therefore presented their findings without explanation, as "facts for a science that is still new, that of the influence of the moral upon the physical." Mesmerism had been the first "great experiment concerning the power of the imagination." Sentimental empiricists, fusing the moral with the physical, sentiment with sensation, had argued that one could neither feel without perceiving nor perceive without feeling. But Mesmer, with his bald reduction of perception to feeling, drove the commissioners to distinguish the two afresh. In the wake of their investigation, the relations between knowledge and sensation were newly problematic. And while the root of the problem was epistemological, its domain was social. Bailly concluded, "What we have learned is that man can act upon man, at any moment and almost at will, by striking the imagination."[119]

To initiate their new moral-physical science—and once again reflecting their assumption of the importance of sensibility as a basis for social union— the commissioners concluded their reports on a political note. Bailly sketched a moral physics of two forces: imitation, by which man "forms, perfects himself," and imagination, by which he "acts, becomes powerful." Imitation was a conservative force, the source of habits, conventions, national character, prejudice, and patriotism. Imagination was a radical force, the "eminently active faculty, author of good and evil," the source of progress. Imitation grasped only what it saw; imagination saw the whole, the "future as well as the present, the worlds of the universe as well as the point where we are." Imitation was stable and imagination, volatile. Imitation held communities together; imagination destabilized them and set loose their collective powers: "When imagination speaks to the multitude, the multitude no longer knows either dangers or obstacles. . . . Nations follow sovereigns, and armies their Generals." In opposing ways, both imitation and imagination had the potential to overwhelm the senses: while imagination focused the mind on its own fancies, imitation could make a person blindly follow others. As long as the two balanced one another, all was well. But when they worked together, they could produce fearsome results.[120]

The commissioners had noticed that the effects of mesmerism were more pronounced in public sessions attended by crowds of patients than in private ones. Their discussion of this crowd phenomenon reflected a new preoccupation with crowds in the wake of such events as the bread riots of 1775–76. Read retrospectively, it also acquires extra resonance from its publication date, five

119. Bailly et al., *Exposé* (1784), 12, 16, 15.
120. Ibid., 12–15.

years before the Revolution. The commissioners proposed that in crowds, the force of imitation combined with imagination to inspire a kind of collective "intoxication." The same phenomenon occurred at theatrical events and in armies on the day of battle. The "sound of the drums, the military music, the noise of cannons" accomplished what Mesmer's eerie harmonica music and impressive costumes did during séances; they "raised the imaginations to the same pitch." Meanwhile imitation instilled the emotions of each in the imaginations of all, overwhelming sensation, for "men united are no longer subject to their senses." Together, imagination and imitation defeated sensibility and erected in its place a "fanaticism" that, the commissioners judged, "presides at these assemblies." Imagination was the original source of "revolts" and "seditions." The imagination, Lavoisier concluded, although "obscure and hidden," was "an active and terrible power."[121]

Mesmerists turned the commissioners' crowd psychology against them. Their "violent" reaction against mesmerism, Servan charged, had arisen from their own *esprit de corps*, for crowds always tended to regard extreme positions as heroic, and moderate ones as cowardly.[122] Meanwhile, Galart de Montjoie—a mesmerist, novelist, and future anti-Jacobin writer—declared the commissioners were guilty of the spirit of system. In a letter addressed to Bailly just after the reports were published, Montjoie protested that "to always cry out against systems, as though there were a science without system," was itself a form of system-building and "a great abuse, Monsieur." One should rather accept imperfect systems, just as one must accept "governments, [though] there is perhaps none that does not offer some abuse." Men were "made in such a way, that there are principles which, bad in themselves, must yet be respected." Someone who did not realize this essential fact of humanity was indeed a "man of System, a very dangerous man."[123]

Thus, in the commission reports and the ensuing dialogue between commissioners and mesmerists, insensibility—the insensibility of crowds, of the popular classes, of royal commissions with their *esprit de corps* and *esprit de système*—became a real and dangerous political force. The commissioners had made mesmerism into the source of a new psychology, a nascent theory of the unconscious that credited the mind with startling powers over the body. This new psychology meant, too, that a person could render others insensible by manipulating their imaginations. Insensibility was no longer simply the de-

121. Ibid., 12–16; Franklin et al., *Rapport* (1784), 64–67; Lavoisier, "Résumé du rapport" (1784), in *Oeuvres*, 3:524. See Rudé, *The Crowd in the French Revolution* (1959), chap. 2, and *The Crowd in History* (1964), chap. 1.

122. Servan, *Doutes* (1784), 95–96.

123. Montjoie, *Lettre* (1784), 57–59.

tachment from one's proper sensations; it could now mean false sensations so powerful as to cause seizures. And insensibility could now have consequences even worse than system-building: it could bring about fanaticism and rioting. Where the imagination was unleashed, a dangerous insensibility threatened the land.

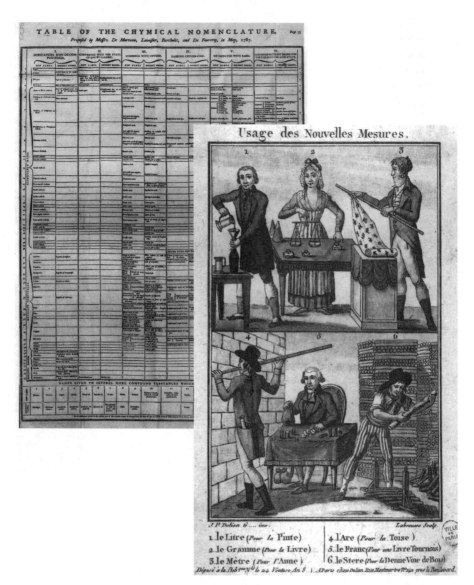

Figures 7.1 and 7.2. Table of correspondences between old and new chemical names, from the 1788 English edition of Guyton de Morveau et al., *Méthode de nomenclature chimique* (1787). Courtesy of the Bancroft Library, University of California, Berkeley. And Revolutionary-era engraving on the use of the new metric system. © Photothèque des musées de la ville de Paris, negative: Svartz.

Chapter Seven

LANGUAGES OF SCIENCE
AND REVOLUTION

Chemists . . . comprise a distinct people, few in
number, having their language, their laws, their mysteries.
—Gabriel-François Venel, "Chymie" (1753)

As men cannot make signs for themselves, except
when they live together, it is a consequence that the basis of their
ideas . . . is uniquely in their reciprocal commerce.
—Condillac, *Essai* (1746)

To be good citizens, it is necessary
to have less science and more virtue; to
talk less . . . and to act better.
—Durand de Maillane, "Opinion sur les écoles" (1792)

The vice of the current age is owing to
an abstract way of thinking.
—Edmund Burke, "Speech on the Report of the Army" (1789)

Before the bloody Terror of 1793, there was, from 1765
to 1780, in the republic of letters, a dry terror, of
which the *Encyclopédie* was the Committee of Public
Safety, and d'Alembert the Robespierre.
—Augustin Cochin, *L'Esprit du Jacobinisme* (1920)

Alongside the faculties of reason and imagination, a third, essential
human capacity seemed dangerously to loosen the senses' grip on the
mind: the power of language. The potentially pernicious action of words was,
once again, equally an epistemological and a moral worry, and the intense controversy it provoked straddled a major scientific debate and a crucial political
one. Here, then, is a new instance of the fusion I have been tracing throughout
this book of epistemological and scientific with moral and political argument.

The proceedings of the Paris Academy of Sciences for February 1776 record major disagreements among chemists on the subject of fixed air (what today is called carbon dioxide). The problem was not so much the substance as the phrase itself, "fixed air": "They generally agree well enough upon the thing, but they are not in accord about the name."[1] This strife over names went deep and escalated to a raging debate a decade later, following Antoine Lavoisier and his collaborators' introduction, as the crucial element of their "chemical revolution,"[2] of a new chemical nomenclature (fig. 7.1). At the same time, certain other disputed programs of systematic linguistic reform became prominent features of the political Revolution (see fig. 7.2). One debate in particular ran alongside and entwined with that among chemists over the proper language for conducting their new science: political revolutionaries bent on establishing a system of civic education argued over the best language for molding a new citizenry. The two disputes about language overlapped, centrally involving some of the same people and many of the same arguments.

Studies of the new chemical nomenclature of 1787 have examined its epistemological function, on the one hand, and have called attention to the moral and political tenor of the surrounding controversy, on the other. But there has been little study of the relation between the two. Bernadette Bensaude-Vincent proposes that the lexicon was a "complex of claims, not just chemical, but also philosophical and even political."[3] The philosophical and political elements of this complex have generally been treated independently, though in close proximity. In a 1992 issue of the *Eighteenth Century* devoted to the chemical revolution, two essays focus primarily upon the nomenclature. Lissa Roberts's interest is epistemological; she assesses the influence of Etienne Bonnot de Condillac's philosophy of language upon the nomenclature and upon new chemists' methods. Jan Golinski's interest is political; he contrasts French and English objections to the nomenclature. More recently, Bensaude-Vincent and Ferdinando Abbri have edited a volume on the dissemination of the new nomenclature in Europe, in which several contributors indicate that the debates accompanying this dissemination were often political.[4]

---

1. Registre des procès-verbaux (24 February 1776), AS. The previous spring, Lavoisier had claimed before the Academy that "fixed air" was a misnomer, for the substance was not released during burning from fixation in a solid but was produced by the combination of "the eminently respirable portion of the air" (oxygen) with charcoal. See Lavoisier, "Principe qui se combine" (1775), in *Oeuvres*, 2:128.

2. Lavoisier, "Mémoire sur la nécessité" (1787), 12.

3. Bensaude-Vincent, "A propos de *Méthode de nomenclature chimique*" (1983), 1. For work on the new nomenclature, see especially Marco Beretta, *Enlightenment of Matter* (1993). Beretta treats the evolution of early modern chemical language and the place of the nomenclature in that evolution. For earlier seminal treatments of chemical language in general, and the new chemical nomenclature in particular, see Crosland, *Historical Studies* (1962); Dagognet, *Tableaux* (1969); Albury, "Logic of Condillac" (1972); Anderson, *Between the Library* (1984).

4. Roberts, "Condillac, Lavoisier" (1992); Golinski, "Chemical Revolution" (1992); Bensaude-

Yet to the authors of the nomenclature and their critics, and to those involved in the interlocking controversy over language in civic education, the two functions of language—epistemological and moral—really formed just one. The moral was inseparable from the epistemological in the chemical controversy, the epistemological inextricable from the moral in the political controversy. Both disputes were propelled, this chapter suggests, by a rivalry between two conceptions of language: a cultural conception, that of the sensibilists, and a competing social understanding of language in which the romanticism and sentimental empiricism of the sensibilists gave way to the idealism and instrumental empiricism of social engineers. While sentimental empiricists argued that understanding and virtue resided in a spontaneously expressive language that directly and vividly conveyed the sentiments, their challengers countered that not spontaneous expressions, but carefully designed linguistic conventions must be the building blocks of knowledge and social unity.

The story of their disagreement has larger implications, both historical and historiographical. Its historical implications derive from the fact that these disputes about the legitimacy of inventing a new language, and thereby remaking the scientific and social worlds, crystallized a general discussion of the origins and nature of language, science, and society. In the general discussion, as in the particular ones treated here, the disputants agreed upon a great deal, making it often difficult to discern their narrow but crucial margin of disagreement. They agreed that language was the medium of natural knowledge and social harmony, and that both of these arose from sensory experience and from an intimate connection between the linguistic and natural orders. Where, then, did they differ? I suggest that their opposing sides were drawn by a clash between two attitudes toward human experience of nature: sentimental empiricism and what I propose to call instrumental empiricism. These two standpoints were equally expressive of Enlightenment values, yet they were irreconcilable. Indeed, it was precisely by their mutual defiance that they helped to define one another. In what follows, I use the disputes over language in chemistry and civic education to show how this disagreement among allies operated at the heart of Enlightenment projects.

Vincent and Abbri, eds., *Lavoisier in European Context* (1995). John McEvoy's essay in this volume, for example, treats the British rejection of the nomenclature, calling upon "competing models of the chemical community" in Britain and France. McEvoy, "Priestley Responds" (1995). Concerning the instrumentalism of the new chemists' use of language, see also Golinski, *Science as Public Culture* (1992), 148; McEvoy, "Priestley Responds" (1995), 138; Levere, "Lavoisier" (1990), 211; and Levere, *Chemists and Chemistry* (1994), 317. Elsewhere, Levere has suggested that Lavoisier's analytical use of language, inspired by Condillac, shaped his analytical procedures in the laboratory: Levere, "Lavoisier" (1990); and Levere, *Chemists and Chemistry* (1994), 317.

The larger historiographical implications of the story told in this chapter arise from another matter upon which all the disputants agreed, which was the inseparability of epistemological and scientific concerns, on the one hand, from moral, social, and cultural concerns, on the other. Social constructivist and cultural histories of science have not, it seems to me, sufficiently taken into account the explicitness and purposefulness with which their subjects often treated the social and cultural engagement of scientific thought.

I. TWO CHEMISTRIES OF LANGUAGE

Gabriel-François Venel, professor of chemistry at the university of Montpellier, early staked out his position in the struggle over a language for the emerging science of chemistry. Venel was the chief writer on chemical topics for Diderot's *Encyclopédie* and the author of its article entitled "Chymie."[5] Here, he called chemists a "distinct people" with their "own independent manner" of conceiving things, a manner "which gives them a language" of their own. Arguing the importance of preserving this separate language for chemistry, Venel appealed to the Lockean epistemological principle that knowledge derives entirely from sensory experience. He praised the "chemical idiom" for expressing "reflections suggested by the immediate exercise of the senses." A distinct language, recording chemists' unique experience of nature's diversity, yoked their science firmly to their sensations.[6]

Similarly, Venel admired the experience-based language of chemical artisans and even respected the enigmatic writings of the ancient alchemists for having been "rich in facts." Indeed, he deemed obscurity itself a virtue in chemical writing. In his view, clarity and order, so admired by "the journalists, the novelists, the Poets," actually distorted scientific research. The artificial clarity of systematic treatises belied the burgeoning confusion of sensory experience. Luckily, the opacity of the chemists' argot shielded their philosophy from the popular tendency to spin philosophical systems. The style of the German chemist Georg Stahl was a model: "difficult" and "dense," it preserved chemistry from becoming "*à la mode*," while at the same time "swarming with the sort of images that spread from the sensible object."[7]

---

5. Venel and Diderot himself wrote most of the chemical articles for the *Encyclopédie* along with Paul-Jacques Malouin and the baron d'Holbach. Venel wrote the ones of greatest theoretical interest, such as "Chymie" and "Principe."

6. Venel, "Chymie" (1753), 408, 409, 416, 419.

7. Ibid., 419, 436–37. The style evolved from these sources comprised what Bensaude-Vincent has called "the natural language of chemists," a natural language that was later supplanted, she implies, by the artificial nomenclature. Bensaude-Vincent and Abbri, eds., *Lavoisier in European Context* (1995), 2. Her phrase echoes Venel's own belief that the obscurity of traditional chemical language reflected the diversity of physical sensation.

These passages of the article "Chymie" appear at first oddly contradictory. Venel argued, in terms that Charles Gillispie has described as proto-Jacobin, that the "high contemplations" of natural philosophy are nothing but the "experience of the worker covered with the varnish of science." But, to defend chemistry against the trend toward system-building, he ultimately concluded it should be reserved for an elite of connoisseurs. Writing for the third volume of Diderot's *Encyclopédie*, the mission of which was to disseminate arcane knowledge, rendering the arts and sciences accessible to a literate public, Venel nonetheless closed with a vindication of "obscurity" in chemical writing.[8] These contradictions no doubt reflect a more general uncertainty about the *Encyclopédie*'s project: at once to celebrate the specialized technical knowledge of craftspeople and to challenge the cloistering of such knowledge within the crafts, two purposes that can fall into conflict.

But Venel's apparent contradictions can also make sense. He did not intend to hide chemistry behind a veil of secrecy. On the contrary, the article opens by regretting the general "incuriosity" about chemists and their work. In its most cited passage, Venel called for a "new Paracelsus," "clever, enthusiastic and bold," who would place chemistry on a par with physics. This new Paracelsus would promote chemistry through powerful language, with a "noisy ostentation . . . a decided and affirmative tone."[9] Venel meant to popularize chemistry; only his was a particular, and somewhat polemical, conception of chemistry and so, too, of chemical language.

He wrote that chemical theories were not rational expositions but "exhibitions" of nature. They were derived not from principles but from unguided "groping," from the "vague signs" chemists were able to divine in their objects, and from what he described as chemists' capacity for "*experimental premonition.*" Physics and chemistry, each with its own "general manner of envisaging" its subjects, must each have a "different language." While the language of physics was rigorous and mathematical, chemistry would always be "vague and approximate." Chemists' knowledge, Venel implied, could be expressed but not articulated; it was a matter of sensation and intuition, not system.[10]

A manner of thinking, a mode of speech, a way of life, all rooted in a natural basis of physical sensation and instinctive response: these were principal ingredients also in the nascent conception of culture.[11] Condillac defined "culture" in his *Dictionnaire des synonymes* as "the care one gives to a piece of land

8. Gillispie, "*Encyclopédie*" (1959), 260; and Venel, "Chymie" (1753), 419, 437.

9. Venel, "Chymie" (1753), 409–10.

10. Ibid., 416. Leon Chertok and Isabelle Stengers characterize Venel's conception of the chemist's craft as "the fruit of . . . an intimate and passionate contact" with chemical phenomena. Chertok and Stengers, *Critique* (1989), 50.

11. Isaiah Berlin traces the "modern concept of culture, and of what may be called cultural pluralism" to Vico and, more generally, to the eighteenth-century "birth of the new belief . . . in the

to render it fertile. It is said figuratively of customs, the mind, the sciences, etc."
By analogy, one did not deliberately invent customs, manners of thought, or
sciences according to first principles, but only fostered their natural growth.
To "cultivate" the "mind, the memory, the arts, the sciences" was a matter of
developing one's responsiveness to this organic process—one's sensibilities
and intuitions—more than one's rational faculty. Candide and his comrades,
cultivating their garden, resolved to "work without reasoning." Though
Venel's recommendation to chemists to shun the systematizing "tone" of contemporary physics, and return to the traditional "chemical idiom," was in that
sense reactionary, his project was, in its effort to preserve a cultural identity,
strikingly modern. In the language of enlightened political philosophy, Venel
defended the "right" of chemists to their idiosyncratic style as at once a "liberty" and a "possession." His chemical philosophy was the one presented in
the central document of the French Enlightenment, Diderot's *Encyclopédie*, to
which the authors of the nomenclature contributed no articles. Venel's ideal
of chemistry and of scientific language was far from old-fashioned; it retained
its vitality through the last years of the century.[12]

Opposing Venel's vision of an organic style of chemical expression was a
different tradition of reformers of chemical language. The reformers in this
tradition followed the model set forth in Carolus Linnaeus's 1735 method of
botanical nomenclature, which inspired new systematic languages throughout
the natural sciences, particularly in anatomy, medicine, and mineralogy as well
as in chemistry.[13] One of the first to apply the model to chemistry was Pierre-

---

value and importance of the singular and unique, of variety as such." Berlin, "Giambattista Vico"
(1991), 59, 56. Natalie Davis writes that in the eighteenth century, "reflections on the history of language as well as on the history of religion . . . produced a theory . . . of primitive culture." She
identifies this theory as an early step toward a modern "anthropological concept of culture" and
indicates this concept at play in an eighteenth-century work on proverbs. See Davis, *Society and
Culture* (1965), 255–56. Raymond Williams writes that "in French, until C18, culture was always accompanied by a grammatical form indicating the matter being cultivated. . . . Its occasional use as
an independent noun dates from MC18." Williams, *Keywords* (1976), 78.

12. Condillac, *Dictionnaire des synonymes de la langue française* (1758–67), in *Oeuvres philosophiques*,
3:171; Voltaire, *Candide* (1759), 101; and Venel, "Chymie" (1753), 409, 419. On the continued vitality
of Venel's ideal of scientific language, see below, §3. "Chymie" continued to be widely read, witness Antoine-François Fourcroy's inclusion of it in his 1796 *Encyclopédie méthodique, chimie*. On the
lasting importance of this "brilliant and influential article," see Thackray, *Atoms and Powers* (1970),
192–98. Venel continued to dominate chemical writing in Diderot's *Encyclopédie* through its first
supplement in 1776, to which he contributed two articles. In 1770 the tireless editor wrote admiringly of Venel, "He is a man of rare merit, an excellent chemist, . . . and the most exacting moralist
I know." Diderot, "Voyages à Bourbonne et à Langres" (1770), in *Oeuvres complètes*, 17:327–61 (quote
on 341).

13. Félix Vicq-d'Azyr, André-Marie-Constant Duméril, and Charles-Louis Dumas advocated systematic linguistic reforms in anatomy, Philippe Pinel in medicine, and René-Just Haüy
in mineralogy. On the "reign of nomenclatures" throughout the natural sciences, see Duris, *Linné
et la France* (1993), 126–32.

Joseph Macquer, who in 1766 published a dictionary of chemistry in which he proposed new systematic conventions governing the names of salts.[14] This proposal was adopted and expanded to include all chemical names, first by the Swedish chemist Torbern Bergman; then by the Dijon lawyer and chemist Guyton de Morveau, who had been a student of Macquer's and was a friend of Bergman's; and, finally, by the tax farmer, academician, gunpowder commissioner, and chemist Lavoisier (fig. 7.3). It was this movement for the instantiation of a systematic chemical language that culminated in the new chemical nomenclature published by Morveau and Lavoisier in 1787, in collaboration with two others, Antoine-François Fourcroy and Claude-Louis Berthollet.[15]

From Macquer to Lavoisier, these natural philosophers sought conventions grounded not in experiment, enthusiasm, tradition, or cultural autonomy, but in "the relation of ideas with words." They defied the Lockean dogma that ideas originate solely in physical sensation, and the corollary drawn by Venel and his followers, that natural philosophical theory and culture must be allowed to emerge spontaneously from sensation and emotion. While Venel disparaged books as artificial sources of chemical knowledge, Macquer claimed the sciences had been conceived with the invention of writing and born with the practice of clear, systematic exposition. Research required a written plan.[16]

Those who adopted Macquer's project of renaming chemical substances believed that language directed as well as recounted experience, shaped as well as described experiments. Rather than growing spontaneously, as a jargon, from accreted experiences and responses, chemical theory was to be engineered, as a lexicon, from first principles. Thus, Bergman observed that technical words, "like coins, owe their currency to prescription."[17] A science

14. Macquer suggested naming salts according to their constituents. Each salt would have a generic name denoting the acid it shared with other members of its group, and a specific name denoting the metal that combined with the acid to form it. Macquer, *Dictionnaire de chymie* (1766), 2:673–74. On Macquer's innovations for the naming of salts, see Beretta, *Enlightenment of Matter* (1993), 137.

15. In their 1787 publication, all four authors of the nomenclature defended a view of scientific language formulated primarily by Morveau and Lavoisier. In what follows, I ascribe this view in its essence to all four authors. For information about "the complexity of Lavoisier's intellectual heritage" in the next decade, and in particular about his collaborators' adherence to and departures from their own new chemical system, see Janis Langins, *La République* (1987), 53–62. Langins analyzes the chemistry curriculum at the new Ecole polytechnique, first called the Ecole centrale des travaux publics, in 1794. Fourcroy, Berthollet, and Morveau were largely responsible for the curriculum and adhered to the new chemical program to varying degrees. See also Dhombres and Dhombres, *Naissance* (1989), 496–504. On Fourcroy's promulgation of the nomenclature, see Siegfried, "Chemical Revolution" (1988).

16. Lavoisier, "Mémoire sur la nécessité" (1787), 5; Venel, "Chymie" (1753), 437; and Macquer, *Dictionnaire de chymie* (1766), 1:xxv, vii.

17. Bergman, *Physical and Chemical Essays* (1784), 3:303.

Figure 7.3. Jacques-Louis David's 1788 *Portrait de Monsieur Lavoisier et sa femme* shows Lavoisier
at the time when he was devising a new scientific language for chemistry. Courtesy of
the Metropolitan Museum of Art, Purchase, Mr. and Mrs. Charles Wrightsman Gift,
in honor of Everett Fahy, 1977. (1977.10).

depended upon a conventional vocabulary just as a society relied upon a conventional system of values. These conventions were formal, expressing no
inarticulate intuitive power. They were deliberately chosen and could be
deliberately altered.

If the conception of language informing Venel's "Chymie" was a cultural
one, the new chemists' model of scientific language might be described as so-

cial. The word "social," we learn from Diderot's *Encyclopédie*, was also a "word newly introduced into the language," used to designate that which suited one "to the commerce of men."[18] Like "culture," "social" emerged as part of the midcentury fascination with the question of the origins and nature of human collective life.[19] Whereas "culture" implied shared intangibles—modes of thought, varieties of experience and intuition—"social" denoted deliberate and rule-governed collaboration. Condillac defined "social" as pertaining to those "qualities that render one suited to society," and "society" as a "body of several persons . . . joined by mutual engagements called laws." Rousseau's "social" meant "corporate and collective." His "social order" was formed by an act of will and "founded on conventions." It resulted, in Jean Starobinski's phrase, from "man's dangerous privilege to possess in his own nature the powers by which he combats that nature and nature itself." The "social" in Rousseau's philosophy is for Starobinski ultimately an "antinature."[20] "Social" differed from "cultural" by connoting deliberately orchestrated, rather than organically arising, human activity.

Dismissing organic tradition as a valid source of chemical names, Bergman applied the binomial system of his compatriot, Linnaeus, to the project of prescribing a new set of linguistic conventions for chemistry. Nevertheless, Bergman preserved a natural basis for scientific language, proposing that denominations should be "in keeping with the nature of things." Newly discovered substances, he said, should be given "names conforming to their characters"—names that directly expressed their observable properties rather than offering metaphorical descriptions ("oil of vitriol") or conjectures about their origins ("fixed air").[21]

18. Anon., "Social" (1765), in Pons, ed., *Encyclopédie*, 2:319. Williams writes that a "development can be seen in *social*, which in C17 could mean either associated or sociable, but by LC18 was mainly general and abstract: 'man is a social creature; that is, a single man, or family, cannot subsist, or not well, alone out of all society.'" Williams, *Keywords* (1976), 246.

19. Daniel Gordon has traced "the invention of the social as a distinctive field of experience" to eighteenth-century France. He writes: "Before the late seventeenth century, the word *société* did not refer to durable and large-scale community. Instead, it referred to small associations and to the convivial life that took place in them. . . . [I]n the eighteenth century, it . . . took on a distinctively new meaning as the general field of human existence." Gordon, *Citizens without Sovereignty* (1994), 5, 51–52. On the eighteenth-century origins of the modern words "société," "sociabilité," "sociable," and "social," see Gordon, *Citizens without Sovereignty* (1994), 51–54. See also Baker, "Enlightenment and the Institution of Society" (1994).

20. Condillac, *Dictionnaire des synonymes* (1758–67), in *Oeuvres philosophiques*, 2:512. "The social order is a sacred right which serves as the basis of all other rights. Nevertheless, this right does not come from nature; it is founded on conventions." "*Each of us puts his person and all his power in common under the supreme direction of the general will.* . . . [T]his act of association produces a corporate and collective body." Rousseau, *Du Contrat Social* (1762), bk. I, chap. I, 174; chap. VI, 192–93. Starobinski, *Jean-Jacques Rousseau* (1971), 305.

21. Bergman, *Opuscules* (1780), xxviii. In the binomial system, a first name denoted the class to which a substance belonged, and a second name identified it as a particular member of that class.

The subsequent abandonment of the principle that a name should reflect the nature of its object marked the origin of the new chemical nomenclature. In 1782 Morveau published a call for a systematic reform of chemical names in which he denied outright Venel's and Bergman's common assumption that names should be founded in facts. Facts alone, Morveau asserted, "say nothing to the mind." He then casually anticipated a concept that would dominate much twentieth-century language theory, that is, the arbitrary relation of the sign to the signified.[22] Morveau claimed that all names were essentially artificial: "Sounds, and the words they represent," he wrote, "in reality have, by themselves, no relation, no conformity with things." So, in the case of an individual substance that one "envisions only for itself," and not in relation to any other substance, Morveau argued, any name that "means nothing" would serve the purpose. In fact, he preferred meaningless names for such independently considered substances and recommended that nomenclators "distance themselves as much as possible from familiar usage." For the purpose, he advised taking roots from classical rather than vulgar languages.[23]

Linguistic conventions were prior to meaning, Morveau believed, and therefore they were themselves meaningless. They alone connected sounds with things. In the case of chemistry, only after convention had "attached a first idea to a word" could denominations be sought that expressed, in a lim-

---

Bergman assigned single names to acids, alkalis, and earths; he then proposed double names for the compounds formed from these substances. These double names would reflect the compounds' compositions: a compound's first name denoted the substance it shared with other members of its group, and its second name denoted the individuating component unique to it. Concerning Bergman's revisions of chemical nomenclature, see Beretta, *Enlightenment of Matter* (1993), 137–47. For a more general discussion of binomial nomenclature in Enlightenment science, see Lesch, "Systematics" (1990), 73–111.

22. Guyton de Morveau, "Mémoire sur les dénominations" (1782); and Guyton de Morveau et al., eds., *Encyclopédie méthodique* (1786–1808), 1:iii–iv. It should be noted that though Bergman had preserved a basic commitment to naturally derived names, he had also begun to move in the direction of arbitrary naming, writing at one point that "as it is not easy to apply names exactly expressive of the thing defined, we are to adopt such as having no determinate meaning may have their sense ascertained by definition." Bergman, *Physical and Chemical Essays* (1784), 3:303. Sylvain Auroux, in his comprehensive study of Encyclopedists' writings on language, writes that the "originality of the Enlightenment was perhaps to have invented linguistic arbitrariness." Auroux, *La Sémiotique* (1979), 47; see especially 48–53, on eighteenth-century theories of the arbitrary relations of words to their objects.

23. Guyton de Morveau, "Mémoire sur les dénominations" (1782), 373; cf. Guyton de Morveau et al., eds., *Encyclopédie méthodique* (1786–1808), 1:v–vi. For the advice about taking roots from classical languages, see Guyton de Morveau, "Hépar" (1777), in Diderot, ed., *Nouveau dictionnaire*, 3:347. From the Enlightenment emergence of this conception of language as fundamentally artificial, Foucault derives his influential argument that the natural sciences transformed in this period into an exercise in classification, the imposition of an artificial taxonomy upon natural phenomena. See Foucault, *Les Mots et les choses* (1966).

ited way, the natures of their objects. For example, having arbitrarily assigned a name to a substance, chemists could give its derivatives names that were derived from its name.[24] These derived names would then express the derivative substances' natures, but only in terms of the original, arbitrary name. The empirical truth of chemical names was thus, to Morveau, necessarily a derivative truth, ultimately founded in arbitrary convention.

He did not confine his arguments to chemical language, nor yet to the technical vocabularies of natural philosophy, but intended them to apply to language in general. Morveau was participating in a broader controversy over the origins of language that had been set off by a philosopher upon whose work he and Lavoisier both drew heavily, Condillac. A crucial element in Condillac's theory of language was the notion that names refer only arbitrarily to their objects.[25] Condillac named John Locke as his chief inspiration. However, Condillac's (and subsequently Lavoisier's) Lockean epistemology had been given a sharp interpretive twist, a twist that transformed the function of language, and of social convention, in natural science.[26]

Locke had warned natural philosophers against "abusing" words by "taking them for Things," and indeed had drawn an example from chemistry to illustrate this pitfall. The word "gold," Locke claimed, referred not to any real thing-gold, but only to a certain complex of sensations and perceptions that people attributed to some essentially unknowable object. When chemists demonstrated, for example, the "Fixedness or Solubility" of gold under certain conditions, they merely expanded the definition of the word "gold" by adding an experience to the complex. Yet chemists were seduced by their own language into thinking that they had perfected their idea of the thing itself. Locke's firm belief in the eternal impossibility of understanding natural objects in themselves led him to conclude that natural philosophy was just "not capable of being made a Science."[27]

But if language was no source of natural knowledge, it was, Locke wrote, the medium of "Society," God having "designed Man for a sociable creature."

---

24. Guyton de Morveau, "Mémoire sur les dénominations" (1782), 373.

25. Beretta stresses the importance of the "fierce debate on the origin of language," triggered by Condillac's writings, to the genesis and reception of the nomenclature, writing that "between 1746 and 1780, the question of language and its origins became the main topic of European philosophical and scientific discussion." Beretta, *Enlightenment of Matter* (1993), 189.

26. Condillac, *Essai* (1746), vi; and Condillac, *Traité des systèmes* (1749), 60–66. On Condillac's transformations of Lockeanism, see Buchdahl, *Image of Newton and Locke* (1961), 36; Yolton, *Locke* (1991), 72–76; Knight, *Geometric Spirit* (1968), 1–3; Rousseau, *Connaissance* (1986), 100–10; Acton, "Philosophy of Language" (1966), 167; Albury, "Logic of Condillac" (1972), 21–25; and Beretta, *Enlightenment of Matter* (1993), 188–206. Smith writes that Condillac "appropriates Locke's *Essay* by redefining it as a theory of language." Smith, *Politics of Language* (1984), 133.

27. Locke, *Essay* (1690), 645–46.

Moreover, moral philosophers did not share natural philosophers' worries about words, not studying natural entities but only social conventions, or what Locke called "mixed modes." Because mixed modes were essentially conventional, they entailed no gap between name and object. The "real Essence" of a mixed mode was a matter of definition and could be fully captured by language. So, for example, a moral philosopher might distinguish among "*Chance-medly, Man-slaughter, Murther, Parricide*" without risk of abusing the terms. The actual things to which these words referred were themselves conventions, identical to the definitions of their names. Words simply dictated the difference between, say, murder and manslaughter. This capacity for precision in moral philosophy meant, to Locke, that morality, in contrast with chemistry, was "*capable of Demonstration,* as well as Mathematicks" and was therefore the "*proper Science . . . of Mankind.*"[28]

Condillac accepted Locke's claim that words were purely arbitrary conventions, but not the corollary that conventions were incapable of generating ideas or indicating natural facts. On the contrary, Condillac granted the arbitrary conventions of language a primary and powerful epistemological function. Experience generated ideas, he said, by means of a process of analysis, like mathematical analysis, in which people decomposed and recomposed their sensations. This process resided in the art of naming. To break a sensation into its component parts was simply to name them. Words were accordingly the prerequisites of ideas and an original source of all knowledge.[29] Concerning the relation of natural and moral science, Condillac thus sharply diverged from his model philosopher. Locke had used what he took to be the essentially social function of language to set moral science apart from natural philosophy. But Condillac used what he took to be the dual purpose of language, social and epistemological, to bring the natural and moral sciences together. A successful science and a successful society came together in a well-made language.

Condillac argued that thought relied upon the use of what he called "institutional signs." The defining features of these were the deliberateness and artifice associated with the word "social." Institutional signs were deliberately

---

28. Ibid., 402, 516–17. Locke defines "mixed modes" as consisting of "several Combinations of simple *Ideas* of different kinds" (288). He refers to mixed modes as belonging to the fields of "Divinity, Ethicks, Law and Politics" (294). His initial examples of mixed modes are "*Obligation, Drunkenness, A Lye*" (288).

29. For the acceptance of Locke's claim, see Condillac, *Essai* (1746), part I, §2, chap. 4, 65. For an argument that Condillac modeled the process of decomposition and recomposition of sensations upon chemical analysis, see Lambert, "Analyse Lavoisienne" (1982), 371. Condillac's comparison of linguistic to mathematical analysis was put to wide use. Concerning Enlightenment attempts at inventing artificial languages on the model of mathematics and on the applications of mathematized language in mechanical communication, see Rider, "Measure of Ideas" (1990).

chosen and had "but an arbitrary relation with our ideas." For Condillac, the purposefulness involved in associating an arbitrary sign with an idea was a feature of social interaction rather than individual reflection. A child raised by bears, he imagined, would not even recognize his own natural cries as signs, never hearing the like of them from any other creature. Only if he "lived with other men" could he perceive the general significance of natural human cries. Institutional signs, without even a natural basis, were all the more reliant upon interaction with others. Condillac's representation of the origins of language presents an interesting contrast with Rousseau's. Rousseau argued that language, being the "first social institution, must owe its shape entirely to natural causes," and characterized the emergence of language as a matter of spontaneous expressions of the "passions" rather than deliberate communication. But Condillac attributed language to social rather than natural causes, believing that people could "make signs for themselves only when they live[d] together." Thus the original basis of their ideas was "uniquely in their reciprocal commerce." [30] Language and, therefore, thinking and, therefore, natural philosophy were all ineluctably social enterprises. This was the central principle of the linguistic philosophy informing Morveau's and Lavoisier's new language.

Venel and Morveau thus represented starkly divergent philosophies of the function of language in natural science. Venel the sentimental empiricist espoused a scientific language steeped in emotional and traditional meaning, an eloquent jargon spontaneously expressive of natural philosophers' sensibilities and of chemists' own *esprit.* In contrast, Morveau the instrumental empiricist proposed a lexicon whose first terms were purely and deliberately meaningless, strictly neutral with regard to both culture and experience. His code's rigor required not only its divorce from traditional language, but the severing of scientific words from their primary grounding in empirical fact. The epistemology Morveau and his collaborators assumed in their use of language was thus neither natural nor cultural; it was social. It arose from what they took to be the essentially collaborative nature of thought—and of natural science.

II. SCIENCE AS CULTURAL EXPRESSION, SCIENCE AS SOCIAL ENGINEERING

These opposing views of scientific language emerged in the context of general disagreement not only about the origins of language, encapsulated in the

---

30. Condillac, *Essai* (1746), part I, §2, chap. 4, 65, 205–6; Rousseau, *Essai* (1781), 46, 50–51. Rousseau objected to Condillac's institutional signs on the ground that their deliberate invention seemed paradoxically to require a preexisting society. For an analysis of Rousseau's disagreements with Condillac's theory of language, see Guilhaumou, *La Langue politique* (1989), 97–98.

contrast between Rousseau's and Condillac's conflicting theories, but also about the genesis of the arts and sciences. Once again, the central question was whether these had arisen spontaneously and owed their progress to organic causes, or had been deliberately constructed and owed their progress to social causes.

A commonplace notion in eighteenth-century France, codified by Montesquieu in *L'Esprit des lois*, held that since the arts and sciences were the product of human sensibilities, and since the human nervous system was affected by its environment, geographic variations in climate provided a natural explanation for the differences in artistic and scientific achievement among nations. Condillac rejected this idea on the ground that differences in climate influenced "only the organs," not the development of talent. He identified a social rather than natural basis for the arts and sciences, and therefore for differentials in national achievement, maintaining that their progress relied "uniquely upon the progress of languages." The poor use of language in a society posed the most serious problem for scientific research, Condillac warned his readers, and the greatest impediment to their scientific education was the fact that "before studying the sciences, you already speak the language, and you speak it badly." He recommended, "Do you want to learn the sciences with ease? Begin by learning your language."[31]

In keeping with this instruction, Lavoisier sought the source of scientific errors in the popular misuse of words. Studying transpiration in animals, for example, and the reason why humidity renders heat more disagreeable, he interposed a rigorous analysis of a common phrase. To say that "the weather is heavy," Lavoisier remonstrated, was to mix several statements into an enunciation so "vague" as to be true only because "it presents no determinate idea." The exact meaning of "the weather is heavy" must be some combination of "*the air just now has no dissolving virtue, it is saturated with water; insensible transpiration is suppressed, and replaced with sweat.*" The mixing of statements always reflected an essential ignorance of causes. In order to discover causes, one had first to identify and label their separate effects—to analyze phenomena, in Condillac's terms, by means of the art of naming.[32]

When Lavoisier and his collaborators later devised a new language for their science, they would assume, following Condillac, that the knowledge of natural causes relied upon the social institution of language. With Condillac, they denied that philosophy arose naturally from either sensation or *génie*. However necessary sensation was to the progress of a science, Lavoisier and

---

31. Montesquieu, *Esprit des lois* (1748), vol. I, bk. XIV. Condillac, *Essai* (1746), part II, §1, chap. 15, 200–1; Condillac, *La Logique* (1780), 109–10, 137, 168; and Condillac, *Traité des systèmes* (1749), chap. 18, 268.

32. Lavoisier, "Transpiration" (n.d.), in *Oeuvres*, 5:386.

Morveau maintained that it was insufficient. To keep themselves on the straight-and-narrow path to truth, chemists needed something else: a "well-made language" expressing neither individual experience nor cultural identity. By its very neutrality, this language would permit philosophical collaboration, the reciprocal commerce of people prerequisite in Condillac's philosophy to rational thought.[33]

This self-consciousness about the importance of institutional change to scientific progress contributes to, and complicates, the picture drawn by current social constructivist and cultural historiographies of science. Historians who have emphasized the force of social and cultural institutions in shaping scientific thought have tended to choose examples in which experimenters and natural philosophers were largely unaware of the influences of their social and cultural contexts, or in which they downright denied these influences. According to these historiographies, the inventors of modern scientific empiricism professed to serve up natural facts unsullied by interest or interpretation.[34] With the new chemical nomenclature, we have a contrary instance in which the new chemists and their critics alike treated chemistry as a morally engaged business. The question was only what form this moral engagement should take. The authors of the nomenclature were not only aware of the dependence of chemical knowledge upon social institutions, but actively embraced this dependence as an integral part of their philosophical program—in fact, as an instrument of research. Their critics meanwhile rejected the social construction of a chemical language, not because scientific languages should be straightforward renderings of natural facts, but because they should be eloquent and organic expressions of a culture.

The new chemical nomenclature of 1787 rested upon two axioms drawn from Condillac's philosophy of language: that names are social conventions, and that all thinking is therefore dependent upon a social institution, the institution of language. These tenets were responsible for the two most interesting and radical features of the new chemical language.

The first of these concerns the names of the elements and Morveau's principle of arbitrary names for the simplest entities. In keeping with this principle, the elements' names ought to have been meaningless. In practice, they were neither arbitrary nor meaningless, except in a technical but significant sense. Only four names in the table were entirely new: caloric (heat), oxygen,

33. Lavoisier, "Mémoire sur la nécessité" (1787), 12–14.

34. An example is Steven Shapin's *A Social History of Truth* (1994), in which he argues that the modern conception of scientific truth was a product of the rules of gentlemanly behavior in early modern England. On the history of objectivity, see Daston, "Marvelous Facts" (1991), "Baconian Facts" (1991), and "Objectivity" (1992); Daston and Galison, "Image of Objectivity" (1992); and Poovey, *A History of the Modern Fact* (1998).

hydrogen, and azote (nitrogen). And even these four had meanings. The nomenclators did follow Morveau's earlier recommendation of using classical roots to distance their technical language from the vernacular. But they coined names whose etymologies reflected salient characteristics of an element (e.g., "hydrogen" to suggest a relation to water, or "oxygen" to acids). Lavoisier explained that these meaningful etymologies were to "relieve the memories of beginners, who retain with difficulty a new word when it is absolutely empty of meaning." Even Morveau conceded that "altogether meaningless words" offered no "hold for the memory." [35]

Still, despite such concessions to practicality, the elements' names were arbitrary in a crucial, technical sense, that is, with regard to the rules of nomenclature. This was because these rules made composition the basis for classing and naming chemical substances. So the compounds were classed and named according to their elementary constituents. But the elementary substances themselves had, by definition, no constituents. Therefore their names were necessarily prior to the rules.[36] This practice represents an important departure from the Linnaean system of botanical nomenclature, in which the same rules applied to naming and classing all specimens. The Linnaean system therefore did not place certain specimens outside its own bounds and name all the rest in terms of these primitives. It was this new practice, of placing the simplest chemical substances, and their names, outside the rules of classing and naming, that applied Morveau's principle of arbitrary naming and was foundational to the nomenclature. The nomenclators' commitment to technically arbitrary names for the elements was tested in 1787, when Berthollet disproved the hypothesis that oxygen caused acidity by producing an acid (prussic) without it. He himself, however, urged the preservation of the name "oxygen," arguing that since it had in theory been meaningless from the beginning, its falsity as to meaning should now be irrelevant.[37]

The second important feature of the new chemical nomenclature involved the names of the compounds. Like the botanical specimens in Linnaeus's binomial nomenclature, each chemical compound had two names in the new system. But whereas the two names in Linnaeus's plant taxonomy denoted directly observable structures, stamens and pistils, the new chemical names referred to hidden, and often hypothetical, components.[38] The first

35. Lavoisier, "Mémoire sur la nécessité" (1787), 18–19; Guyton de Morveau, "Mémoire sur la nomenclature" (1787), 34–35.
36. See Lavoisier, "Mémoire sur la nécessité" (1787), 17; Lavoisier, Traité (1789), 6–7.
37. Berthollet, Essay (1803), 450, 455.
38. See Stafleu, Linnaeus and the Linnaeans (1971), 49–55; Blunt, Compleat Naturalist (1971), 34–35. Lesch writes that Linnaeus's "system gives a privileged place . . . to the visual sense." Lesch, "Systematics" (1990), 77. External form and internal composition were of course not unrelated in eighteenth-century natural philosophy. In contemporary crystallography, indeed, a central proj-

name of a compound indicated the group it belonged to and was formed from the name of the simple substance that the compound ostensibly shared with all other members of the group. A second name identified the individuating simple substance, or "radical," that was unique to the particular compound. So, for example, the nomenclators defined a group of compounds each formed of oxygen and a metal. Each member of this group had the first name "oxide" and a second name of the individuating metal, as in *"oxide de zinc."* [39] This central rule for naming compounds was responsible for a striking characteristic of the resulting table: many, indeed *most*, of the substances in the nomenclature had predictive names—that is, names that referred to substances not yet isolated and compositions not yet determined.

Consider the acids. Lavoisier believed that each acid was the compound of an "acidifying principle," which he mistakenly identified with oxygen, and a radical or "acidifiable base." The nomenclators assumed "by analogy" that acids whose radicals had not yet been discovered nevertheless did contain acidifiable bases. Acting on this assumption, they named twenty-two simple substances that they had not yet isolated, forming names from the word "radical" and the relevant acid—for example "citric radical." The first column of the nomenclature listed the simplest substances, and of these, numbers nine through thirty (out of fifty-five) were the undiscovered bases of undecomposed acids. Because the new chemists were committed to a language founded on the assignment of arbitrary names to simple objects, they rested their entire nomenclature upon the "meaninglessly" named elements. Yet almost half of these were unknown substances. In turn, nearly all the acids were named in terms of these unknowns. And likewise for the salts, which were named in terms of the acids, and therefore in terms of the elements. This domino effect meant that more than half of the new chemical names, compounds and elements alike, involved references to substances not yet discovered.[40]

---

ect was to relate visible structure to hidden composition, a task that the crystallographer René Haüy achieved by means of his hypothetical *molécules intégrantes*. See Metzger, *La Genèse* (1918), 197. But for our purposes, the difference between classification by visible structure and by hypothetical components was the difference between descriptive and predictive naming.

39. Lavoisier, "Mémoire sur la nécessité" (1787), 19–21; and Guyton de Morveau, "Mémoire sur la nomenclature" (1787), 55–57, 38 ("radical"). Morveau had inaugurated the use of the term "radical" in its modern chemical sense the previous year, in the *Encyclopédie méthodique* (1786), 1:142. See Crosland, *Historical Studies* (1962), 302. The word "radical"—developed from sixteenth- and seventeenth-century botanical usage denoting that which pertains to the root—had been used to signify the crux or basic principle of an entity in linguistics as of 1754 and in mathematics as of 1762. *Trésor* (1990), 14:241.

40. Guyton de Morveau, "Mémoire sur la nomenclature" (1787), 46–50; and Guyton de Morveau et al., "Tableaux de la nomenclature chimique," in *Méthode* (1787), 101. Beretta writes that "Lavoisier often named substances before he could isolate them chemically." Beretta, *Enlightenment of Matter* (1993), 267.

To say that the nomenclature was an artifice is not to say it was a fiction. On the contrary, in the eyes of its authors, the nomenclature's whole value lay in the empirical truths it would generate. But, they held, language could only be a "faithful mirror" of nature through artifice, the conventional manipulation of arbitrary signs. The result was a scientific language conceived as instrumental, not expressive. The new nomenclature was engineered to precede and direct experiment, even more than to follow and describe experience. Lavoisier said that it "mark[ed] in advance the place and the name [of] new substances yet to be discovered." In fact, he called the nomenclature "more a method of naming than a nomenclature," because it would "adapt itself naturally to the work that will be done in the future." The nomenclature and its accompanying system of signs would present "at once what has been done in chemistry, and what remains to be done." The new chemists offered their system of names as a portrait of their science's future.[41]

These claims were not window dressing. The nomenclators' attitude toward language was integrally involved in the internal development of their science and in its practice.[42] Word changes were foundational to Lavoisier's chemistry, indeed even before the nomenclature. For example, he credited the word "expansibility," coined by Turgot, with having figured centrally in his understanding of combustion and conception of gases. Turgot had defined "expansibility" as the mutual repulsion caused by heat among the parts of a fluid or air. Lavoisier never suggested that Turgot had identified a new phenomenon. Mutually repulsive fluids were common in contemporary natural philosophy. Rather, Lavoisier credited Turgot with having "fixed the sense" of a new word, thereby making available "the most vast and the newest views." For instance, the word "expansibility" helped Lavoisier to see airiness as a state of matter rather than a kind of matter. That is, he was now able to surmise

---

41. Lavoisier, "Mémoire sur la nécessité" (1787), 14, 16–17; and Lavoisier, "Rapport sur les nouveaux caractères chimiques" (1787), in *Oeuvres*, 5:378. There was a recent precedent for the predictive use of names in chemistry. Macquer's proposals for the binomial naming of salts according to their constituents, presented in his *Dictionnaire de chymie* (1766), included names of unknown salts that he assumed could be made from familiar metals, like gold, tin, zinc, and antimony. See Crosland, *Historical Studies* (1962), 137. I am grateful to Larry Holmes for bringing this precedent to my attention.

42. Whether the nomenclators were correct in assigning such importance to language in the progress of their science is a controversial matter. Holmes has objected that the theories of composition expressed in the nomenclature were virtually complete before the language reform was begun; thus the nomenclators' own previous research belied their claims for the necessity of linguistic change to scientific progress. In the following discussion, I respond to this objection by suggesting that Lavoisier's attention to language long predated his work on the nomenclature itself. More generally, however, I am interested in the nomenclators' theory of scientific language. That theory informed their transformation of their science and its language, whether or not it presented a strictly accurate representation of their own trajectory.

that any common body "in a state of expansibility" would make a gas. Moreover, if aerial fluids were not intrinsically airy, but could pass from a state of "expansibility" to one of solidity, then oxygen might pass from the air to combine with roasted metals. Thus, the word "expansibility" also helped Lavoisier see combustion as the combination of a substance with oxygen, rather than the release of a hypothetical fiery substance that contemporary chemists called "phlogiston."[43]

The chemical nomenclature made manifest Lavoisier's principle of the reliance of views upon words. It was intended, and was treated, as a research program, even by those who emended it. Humphry Davy, for example, while engaged twenty years later in decomposing the undecomposed acids to discover their radicals, determined that what had been called "oxymuriatic acid," ostensibly an oxygenated form of muriatic (hydrochloric) acid, in fact contained no oxygen. Davy renamed the substance "chlorine," deliberately choosing a simple, "arbitrary designation" in keeping with its newly elementary status. He wrote that such changes of terminology were intrinsic to the progress of chemistry, a process of naming, experimenting, and renaming. Echoing Lavoisier on "expansibility," Davy thought the word "chlorine" could help in "unfolding just views."[44]

This engagement of naming in the conduct of chemical experiments and the elaboration of chemical theory has led several historians—notably Golinski, Roberts, and Trevor Levere—to liken the nomenclature to a laboratory instrument.[45] The particular sort of philosophical instrumentalism that Morveau and Lavoisier assigned to language had as much in common with social engineering as it did with experimental manipulation. Condillac's deliberately chosen institutional signs, the elements of language, required the "reciprocal commerce" of people, and in turn directed that commerce by shaping its participants' ideas. The instrumentality of words in his philosophy arose from the inseparability of thought and communication, and the reliance of rational reflection upon social collaboration. Condillac himself had called chemistry a hybrid of physics, which explained effects, and politics,

43. Turgot, "Expansibilité" (1756), in Diderot and d'Alembert, eds., *Encyclopédie*, 6:277–78; and Lavoisier, "Principe qui se combine" (1775), in *Oeuvres*, 2:122, 127. This incorporation of oxygen during burning made sense of the weight gained by calcinated metals, a phenomenon that had stumped phlogistonists since they believed burning involved the loss of a fiery matter. Crosland has emphasized "the importance that Lavoisier, early in his career, attached to the names of substances." Crosland, *Historical Studies* (1962), 169.

44. Davy, "On Oxymuriatic Gas" (1810), 62, 61n.

45. Roberts writes that language, for the new chemists, was "as much an instrument of experience as the material instruments of the laboratory." Roberts, "Condillac, Lavoisier" (1992), 260. See also Golinski, "Chemical Revolution" (1992), 244; Levere, "Lavoisier" (1990), 211; and Levere, *Chemists and Chemistry* (1994), 317.

which created them.[46] Correspondingly, the authors of the nomenclature conceived their project as a compound of epistemological and social reform. Their sentimental-empiricist critics, treating the sciences as cultural expressions and believing that scientific language should describe rather than transform experience, found the nomenclators' social engineering not only misguided but downright evil.

Much of the controversy following the publication of the nomenclature in 1787 focused upon the assignment of arbitrary names to simple objects envisioned uniquely "for themselves." This practice distanced chemical language from common experience, the objects of which are seldom perceived in isolation. New chemists invoked an uncommon kind of experience, generated by a hybrid of linguistic and chemical analysis. They decomposed common substances like air, water, and the acids, into substances so uncommon as to be found only in the laboratory. By Condillac's art of naming, they bestowed upon the rare products of these experiments the status of elementary substances. Their opponents did not object to the experiments themselves but to the practice of naming the arcane results as basic substances, which they judged an inversion of the natural order of experience.[47]

So, for example, the commission appointed by the Academy of Sciences to evaluate the new nomenclature doubted whether it was "more natural" to consider sulfur, composed in a laboratory from vitriolic (sulfuric) acid and hydrogen, as a simple substance, than to treat common air as elemental. In general, they hesitated to credit a "crowd" of simple substances that "all analogy" seemed to suggest were actually composed, being the end products of laboratory procedures, in place of the traditional four elements, found everywhere in nature.[48]

Jean François De Machy, one of the most outspoken opponents of the new nomenclature, had the same quarrel with substituting elements of laboratory analysis for elements of common experience. He found it unthinkable that simple bodies might be derived from operations upon composed bodies. "Nothing will persuade us," De Machy wrote, "that carbon, the result of the combustion of a body as composed as wood, could be a simple body." As for calling sulfur an element, the substance was "not even natural in a volcano … too many complicated conditions are necessary to obtain it." He found it foolish to "take the products of art" for nature's own elements. To name the acids according to the products of their decompositions, rather than naming

46. Condillac, *Traité des systèmes* (1749), chap. 14, 247.

47. On the contrast between the direct, unmediated experiences of nature that Venel and Diderot advocated for chemists and the instrumental mode of experience that Lavoisier and the new chemists introduced, see Roberts, "Death of the Sensuous Chemist" (1995).

48. Baumé et al., "Rapport sur la nouvelle nomenclature" (1787), 246, 248.

the products according to the acids from which they were derived, seemed to De Machy sheer perversity: "Products, the last efforts of analyzing subtlety, taken for elements, for principles!"[49]

He and his fellow sentimental-empiricist chemists described a Heraclitean world in which philosophical analysis ranged from futile to dangerous. All things in nature existed in a burgeoning, chaotic state of "constant activity" and "more or less intimate combination." Matter itself was "endowed with an active force" that tended toward composition and disorder, and resisted the tidy regularity of decomposition.[50] In their analytical fervor, new chemists were insensitive to "composing nature" with its "phenomena of disorganization." De Machy wrote that "the more living nature tends to compose, the more dead nature tends to simplify." The new chemists with their analytical methods could only render nature inanimate. They were blind to the irreducible wonders of the living:

> You would be hard put to tell us, which gases constitute the germ of a plant, organize it, give it the property of making a *sui generis* plant. You obtain them by analysis; but use them to reproduce an atom of substance analogous to a flower or a fruit, and maybe we will begin to believe you.[51]

Synthesis was the process of life; analysis could capture only death. The view that chemical analysis was destructive gained a wide currency. Frederick II lamented in a letter to Voltaire, for example, that chemists had come to resemble geometers in their deplorable tendency to reduce the world to first principles: "Rather than creating gold, [they] send it up in smoke with their operations."[52]

Throughout the natural sciences, sentimental empiricists objected to systematic technical languages on the ground that such languages, as rational constructs, were artificial and untrue to nature. Buffon, in particular, vigorously opposed the use of Linnaean taxonomy in natural history.[53] But it was in chemistry that this methodological debate about language became most overtly moral. In particular, Lavoisier and his collaborators' linguistic replacement of "phlogiston" by "caloric" epitomized the evils of their science.

49. De Machy, "Suite" (January 1791), 110–11, and see also 189, 351–52; and De Machy, "Suite" (July 1791), 97, 95.

50. Darcet, "Mémoire sur la calcination" (11 November 1782), 29–30.

51. De Machy, "Précis" (January 1792), 42–43, 192–93.

52. Frederick II to Voltaire, 17 December 1777, in Voltaire, *Correspondence*, 97:233.

53. On the opposition to Linnaean taxonomy, see Duris, *Linné et la France* (1993), chap. 9; Stafleu, *Linnaeus and the Linnaeans* (1971), 291–336. For Buffon's dislike of systematic nomenclature, see Buffon, "Discours sur la nature des animaux" (1753), in *Oeuvres philosophiques*, 347; Buffon, *De la manière d'étudier* (1749), 19–20, 22, 65; and Roger, *Buffon* (1989), 122–23, 125, 255–57, 317–20.

Phlogiston was the matter of heat in eighteenth-century chemistry before Lavoisier replaced it with his own igneous fluid, caloric. In addition to being the substance of heat and fire, phlogiston performed "the office of composing." It was the basis of all organic bodies and an essential part of mineral ones as well. Material bodies owed their integrity to phlogiston. This was evident because when they burned and gave off phlogiston, they were reduced to ashes. Or in the case of roasted metals, when they lost their phlogiston, they became brittle and crumbly calxes. These could be restored to their former integrity by having their phlogiston replenished. (This occurred when calxes were heated with carbon, which phlogistonists regarded as highly phlogistic).[54]

While it gave vegetable and mineral bodies their wholeness, phlogiston was the spark of life in living creatures, warming and animating them. Plants grew by absorbing phlogiston from light. The Scottish chemist James Hutton gave a lyrical description of the vital importance of phlogiston to vegetation: "The resinous fir grows in a soil of sand," he wrote, "with nothing but what the heavens may bestow." This view of nature derived its support from "moral argument," Hutton was careful to say, and not vice versa: "It is not to evince . . . metaphysical and moral truths that the physical system of things is here inquired into; but it is the physical system that I would now support by metaphysical and moral argument."[55]

Lavoisier was cold to moral arguments for phlogiston. He replaced this source of life, warmth, and integrity with the word "caloric," which he coined for the sake of analytical rigor: in order to avoid the phlogiston theorists' pitfall of giving the same name to two distinct things. He wrote, "We cannot, in a rigorous language, call [the substance that causes heat] by the name of heat, because the same appellation cannot express both the cause and the effect." Initially Lavoisier referred to the cause of heat as the "igneous fluid" or the "matter of heat." In the chemical nomenclature of 1787, he renamed the igneous fluid "caloric." This new name had advantages beyond convenience. "Caloric" did not, "rigorously speaking," oblige its users to suppose that it referred to a real substance. It might be any "cause" of mutual repulsion among the molecules of a material. As long as one used the term consistently, one was free to envisage its effects "in an abstract and mathematical manner."[56]

From the phlogistonists' point of view, the word "caloric" was a meager replacement for phlogiston, a formal placeholder that performed none of phlogiston's life-giving functions and had none of its visceral tangibility. After one of Lavoisier's performances before Joseph II, Georges Balthazar Sage, a prominent detractor, claimed to have lingered in order to revive an as-

54. De Machy, "Suite" (January 1791), III.
55. Hutton, "Chymical Dissertation" (1792), 222, 261. See also Carra, *Dissertation* (1787), 15–16.
56. Lavoisier, *Traité* (1789), 19.

Figure 7.4. Joseph Wright of Derby's 1768 *Experiment on a Bird in the Airpump* depicts the violence of experimental chemistry in the figure of the dying bird and in the visible emotions of sensitive onlookers. Copyright Victoria and Albert Museum, London/Art Resource, NY.

phyxiated bird. He said he had been rewarded by praise from the German emperor, who told him it was much more impressive to give life than death.[57] Living nature, an ever-blossoming paradise, could not be decomposed other than violently in the eyes of sentimental-empiricist chemists. Properties and substances were mixed into a great living soup. The new chemists' linguistic and analytical abstractions were precisely as deadly as the asphyxiation of birds and mice implied (fig. 7.4).

Instead of hypothetical substances like caloric and rare and contrived substances like carbon and sulfur, an element must be something vital and common, like water. But Lavoisier and his collaborators claimed that water was made of airs. Indeed, they took the compound nature of water as given: their experimental synthesis of water rested upon the prior assumption that water was a compound whose weight was equal to the sum of the weights of its components. In the apparatus Lavoisier used to compose water, hydrogen and oxygen traveled through tubes into a glass bell, where they were burned together. Because these tubes were made of permeable leather, it was impossible to be certain of the weights of the ingredients, but, Lavoisier wrote,

57. Sage, *Opuscules* (1813), 235–36.

"since it is no less true in physics than in geometry that the whole is equal to its parts . . . we felt justified in concluding that the weight of this water was equal to that of the two airs that formed it."[58]

Another substance made elemental by daily experience was air. De Machy insisted that atmospheric air was a "*sui generis* fluid." In the same way that they assumed water was a compound, Lavoisier and his collaborators assumed air to be a mixture. They considered the atmosphere as a reservoir containing all substances that would vaporize within the range of temperatures characterizing our environment. De Machy maintained that even if substances could enter and leave the atmosphere, there must be a pure substance differing from the mingled atmosphere "as pure water . . . from composed water." He rejected the divorce of sensible quality from essential substance implicit in the new chemists' claim that all sorts of matter could travel among all states of matter. "Go find us some bodies that are essentially solid," he challenged Lavoisier, "and that you will liquify . . . [or] render gaseous at your will."[59]

The new chemists' explanations of chemical phenomena in terms of their analytical elements contradicted the most basic and obvious dictates of common experience. Outrageous examples included the notion that air dissolved in water rather than vice versa, water being made of two elemental airs; that metal was combustible, calcination being essentially the same as combustion, the combination of the metal with oxygen; and that inflammable air (hydrogen), the most combustible of substances, was the principle component of the least combustible body, water. De Machy concluded that the new chemical language was nothing but a misguided "jargon, swollen with Greek, bristling with Latin, worthy of the Theban Sphinx."[60] He and other critics condemned the nomenclators' estrangement of scientific from traditional language, along with their distancing of scientific from common experience. The two went hand in hand, as an anonymous critic pointed out. The objects of common experience had names furnished by "popular idiom," so he proposed that the elements of a proper nomenclature would be the most composed (and therefore most common), rather than the simplest bodies.

Jean-Claude de La Métherie, editor of the *Journal de Physique* and foe of the new chemistry, likewise denied there was any valid distinction between technical and vernacular vocabularies. Words must arise "tacitly" out of "usage" and could not be imposed by artificial "convention." La Métherie argued for the organic unity of a society's language in aesthetic terms. Making "euphony"

58. Lavoisier, "L'Eau n'est point une substance simple" (1783), in *Oeuvres*, 2:338–39. Holmes has discussed Lavoisier's assumption, rather than experimental confirmation, of the principle of conservation of weight, in Holmes, *Lavoisier* (1985), 269, 283.

59. De Machy, "Examen" (July 1790), 310, 257–59.

60. Ibid., 122; De Machy, "Suite" (July 1791), 48, 50.

a cardinal rule of nomenclature, he held that technical terms should adhere to the *génie* of the French tongue. He then objected, term by term, to the nomenclature's music: "I prefer *lactic salt* to *lactate*, which is *hard* and *barbaric*."[61]

Some critics had recourse to ridicule. One anonymous reviewer corrected Lavoisier's Greek, claiming he had reversed his active and passive inflections. "Oxygen," he reprimanded, properly meant "engendered by acid," not "engendering acid," and "hydrogen" similarly meant "engendered by water." Sage riffed on this theme. Claiming that "oxide," translated correctly, meant "vinegar," he construed "oxygen" as "the son of a vinegar-maker." To such frivolities, Morveau responded seriously. He cited the article "Etymology" that Turgot had written for the *Encyclopédie*, in which he favored the mixing and changing of languages. For those who adopted classical roots to create French words, Turgot had recommended studying the ancient languages "not just in [their] purity and the works of good authors, but also in [their] most corrupted forms," as a lesson in inventiveness and flexibility. Words, Turgot had said (and, as we know, Morveau agreed) had no necessary relations with the things they expressed and should therefore be changed as needed.[62]

The new chemists assumed, in keeping with Turgot's article, that words were deliberately chosen social conventions. They departed from Venel's cultural understanding of scientific language by declining to privilege custom as the natural expression of sensations and intuitive responses. In return, critics objected to what they took to be the nomenclators' cavalier attitude toward experience—both the empirical experience of the senses and the historical experience invested in tradition. Venel's exclusivism had disappeared; no one on either side of the 1780s debate over chemical language suggested that chemists' vocabulary should insulate them from outside meddling. But exclusivism had never been Venel's essential purpose. Rather, he had intended to popularize a particular view of chemistry, according to which chemical knowledge was so firmly rooted in sensation and intuition as to be essentially inarticulable. The job of a chemical language, in Venel's view, was to express rather than to articulate.

Like him, the critics of the new nomenclature took a cultural view of language. They maintained that a scientific vocabulary should reflect the tradition that produced it—not, as Venel had argued, the chemists' own tradition,

---

61. Anon., "Lettre aux auteurs du *Journal de Physique*" (1787), 421; and La Métherie, "Essai sur la nomenclature chimique" (1787), 270–72, 274–75, 281. Cf. Rousseau, *Essai* (1781), 78–79. Rousseau argued that eloquent languages, suited to persuasion, are "favorable to liberty: these languages are sonorous, poetic, harmonious."

62. Anon., "Lettre aux auteurs du *Journal de Physique*" (1787), 423; Sage, *Exposé des effets de la contagion nomenclative, et réfutation de paradoxes qui dénaturent la physique* (1810); cited in Beretta, *Enlightenment of Matter* (1993), 288; Guyton de Morveau et al., eds., *Encyclopédie méthodique* (1786–1815), 1:637–38; and Turgot, "Etymologie" (1756), in Diderot and d'Alembert, eds., *Encyclopédie*, 6:100, 98.

but the national traditions to which chemists belonged. The transition from Venel's culture of chemists to the later notion that chemists participated in national cultures is part of a general development that historians of eighteenth-century Europe have detailed extensively, the formation of a concept of national cultural identity. The question of the epistemological role of words was crucial in this larger development as well. Discussion of French culture involved competing assessments of the function of the French language. The philosophes, unsurprisingly, agreed upon the supremacy of French, but marshaled conflicting epistemological arguments in its support. Diderot and other sensibilists argued that French was the most natural of languages because French syntax followed the primitive order of perception.[63] In contrast, Condillac judged French to be the clearest rather than the most natural of languages, exactly because it departed from the order of perception to impose its own rational order upon one's experiences.[64]

The authors of the nomenclature followed Condillac in their judgment of what makes a language good and promised their predictive names would conduct chemists' ideas toward yet unperceived truths. Opponents objected on sensibilist grounds that only sensory experience could generate authentic knowledge. The linguistic reformers had placed "words . . . *but Daughters of Earth*," on the throne of "things . . . *the Sons of Heaven*." To those who conceived of scientific language in cultural terms, as the spontaneous expression of natural experience, the nomenclature's social engineering appeared flatly coercive. Lavoisier had advertised that the new language would "bring about a prompt and necessary revolution" in the teaching of chemistry by forcibly preventing teachers from straying from the theoretical and pedagogical path it defined. Henry Cavendish believed him, for what could more effectively "rivet a theory in the minds of learners than to form all the names which they are to use upon [it]?" He pronounced this feature of the new chemical language "very mischievous." Sage, in an angry letter to Lavoisier, likened the right to choose one's technical vocabulary to religious and political freedom: "Allow me, my dear colleague, to have my religion, my doctrine, my language."[65]

Believing that experience and tradition were the only legitimate sources of scientific language, Sage and others were conservatives in the matter of linguistic reform. The combined principles of meaninglessness and predictive-

---

63. Diderot, *Lettre sur les sourds et muets* (1751), 136–38.

64. Condillac, *Essai* (1746), part II, §1, chap. 12 (see esp. 174). On the debate over the relation between the French language and the order of perception, see Higonnet, "Politics of Linguistic Terrorism" (1980), 50–51; Mah, "Epistemology of the Sentence" (1994), 66–67.

65. James Keir, *The First Part of a Dictionary of Chemistry* (1789), xviii; cited in Beretta, *Enlightenment of Matter* (1993), 292. Lavoisier, "Mémoire sur la nécessité" (1787), 12; Henry Cavendish to Charles Blagden, 16 September 1787, in Cavendish, *Scientific Papers*, 2:325–26; Sage to Lavoisier, 17 May 1787, in Lavoisier, *Oeuvres de Lavoisier—Correspondance*, 7, fasc. 5:41.

ness, a vocabulary denying past meanings while claiming to shape future experience: these held sinister implications that were reinforced by Lavoisier's boasts of bringing about a "revolution," a word that was gathering new and powerful reverberations.[66] In response, La Métherie made conservatism both the first and second of his rules of linguistic reform: "1. these changes must be affected little by little"; and "2. one must ... distance oneself as little as possible from the old words." The message of the Academy of Sciences commission appointed to evaluate the new nomenclature was also essentially conservative. They wrote that it was not "in a day that one reforms, that one practically annihilates a language." In 1788 Jean-Antoine Chaptal reported to Lavoisier, "I have read everything written against the new nomenclature, I have even become acquainted with the jokes they permit themselves." He permitted himself a small linguistic joke: "They want the revolution to happen gradually."[67]

During the political revolution that was brewing when Chaptal made his quip, as during the chemical revolution, the question of whether language records past experience or directs future experience, whether it codifies knowledge or creates it, was intensely controversial, in particular in the debate surrounding the issue of public civic education: how to create an intellectually and morally fit citizenry for the new republic by means of instruction. This debate featured conflicting conceptions of the proper use of language in teaching and in natural science familiar from the controversy surrounding the new chemical nomenclature. Indeed, Lavoisier was a central participant in the discussion of public instruction, as a prominent member of the Academy of Sciences; as a close associate of the members of the Revolutionary Committee of Public Instruction, notably of the marquis de Condorcet; and as the author of a treatise on public instruction and of a proposed law governing it, delivered to the National Convention in 1793. Meanwhile his *Traité élémentaire de chimie*, presented as the textbook of a science revolutionized by language, first appeared in March 1789 and went through four editions by 1793.[68] Condorcet, in his reflections on public instruction, understandably looked to the new chemical nomenclature as an example of the pedagogical function of language. There were those, however, who saw the nomenclature as an example to be avoided at all costs.

66. On the meaning of "revolution," and the emergence of "revolutionary," in political and scientific rhetoric during the Revolutionary decade, see Langins, "Words and Institutions" (1987).

67. La Métherie, "Essai sur la nomenclature chimique" (1787), 273; Marivetz, "Lettre à M. de la Métherie" (1788), 61–63; Baumé et al., "Rapport sur la nouvelle nomenclature" (1787), 244; and Jean-Antoine Chaptal to Lavoisier, 6 February 1788, in Lavoisier, *Oeuvres de Lavoisier—Correspondance*, 7, fasc. 5:122.

68. "One cannot perfect the language without perfecting the science, nor the science without the language." Lavoisier, *Traité* (1789), 2. For information on the editions to 1793, see Douglas McKie, "Introduction," to Lavoisier, *Elements of Chemistry*, xxvi.

III. LA SCIENCE PARLIÈRE VS.
LA SCIENCE SANSCULOTISÉE

Rousseau recommends in his didactic novel *Emile* that one should "limit as much as possible the child's vocabulary. It is a great disadvantage for him to have more words than ideas." Later, he advises, "keep the child dependent on things only; you will have followed the order of nature in the progress of his education"[69] (see fig. 7.5). The Revolutionary debate over civic education turned largely upon such oppositions between things and words, nature and convention, experience and language. This debate's guiding philosopher was, however, not Rousseau but Locke.[70]

Locke's tabula rasa supported much Revolutionary programming. A typical manifesto on public instruction published during the Jacobin ascendancy recommended that educators seek "to give rise to sensations before ideas." It was "clearly demonstrated," the author announced, "that all ideas derive from sensations." Therefore, the "whole art of instruction" must be in the "linking of sensations." Joseph Lakanal made the same argument several years later in a proposal to the Council of Five Hundred for a program of instruction in the technical arts. Rejecting the "sterile" teaching methods of the Old Regime, Lakanal affirmed that "experience is the only candle that can illuminate genius." Jean-Paul Marat in the interim dismissed Old Regime academicians as "little talented in the spirit of observation, ignorant of the art of experiments." An anonymous correspondent of the National Assembly similarly condemned the Academy of Sciences not just for its "aristocracy," but, more damningly, for its detachment from empirical experience, its spirit of "Systematic ignorance." The abbé Grégoire, reporting upon the establishment of the Conservatoire des arts et métiers, promised that, in this haven of handiwork, "experience alone, speaking to the eyes, will have the right to command assent."[71]

From the early days of the Revolution, the Lockean notion of the "abuse of words" was a recurring theme in the National Assembly. Patriots and pam-

---

69. Rousseau, *Emile* (1762), 298, 311. On Rousseau's and, in particular, *Emile*'s profound effect on attitudes toward teaching and childhood throughout Europe, see Py, *Rousseau et les éducateurs* (1997). Py writes that *Emile* became "a model of familial education," emphasizing the child's "hygiene, his physical and mental development, his moral education" (572).

70. Harvey Chisick has argued that throughout the Enlightenment, particularly in France, Lockean sensationalist psychology served as "the basis of a theory of education." Chisick, *Limits of Reform* (1981), 39.

71. Manuel, *Etude de la Nature* (1793), 13, 15–16; Joseph Lakanal, "Rapport fait au Conseil des Cinq-Cents" (23 messidor an 4 [11 July 1796]), 4–5, AN, AD VIII/29, pièce 11; Marat, *Les Charlatans Modernes* (1791), 14. Anon., "A Messieurs de l'assemblée Nationale de France, ou, à son comité d'instruction publique" (24 June 1790), AN, DXXXVIII/2/XIX, pièce 1; and Henri Baptiste Grégoire, "Rapport sur l'établissement d'un Conservatoire des arts et métiers. 8 vendémiaire, an 3 [29 September 1794]," in *Convention Nationale, vendémiaire, an III* (Paris, n.d.), 8.

Figure 7.5. Emile and Jean-Jacques find their way home from the woods, and a practical use for astronomy: "Let us run quickly: astronomy is good for something." Illustration by Moreau le Jeune, engraved by N. Le Mire. Bibliothèque nationale de France.

phleteers invoked what Jacques Guilhaumou has characterized as a "language of facts," an "empiricism of words." Such a revolutionary language would eliminate, as one journalist put it, all words for "rational entities" and leave only "realities." Syntax as well as semantics had revolutionary import; linguistic activists argued that the ordering of words, as well as their individual meanings, could influence citizens' ideas and, therefore, their political judgments. Hence the emergence of a "patriot-grammarian" and a "Sans-culottes Grammar," as well as a "Sans-culottes Alphabet."[72]

Inside the debate over civic education, a distrust of specialized language accompanied the sensationist view of pedagogy. The language of the sciences and the use of language in science teaching under the Old Regime were favorite culprits of Revolutionary pedagogical reformers. An example is a pamphlet published in the year 2 of the Republic, entitled *La Science Sansculotisée: first essay on the means of facilitating the study of astronomy . . . to effect a revolution in teaching*. The author of this manifesto, a Citizen Decremps, found fault chiefly with the language of astronomy, which he deemed too far removed from common experience and practical application. A sailor, having a homely knowledge of the sky's configuration and an urgent need for astronomical results, would want these conveyed in "living languages," not Greek or Latin[73] (see fig. 7.6).

Advocates of an experience-based pedagogy generally disparaged technical vocabularies and saw language as the enemy of empiricism. A plan for a "Republican secondary school" proclaimed, for example, that the "language the *Savants* spoke was not that of nature; that of observation and experience, for which it suffices to have senses." And a report on the new Conservatoire national des arts et métiers urged, as the proper ideals for the new institution, teaching students "la science des faits" rather than "la science parlière," and "[making] them see" rather than "making them speak."[74]

The moral implications of sensationism were constantly invoked during the Revolutionary debate over public instruction and became overtly political. Previous chapters have suggested that since midcentury sensationism had been not only an epistemological doctrine of the sensory origins of ideas, but also a moral doctrine of the sensory origins of virtue. It now became a political doctrine of the source of good citizens. So the comte de Lacépède argued that political stability would grow from a national program of education founded in "facts." Another theorist of civic education, preaching to the Committee of Public Instruction, proclaimed: "It is by way of the senses that

72. See Guilhaumou, *Langue politique* (1989), 52–58, 62–63, 131–50, 70–71; and Anon., *L'Alphabet des Sans-Culottes, ou, premiers élémens de l'education républicaine* (Paris, l'an II [1793/94]).

73. Henri Decremps, *La Science Sansculotisée; premier essai sur les moyens de faciliter l'étude de l'astronomie, tant aux amateurs et aux gens de lettres, qu'aux marins de la République française, et d'opérer une REVOLUTION dans l'enseignement* (Paris, an 2 [1793–94]), 9, 12, 34–35, 20. Among his other recommendations for reintroducing vitality into the language of astronomy, Decremps banned all but round numbers from astronomical texts. On Decremps, see Darnton, *Mesmerism* (1968), 28; and Stafford, *Artful Science* (1994), 56, 95–96, 123–24.

74. Anon., "Programme du Lycée Républicain" (n.d.), 6; and Charles Jean-Marie Alquier, "Rapport sur une résolution du Conseil des Cinq-Cents" (27 nivose an 6 [17 January 1798]), 10, AN, AD VIII/29, pièce 12. (Alquier contrasts *"the science of facts"* to *"the science of chatter."*)

Figure 7.6. The cartoon criticizing Bailly, who served as mayor of Paris in the first two years of the Revolution, drew upon the commonplace that natural philosophers, especially astronomers with their mathematical methods and distant subject matter, had a dangerously abstract cast of mind: "The astronomer B. . . . Observing the stars, allows himself to fall into a well." (The reference is to Thales, who is said to have fallen into a well while observing the stars.) Bibliothèque nationale de France.

the virtues enter the heart . . . [and] vices enter by the same door." A third suggested that moralists seek the source of social ills in the physical sensations that first gave rise to harmful ideas. He decreed that "physics [should] be always the guide of morality" and demanded, "Let a course of experimental physics . . . serve as an introduction to moral education."[75]

The purported sensory origins of ideas had various effects upon the shape of Revolutionary educational programs. One problem logically indicated by sensationist pedagogy concerned the instruction of the sensory-impaired. The implications of Lockean epistemology for the mental processes of blind and deaf people had been tested over the preceding century (see chapter 2). The Revolution transmuted these philosophical exercises into matters of policy. Many correspondents of the National Convention and its Committee of Public Instruction treated the subject of schooling for blind and deaf students.[76] An exemplary petitioner for special institutions of blind and deaf education implored the Revolutionary government to consider the difficulties encountered by those without sight or hearing "if, in fact, our sensations are the only channel for our ideas." Another marshaled the Lockean axiom against those who held that deaf students in schools would suffer from loneliness for their families. Such an objection, this good associationist reasoned, falsely attributed "to deaf-mutes, in whom the minds are absolutely inert, moral sentiments . . . and this supposition is certainly inadmissible, as I have never believed in innate ideas."[77]

The question of whether a well-designed language could compensate for curtailed physical sensation was paramount in discussions of instruction for the blind and deaf, though with miscellaneous results. In one case, deafness implied a need for perfect linguistic rigor, since the "mores-or-lesses of bad definitions" would not be corrected in the daily course of a deaf student's experience. In another, a reporter upon the recent invention of sign language eagerly assured the Assembly that this new way of speaking transcended mere rigor, the application of "cold, purely conventional signs." Instead, like vocal language, signs permitted the empathetic expression of sentiment, exploiting

---

75. Lacépède, *Vues sur l'enseignement* (1790), 2:25; Nicolas Raffron, "Troisième discours" (1793), in Comité d'Instruction Publique, *Procès-verbaux*, 2:233; and Manuel, *Etude de la Nature* (1793), 15–16, 36. For an argument that Revolutionary programs of civic education, in their emphasis upon practical knowledge, crafts, moral development, and fostering sociability, were continuous with Old Regime educational reforms, see Chartier, Compère, and Julia, *L'Education en France* (1976), 293–97.

76. On blind and deaf education during the Revolution, see Weiner, "The Blind Man and the French Revolution" (1974); and *Citizen-Patient* (1994), chap. 8.

77. Jean-Baptiste Massieu, "Rapport sur l'établissemens des Aveugles-nés et sur sa Réunion à celui des Sourds-Muets" (1791), AN, AD VIII/43/C, pièce 3; and Roger Ducos, "Rapport et décret sur l'organisation des établissemens pour les sourds-muets" (n.d.), AN, AD VIII/43/C, pièce 7.

"the organs, and particularly the eyes" to convey "the most secret affections of the soul."[78]

More generally, the sensory origin of ideas was frequently invoked to justify a classical emphasis upon physical training and a preference for the technical arts. Projecting an image of the naturalist's work as an essentially aesthetic exercise, an immersion in the evidence of the senses, theorists of public instruction also tended to make natural history their chosen science. "Citizen Legislators," declared Lakanal, "if there is a science that adds to the beauty of the countryside . . . it is Botany." Often these different emphases traveled in pairs, as when Félix Vicq-d'Azyr argued before the National Convention that natural history must be regarded as "one of the most important branches of human knowledge, for its direct and continual application to the useful arts." In a similar combination of winning arguments, an advocate of establishing botanical gardens throughout France argued that "as a science, natural history has the advantage over the others, of exercising the body."[79]

Lavoisier, defending the Academy of Sciences against these trends during the Convention's period of anti-academic activism, argued for the other side, emphasizing the inadequacy of sensory experience alone in education. He affirmed the reigning pedagogical consensus that young children's bodily sensations naturally guided them to their first discoveries about the physical world, confirming their hypotheses by pleasure or correcting them by pain. But Lavoisier distinguished formal schooling from the natural learning of infancy. Scientific ideas, in particular, "neither affect our existence nor our welfare; and we are not forced by any physical necessity to correct them." Pedagogy must then compensate for the insufficiency of physical experience in formal scientific education.[80]

In forming his pedagogical philosophy, Lavoisier drew upon his own experience as a pupil. He had studied chemistry with Venel's own master,

78. "Compte rendu à la Convention Nationale de ce qui s'est passé a l'établissemens des Sourds-Muets" (n.d.), AN, AD VIII/43/C, pièce 10; and Prieur, "Rapport sur l'établissement des Sourds-Muets fait à l'Assemblée Nationale" (1791), AN, AD VIII/43/C, pièce 2.

79. Joseph Lakanal, "Discours sur l'Education nationale" (n.d.), AN, AB XIX, 333; Félix Vicq-d'Azyr, "Mémoire de la Société d'histoire naturelle," AN, F17/1005/B/932; and Boisset, "Rapport et projet de decret relatifs à l'établissement des Jardins des plantes," AN, AD VIII/43/A/15, pièce 2. Langins has characterized "Revolutionary" pedagogy, as practiced in the early Ecole centrale des travaux publics and Ecole normale, as having had "two aspects," one encyclopedic and utilitarian, the other emotional: "the practical aspect and that of principles or intuitions. . . . [The latter] finds its true sense in the realm of sentiment and of vague enthusiasms for a new future." See Langins, La République (1987), 95. See also Gillispie, "Encyclopédie" (1959), 274–79.

80. Lavoisier, "Réflexions sur l'instruction publique" (1793), in Oeuvres, 6: 516–58; Traité (1789), 3. Concerning Lavoisier's attempts to rescue the Academy of Sciences during the Jacobin period, see Hahn, Anatomy of a Scientific Institution (1971), 230–38, 245–51; and Dhombres and Dhombres, Naissance (1989), 13–22.

Guillaume-François Rouelle, pharmacist and charmingly unruly chemical demonstrator at the Jardin du Roi. Rouelle had transformed French chemistry by reintroducing Stahl's theory of phlogiston in France, along with a thoroughgoing empiricism. His students learned that chemical transformations were due to combinations of material substances rather than occult forces, and the practical art of producing "new, more perfect" combinations by means of decomposition and recomposition.[81]

Rouelle's courses were attended by Rousseau, Buffon, and Diderot, the last for three years running, and met with almost universal approval. Diderot's enthusiasm was such that he had his notes copied for distribution, and Venel closed his article "Chymie" with a plug for Rouelle's lectures. Lavoisier, however, had some serious misgivings about the master's methods, misgivings that focused on Rouelle's use of language. Many years later, in some notes on the teaching of chemistry, he recalled his confusion during Rouelle's course. "I had been accustomed," he wrote, "to that rigor of reasoning which mathematicians put into their work." In chemistry it had been "another world . . . they presented me with words that they were not at all in a position to define."[82] Like Condillac, Lavoisier likened linguistic to mathematical analysis and found rigorous language essential to good teaching.[83]

Lavoisier's treatise on public instruction, delivered to the National Convention in 1793, opened with the requisite avowal of orthodox Lockean associationism: "Man is born with senses and faculties," it began, "but he brings with him not a single idea: his brain is a blank slate." This profession of faith once accomplished, though, Lavoisier argued that the education of a nation through scientific research required not just sensations and experiences, but proper "institutions"—by which he meant the Academy of Sciences, which he was then struggling to preserve. His description of the importance of institutions to scientific progress also, however, evoked another kind of institution: the nomenclature. Research, he said, required institutions that "by their essence, by the very mechanism of their organization, [were] continually expanding the limits of our knowledge." Such institutions were intended to foster mental rather than physical industry, the "employment of the faculties of the mind." Their success relied not just upon empirical methods, but upon "independence" and "liberty"—which meant state support.[84]

81. Rouelle, *Prospectus* (1759); and Denis Diderot, "Partie du règne végétale redigée par M. Diderot," AS, Fonds Chabrol, carton 3, dossier 24.

82. Venel, "Chymie" (1753), 437; Lavoisier, "Sur la manière d'enseigner la chimie" (n.d.), AS, Fonds généraux, MS 1259; and Lavoisier, *Traité des systèmes* (1789), 1:5.

83. On the pedagogical impetus behind the new nomenclatures throughout the natural sciences, see Duris, *Linné et la France* (1993), 132.

84. Lavoisier, "Réflexions sur l'instruction publique" (1793), in *Oeuvres*, 6:516, 527–29.

This support was justified by the importance of public instruction and its institutions, often linguistic, to political and economic improvement. For example, once again following Condillac's example, Lavoisier placed political economy under the rubric of civic education, attributing France's economic difficulties to widespread agronomic ignorance. And he attributed this ignorance, in turn, to linguistic poverty. Condillac's treatise on political economy had begun with his accustomed formula: "Each science requires its particular language." With the introduction of linguistic precision, "the science develops of itself." Economic problems came from sloppy speaking, and their solutions lay in a system of public instruction in the proper employment of economic terms. Clarity of expression would bring economic health. As discussed in chapter 4, Condillac's recommendations for public economic instruction rested upon his understanding of the word "value" as denoting a conventionally determined quantity. Not grain and wine in themselves, but deliberate judgments of their importance, were the original source of a nation's wealth.[85] Like chemical nomenclature, economic words derived their meaning, according to Condillac's philosophy, neither from natural fact nor from cultural expression, but from social prescription.

Lavoisier, too, sought economic well-being in strictly enforced linguistic clarity. In a study of France's agricultural wealth presented to the National Assembly's Committee on Taxation in 1791, he argued the importance of rigorous definitions. Writers on agricultural wealth had, he claimed, mistreated their subject through an inexact use of words, making "a host of double and triple uses; they counted the same value two or three times, and arrived at false and exaggerated results." He gave the example of tallying the costs of a farm by entering the prices of straw and wheat separately. This was a mistake, for since straw was converted into manure and used in the production of wheat, its value was "implicitly mingled into that of the wheat." The same principle applied to horses' fodder and oats: their value comprised part of the value of the final product.[86]

To distinguish intermediate from final products, Lavoisier named three separate categories: the "natural agricultural product," which included products consumed in the making of other products as well as those ultimately convertible into money; the "real revenue" in money or products convertible into money; and the "net revenue," the amount claimed by proprietors and taxes once the "expenses and charges of farming" had been subtracted. Agronomic

85. Ibid., 6:529 (see also Lavoisier, "Mémoires presentés à l'Assemblé provinciale de l'Orléanais" [1788], in Oeuvres, 6:262); and Condillac, Le Commerce (1776), 247, 444–45, 249, 254–59.
86. Lavoisier, "Résultats extraits d'un ouvrage intitulé de la richesse territoriale du royaume de France" (1791), in Oeuvres 6:406–7.

reform, like the advancement of chemical knowledge, rested upon the inculcation of a new terminology, and so Lavoisier made enlightened agriculture a "part of public instruction." It was primarily mental industry, not "working with one's arms in the material use of force," that would increase the nation's wealth. Government sponsorship should free *savants* to perform their "laborious meditations" and educate the public.[87] Unfortunately, compensation for mental industry smelled to members of the National Convention like a restoration of corporate privileges, and Lavoisier fought a losing battle. So did another leading advocate of a program of civic education founded in linguistic reform, upon whose proposals Lavoisier had modeled his own plan, Condorcet.

The regard was mutual. Condorcet implicitly invoked Lavoisier's chemical nomenclature when he argued that both research and education, philosophy and pedagogy, should be based upon two principal sources of philosophical progress. One was the "art of uniting a great number of objects under a systematic disposition." The other was the "institution of a universal language" for each science. Echoing Condillac's and Lavoisier's equation of linguistic with algebraic analysis, Condorcet said the language would be "learned with the science itself, like algebra."[88]

In the weeks leading up to the purge of the Girondins in June 1793, Condorcet made a last attempt to rescue the Revolution from the "false politics" he saw overwhelming it. As his biographer, Keith Baker, has emphasized, Condorcet's instrument was language. He enlisted linguistic reform as a social tool. His collaborators were the abbé Sièyes and Jules-Michel Duhamel, who had previously acted upon his belief in the constructive power of language by affiliating himself with the Institution for the Deaf-Mute. Together, Condorcet, Sièyes, and Duhamel created the *Journal d'instruction sociale*, which, they hoped, would teach a citizenry to think not through sensations and emotions, but through a carefully designed vocabulary. The journal's prospectus claimed that common language and its failings presented a primary obstacle to moral and political progress.[89]

Condorcet identified a cause of errors in the moral sciences that Condillac had already indicated in the natural sciences and that the new chemists had highlighted in their nomenclature. It was a linguistic problem, the use of

---

87. Ibid., 6:406–7; and Lavoisier, "Réflexions sur l'instruction publique" (1793), in *Oeuvres*, 6:527–29. Jean-Pierre Poirier identifies a similar attitude at work in Lavoisier's activities as National Treasury Commissioner in 1791: "He wanted to invent a new . . . method of financial nomenclature similar to the one he had devised for chemistry. For example, the word *comptabilité*, accounting, had been used for all paying operations. Thereafter, one would use the word 'verification' for the control of actual payments and the word 'accounting' for the control of their validity." See Poirier, *Lavoisier* (1993), 283.

88. Condorcet, *Esquisse* (1793), 289–93.

89. Condorcet et al., *Journal d'instruction sociale, Prospectus* (1793), in *Oeuvres*, 3:605–13; see 606.

words that had, "in the vulgar language, different meanings than their philosophical senses." Condorcet charged public instruction in the moral sciences with the task of transforming these words for public usage, supplying people with "the rigor and precision of their philosophical senses." Instruction must collapse the linguistic distance between "the man and the *philosophe*," because, Condorcet believed, justice depended "uniquely upon precision of ideas" and, therefore, of language.[90]

Condorcet, Sièyes, and Duhamel's ideal of social education rested on a recognition of the shortcomings of dogmatic empiricism. Sièyes denied that the science of society should be based upon an empirical study of the historical record. Nor, according to Condorcet, should it be founded in natural facts. He called public instruction an "eternal battle . . . between nature and genius, between man and things." Education should counteract nature. The purpose of social institutions, Condorcet judged, was to "diminish . . . natural inequality." The innate superiority of some would not engender dependence in others if all were made conversant in the essential conventions and institutions governing social life: the three R's and the elements of law. The first year of Condorcet's proposed educational program was thus taken up with reading and writing and focused heavily upon the acquisition of vocabulary. Students should be taught, Condorcet said, that language is the guide of sensory experience, and that books would train them to "see better." He also recommended the use of technical terms in teaching children. Scientific language was preferable to vulgar language, he said, because its "convention" was "less tacit."[91]

Condorcet was criticized for his departure from the principle that all ideas originate in sensations. A reviewer of his memoirs on public instruction derided him for stating that men were born with virtuous "habits" and an innate moral sense. How, the reviewer asked, could "a being who has never acted have habits, when its memory is a blank slate?"[92] Within the Committee of Public Instruction, too, Condorcet defended an increasingly unpopular position. The division of sentiment was encapsulated in a distinction drawn by Jean-Paul Rabaut, a speaker before the National Convention who so impressed the

90. Condorcet, "Second mémoire" (1791), in *Ecrits*, 1:94–95.

91. Condorcet, "Cinquième mémoire" (1791), in *Ecrits*, 1:237; Condorcet, "Premiere mémoire" (1791), in ibid., 1:35–37; and Condorcet, "Seconde mémoire" (1791), in ibid., 1:85–87, 93. "Sièyes had frequently insisted that to base the social art on an appeal to history was to acquiesce in the tyranny of fact. . . . Society, for Sièyes, was an artificial construct, an edifice; and the science of society was therefore to be truly an *architecture sociale*." Baker, *Condorcet* (1975), 354. On Condorcet's "intellectual and philosophical conception of instruction," see Kintzler, *Condorcet* (1984), 213–20.

92. Anon., "Analyse des réflèxions sur l'instruction publique du M. de Condorcet," in *Tribut de la société des neuf soeurs*, 1792, 4:426 (the reviewer referred to Condorcet's section entitled "L'instruction morale doit avoir pour but de fortifier les habitudes vertueuses," in "Troisième mémoire" [1791], in *Ecrits*, 1:156).

Figure 7.7. "A Revolutionary Committee under the Terror," anonymous engraving after Aléxandre-Evariste Fragonard. Bibliothèque nationale de France.

members of the Committee of Public Instruction that they invited him to join their ranks. (For a sense of the scene, see fig. 7.7.) Rabaut distinguished between instruction and education: while instruction sought to enlighten the mind with "books, instruments, calculations, methods," education was a matter of cultivating the "body and heart" by means of "circuses, gymnasia, weapons, public games." He concluded that a state should educate its citizens but never instruct them.[93]

Ultimately, Condorcet lost the Revolutionary struggle over public instruction. The decisive moment came in December 1792, eight months after he had presented his pedagogical program to the Legislative Assembly on behalf of its Committee of Public Instruction. In the interim, a new committee on instruction had been formed, following the replacement of the Legislative Assembly by the National Convention and the declaration of the Republic. Pierre-Toussaint Durand de Maillane, a member of the new committee, criticized Condorcet's plan before the National Convention. Durand de Maillane appealed to what had by then become the dominant popular ideal of public in-

93. Jean-Paul Rabaut Saint-Etienne, "Projet d'Education Nationale" (21 December 1792), in Condorcet, *Ecrits*, 2:232–33. On the distinction between education and instruction, or pedagogy, see Py, "Introduction" to *Rousseau et les éducateurs* (1997).

Figure 7.8. "The Arrest of Lavoisier," engraving by Jean Duplessi-Bertaux.
Bibliothèque nationale de France.

struction. "To be good citizens," he claimed, "it is necessary to have less science and more virtue; to talk less, to write less and to act better." Just as the Academy of Sciences had rejected the new chemical nomenclature, and for reasons that were strikingly similar, the Committee of Public Instruction set aside Condorcet's plan.[94] (Neither Condorcet nor Lavoisier would survive the Revolution. They died within a few weeks of one another in the spring of 1794, at the height of the Terror, Condorcet in prison and Lavoisier, despite his powerful friends and colleagues in the Jacobin government, arrested as a tax farmer [fig. 7.8] and guillotined.)[95]

The question of the epistemological function of language brought natural and moral science together in a conflict that penetrated to the very core of the project of "Enlightenment." In arguing about language, chemists pitted cultural and social conceptions of the practice of natural science against one another. Philosophes and Revolutionaries expanded upon these rival visions of the relations of words to natural things to support competing conceptions of political revolution—as the invention of new social institutions or as the generation of a modern political culture.

A common pattern emerges in both the chemical debate and the quarrel over public instruction. In both the scientific and the political controversy, those who distrusted systematic linguistic conventions, and emphasized the

94. Durand de Maillane, "Opinion sur les écoles primaires" (12 December 1792), in Condorcet, *Ecrits*, 2:223. On the hostility engendered by Condorcet's "elitist" program of instruction, see Hahn, *Anatomy of a Scientific Institution* (1971), 207–15, 228–30; and Palmer, *Improvement of Humanity* (1985), 129–39. For the story of the struggle and its outcome, see Baker, *Condorcet* (1975), 318–39; and Badinter and Badinter, *Condorcet* (1988), 445–50, 610–11. For an analysis of Condorcet's growing isolation in the National Convention, see Barthélemy, *Les Savants* (1988), 163–85. For an argument that Robespierre rejected Condorcet's and Lavoisier's theories of public instruction out of a belief that specialized languages bred linguistic inequality, see Anderson, "Morale de la Terreur" (1995).

95. On Lavoisier's death and the question of why his influential friends did not save him, see Poirier, *Antoine-Laurent Lavoisier* (1992), 382–85.

importance of raw sensory experience, joined an epistemological theory with an evolving conception of culture. They sought a natural basis for culture in sensibility and in a philosophical language directly expressive of sensation and emotion. These sentimental-empiricist theorists of language wanted to rediscover a world unsullied by human systems of thinking and governing, and to grow their philosophy organically from it.[96] They sought therefore to rid philosophy of system and to rid civic education of instruction.

In contrast, those who emphasized the importance of language in the formation of ideas promoted a social rather than a cultural epistemology. In their view, the language of a science or a society was no organic product, received from nature as an accretion of spontaneous expressions of physical sensations and answering emotions. Languages, scientific and common, were deliberately molded things. These instrumental empiricists sought to understand a world that they believed was inevitably shaped by their own efforts, whether to understand, teach, or govern. They made no pretense; they wanted not only to understand the world, but to change it and to transform the thinking of its other inhabitants. They wanted to effect this transformation by shaping not experience or emotion, but understanding. As pedagogues, they believed that human nature could be much improved by social intervention. And as philosophers, they believed they could actively invent rather than merely passively receive truths about nature.

It would be a mistake to treat this divergence as a struggle between tradition and progress. The interlocking chemical and pedagogical disputes about language resist the standard divisions between pre-modern and modern science, or tradition and Enlightenment, or revolution and counterrevolution. Instead, the competing cultural and social conceptions of language express conflicting views of how to carry out the projects of modern science, enlightened philosophy, and political reform.

Both Venel's and Morveau's philosophies of language expressed central Enlightenment preoccupations: cultural diversity and empiricist openness, on the one hand, social universalism and epistemological rigor, on the other. Similarly, in the pedagogical conflict, Condorcet and his opponents were united in their rejection of traditional methods, and their call for modern — even republican — ones.

---

96. Mona Ozouf has called attention to the word "regeneration" as it "looms up in the tide of brochures, lampoons, pamphlets ... that accompany the meeting of the Estates General" and displaces the word "reform." "Regeneration," she writes, primarily connoted physical renewal, to some, and spiritual rebirth to others. Ozouf, "Régénération," in Furet and Ozouf, eds., *Dictionnaire critique* (1992), 373–89, 373–75. Lucien Jaume has described the central theme of Jacobin political philosophy as "the necessity of ... reinstituting a lost *nature*. It is no longer a question here of natural right. ... The 'nature' now evoked is that of the heart, of the sentiments." See Jaume, *Le Discours Jacobin* (1989), 246–47.

The quarrel over language use in teaching and scientific research persisted precisely because its two sides presented no easy contrast between modernity and antiquity, progress and tradition, revolution and counterrevolution. Both sides professed an earnest wish to move beyond the mistakes and constraints of the Old Regime into a remodeled modern world. Beneath their divergent theories of how to carry out the remodeling, they shared a common project. And so they remained locked in dispute.

The Idéologues, practitioners of a new science of ideas that emerged during the Revolution and dominated academic moral and political sciences for almost a decade, would seek the widespread application of Condillac's philosophy of language. They would invoke Lavoisier's chemistry as evidence of their linguistic program's validity.[97] Discussion of civic education would return to the problem of the relative roles of sensations and words in shaping pupils' intellects and moral faculties. The nomenclators' chemical language would gain an ever-wider acceptance, and their conception of chemical research as the elaboration of a table of names, an ever-mounting success, in the first decades of the next century. Even so, their assumption that a science's language does not just express its practitioners' experiences, but actively shapes its progress, would remain controversial. Georges Cuvier, in his Napoleonic retrospective of French science, could yet declare that "to intend [the nomenclature] as an instrument of discovery" was "ridiculous."[98]

As it has emerged over the course of this book, sentimental and instrumental empiricists, both by mutual opposition and by the very proximity of their views, helped to develop and to propel one another's programs. In chapter 2, for example, sentimental-empiricist assessments of the blind as essentially

97. On the role of the new chemical nomenclature, and the method of philosophical and chemical analysis associated with it, as models for the Idéologues' pedagogical reforms, see Gusdorf, *La Conscience Révolutionnaire* (1978), 369–83.

98. Cuvier, *Histoire des progrès des sciences naturelles* (1826), 81. On the application of Condillac's philosophy of language by the post-Thermidorian Idéologues, see Albury, "The Order of Ideas" (1986), 203–25. On the importance of linguistic and sensationist epistemologies in post-Revolutionary Idéologues' educational programs, and specifically on the influence of the new chemical nomenclature evident in these programs, see Hordé, "Les Idéologues" (1977), 55–58. See also Anderson, "Scientific Nomenclature and Revolutionary Rhetoric" (1989). Sensationist and linguistic educational theories reemerged with particular force in discussions of child psychology and lay moral education at the close of the nineteenth century. For a discussion of the emergence of a theory of lay morality in education, relating this development to earlier Enlightenment epistemological and pedagogical theories, see Stock-Morton, *Moral Education for a Secular Society* (1988). Stock-Morton describes the curricular innovations of the 1880s called "*leçons de choses*, also called *enseignement intuitif*, in which children learned through their senses." She cites as another example Emile Durkheim's 1911 article on pedagogy, in which he emphasized the importance of "training the senses" (157). I am greatly indebted to Katharine Norris for alerting me to the recurrence of these themes in debates of the late nineteenth and early twentieth centuries, and especially for allowing me to use her work-in-progress on the subject.

solipsistic, turned on their head by instrumentalist reformers, yielded a program to socialize blind children. The story told in this chapter is the reverse: instrumentalists' invention of a new scientific language provoked sentimentalist reformers to ban systematic linguistic reform and technical vocabularies from their program of civic education. New educational programs represented some of the most direct political applications of Enlightenment thinking. If one can generalize from Haüy's school for the blind and the debates of the Committee of Public Instruction, such political applications of the Enlightenment were shaped by a crucial division at its heart. To be sure, all these reformers called themselves empiricists and professed to be remaking the political world according to the dictates of nature. The division between instrumentalists and sentimentalists was a clash among allies, and all the more powerful for it.

IV. CODA: THE SENTIMENTAL-EMPIRICIST HISTORIOGRAPHY OF THE TERROR

The clash among allies also, however, generated something new after 1789, and with gathering speed after 1793: a particular, critical account of the Revolution. This development began when the nomenclature's opponents developed their moral indictment of the new chemical language by associating it with Revolutionary radicalism and, after Thermidor, with terrorism. These associations provided critics of the Revolution with a ready-made weapon. The sentimental-empiricist indictment of the new chemistry, an indictment that had shaped important Revolutionary programs, notably the plans for a new system of civic education, now began ironically to inform repudiations of the Revolution.

Those who sought links between the chemical and political revolutions were helped along by the prominence of systematic linguistic reform as a feature of the political revolution, from the Babel of Greek prefixes for the new decimal system of weights and measures (fig. 7.2); to the ten bucolic month names of the Revolutionary calendar; to the standardization of the French language; to the renaming of all variety of objects, among which Sage cited pears, hitherto known as *bon chrétiens*, which became *bon citoyens*, while others called *cuisses madames* became *cuisses citoyennes*.[99] In 1793 the geologist Jean André Deluc wrote to a friend:

---

99. I.e., "good christians" became "good citizens," "lady's thighs" became "(female) citizen's thighs." See Sage, *Opuscules* (1815), 78. Golinski calls attention to the post-Thermidor tendency to compare "the chemical nomenclature with such authoritarian acts of the French revolutionary regime as changing the names of institutions and measurements." Golinski, *Science as Public Culture* (1992), 150. McEvoy presents the example of John Robison, who "linked Lavoisier's chemistry, the metric system, and the Revolutionary calendar" as different instances of the Revolutionary "obliteration of the past." McEvoy, "Priestley Responds" (1995), 134. For a discussion of the popular reception of the metric system and Revolutionary calendar, see Heilbron, "Measure of En-

The Néologues set up as despotic legislators, advocate their doctrine by words, and crush those who dare disagree. You had it right in calling them a "Jacobin Chemical Club." . . . Also have you seen these Dictators pass quickly from the little chemical domination to the goal of dominating the whole society by means of words; for all these first Néologues soon became Jacobins. . . . This spirit of domination of the French began, as I say, with chemistry, and it is there that we must demolish their edifice . . . in unveiling their arrogant ignorance on one point, we will stop [their] prestige.[100]

Three years later Joseph Priestley addressed the surviving members of the alleged chemical cabal: "As you would not, I am persuaded, have your reign resemble to that of *Robespierre*, few as we are who remain disaffected, we hope you would rather gain us by persuasion, than silence us by power."[101] He meant that the new language had been an instrument of philosophical tyranny, serving to silence rather than to persuade.

Long afterward Sage would continue to look back upon the chemical and political revolutions and link them with one another by what he took to be their common attitude toward language. One diatribe mentioned a play entitled *Neologomania* in which a main character, "Oxiphile," rallied his co-conspirators thus: "Messieurs! you know that we aim to have a new physico-chemical nomenclature adopted! Messieurs, seconded by you, we will be able to force public opinion." Into old age, after the fall of Napoléon (and after the emperor himself had privately confessed to Sage that he was a "pneumatist," an adherent of new chemical doctrine), Sage stuck to his guns. "The revolution was fecund in neologists," he would write, and many of them had been vengeful chemists. According to Sage's history of the Revolution, Morveau, still stinging from the Academy's rejection of his nomenclature, had avenged himself as a member of the Committee of Public Safety in appropriately neological style: he had eliminated the Academy by renaming it the "Institute."[102]

Historians of chemistry have continued to associate the chemical and political revolutions, but in different alignments. In a 1959 essay entitled "The *Encyclopédie* and the Jacobin Philosophy of Science: A Study in Ideas and Con-

---

lightenment" (1990). On the political and economic interests motivating metric reform, see Alder, "A Revolution to Measure" (1995). On the project to render the French language uniform, see Lartichaux, "Linguistic Politics" (1977); Higonnet, "Politics of Linguistic Terrorism" (1980); and Certaux, Julia, and Revel, *Une Politique de Langue* (1975).

100. J. A. Deluc to G. C. Lichtenberg, 18 November 1793, in Lichtenberg, *Briefwechsel*, ed. Ulrich Joost, vol. 4 (Munich: C. H. Beck, 1992), no. 2317. I am grateful to John Heilbron for this citation.

101. Priestley, *Considerations* (1796), 17.

102. Sage, *Opuscules* (1815), 79; Sage, "Notes particulières" (n.d.), AS, Dossier Sage; Sage, *Opuscules* (1813), vii; and Sage, *Opuscules* (1815), 74–78 (on Morveau's revenge).

sequences," Gillispie draws exactly the opposite connection from Sage's, Deluc's and Priestley's. All three had attributed Revolutionary radicalism to what they saw as the abstract, dogmatic spirit of the analytical new chemistry. Gillispie, on the contrary, links Jacobinism not with Morveau's and Lavoisier's new nomenclature, but with Venel and the sentimental-empiricist chemistry of the *Encyclopédie*. He argued that the Jacobins' "resentment for the new chemistry," their broader attack on academic science, and even, by implication, their violent radicalization of the Revolution were the outcomes of a volatile combination that had emerged around midcentury in *Encyclopédie* articles such as Venel's: an antimodern, romantic natural philosophy coupled with a politics more populist than liberal. In Venel's view of chemistry and its language, Gillispie discerns "the authentic voice of the *sans-culotte*."[103]

Recently, Golinski and John McEvoy have again redrawn the connection between the scientific and the political debates. McEvoy presents two competing conceptions of community: the French model, a "private" and "specialized" meritocracy, reflected in the nomenclature; and the British model, the open "commonwealth." He suggests that Priestley's appeal to "common usage" as the sole legitimate basis for a chemical language was an attempt to apply the commonwealth model to the community of chemists. Similarly, Golinski distinguishes French and English critics of the nomenclature by their divergent political motivations. He sees English objections to Lavoisier's abandonment of traditional usage as largely expressing a communitarian approach to language, and the objections of French critics as expressing their conservatism.[104]

What can be made of these divergent and sometimes contradictory inter-

103. Gillispie, "*Encyclopédie* " (1959), 257–58, 265, 270. Elsewhere, Gillispie has suggested that science is the prerequisite of a liberal polity because "though created out of personality and in culture, it is not then bound by personality or culture." A society that respects the truth of a scientific fact thus has the modicum of independence from tradition, superstition, and authority necessary to govern itself as a democracy. See Gillispie, "The Liberating Influence of Science in History" (1977), 41. Accordingly, an attempt such as that Gillispie associates with Denis Diderot in the 1959 article, to reinvest natural science with moral and emotional meaning, is necessarily illiberal. These arguments are refined and expanded in Gillispie, *Science and Polity at the End of the Old Regime* (1980). Gillispie's proposal generated much controversy, driven by the difficulty of demonstrating links between particular scientific programs and political parties. For responses to Gillispie's 1959 essay see, e.g., Williams, "The Politics of Science in the French Revolution" (1959); and Hill, "On the Papers of Charles Coulston Gillispie and L. Pearce Williams" (1959), both in the same volume as Gillispie's article; and, more recently, Baker, "Science and Politics at the End of the Old Regime," in *Inventing the French Revolution* (1990), 153–66. On the difficulties of demonstrating links between programs and parties, see below, the end of §2 and §3.

104. McEvoy, "Priestley Responds" (1995), 131, 133–35; and Golinski, "Chemical Revolution" (1992). See also Golinski, *Science as Public Culture* (1992), 149–50. In yet a further twist, although Priestley himself likened the French new chemists to Robespierre, Edmund Burke, in *Reflections on the Revolution in France* (1790), lumped Priestley together with the French new chemists and warned that their chemistry sowed the seeds of republican violence. See Crosland, "Image of Science" (1987); and Kramnick, "Eighteenth-Century Science" (1992).

pretations? One thing is clear: Among both the French and the English critics of the nomenclature were people with very different political affiliations. Detractors of the nomenclature held views across the spectrum, prominently including a Mason, a Jacobin, and a constitutional monarchist, as well as both Sage and Marat, on the French side, and Edmund Burke as well as Joseph Priestley among the British.[105] This diversity may account for the remarkably different ways in which contemporaries and historians have drawn the connection between the scientific and the political argument. While Sage condemned the new chemical language as Jacobin despotism, Marat cursed it for academic smoke and mirrors. Of Lavoisier, Marat wrote scathingly, "He changes systems as he changes his shoes. . . . [H]e changed the term *acide* to *oxygine*, the term *phlogistique* to *azot*, the term *marin* to *muriatique*, the term *nitreux* to *nitrique* and *nitraque*. *Voilà* his claims to immortality."[106] The argument about the nature and function of scientific language had moral and political resonance from the start—but the political implications of this debate were as ambiguous as they were apparent.

The variety in political motivations and larger political affiliations among the nomenclature's critics makes their considerable common ground even more significant. They agreed upon both the epistemological principle that experience always precedes language and the moral injunction against denying this precedence. If, as McEvoy and Golinski suggest, some of these critics were meritocrats and others democrats, some communitarians and others traditionalists, these opposing political categories shared a common logic, the logic of sentimental empiricism. Critics unanimously rejected the nomenclators' instrumental attitude toward experience, tradition, and language. The distinctions I have been proposing, between sentimental and instrumental empiricism, and between cultural and social conceptions of scientific language, may be useful here. They run orthogonally to broad political conflicts between Enlightenment and counter-Enlightenment, or Revolutionary and counter-Revolutionary, and so turn up a quieter but no less effective set of differences, which go to the heart of how contemporaries understood the programs of Enlightenment and Revolution.

105. De Machy was a Mason; Jean Darcet, a Jacobin; and La Métherie, a constitutional monarchist. It is notoriously hard to correlate even political attitudes with well-defined groups of political actors during the Revolution. Patrick Brasart has emphasized the difficulty of associating clear political positions with parties in the National Convention, in *Paroles de la Révolution* (1988), 134–37.

106. Marat, *Les Charlatans Modernes* (1791), 36–37. I have italicized the chemical terms to indicate that they are untranslated. For a discussion of Marat's animosity toward both Lavoisier and Condorcet, see Barthélemy, *Les Savants* (1988), 168–76, 192–93. Gillispie has invoked Marat's attacks on Lavoisier as evidence of a Jacobin philosophy of science so far removed from the analytical, systematic one that Burke alleged as to closely resemble Burke's *own* philosophy, in what Gillispie describes as its hostility to abstract systems. See Gillispie, *"Encyclopédie"* (1959), 255–90.

Precisely because these differences penetrated to the core of Enlightenment and Revolutionary thinking, they offered critics of the Enlightenment and the Revolution, especially retrospectively, a powerful story about what had gone wrong. These critics had at their disposal an expedient villain, one that had been widely and intensively discussed, a particular style of natural science epitomized by the new analytical chemistry.[107] The interwoven controversies over the new chemical nomenclature and the role of language in civic education had made technical languages emblematic of a particular, misguided approach to both science and government: abstract, analytical, instrumentalist, systematic.

Consider, for example, a treatise on the causes of political fanaticism published three years after the fall of Robespierre by a prominent counter-Revolutionary. Antoine Rivarol blamed the Terror not on Robespierre, but on the reductive spirit of modern philosophy, and included a sketch of the history of French chemistry, which he took to epitomize the dangerous, analytical tendency. "It is the spirit of analysis which dominates," he lamented. "Its disciples have everywhere employed solvents and decomposition," seeking nothing less than the "reconstruction of all by the revolt against all." Rivarol explained that while nature "links, assembles, composes always," analysis divides, separates, and decomposes; while nature creates, analysis destroys. Analytical thinkers would "dissect living men to better understand them."[108]

"Absolute equality among men" and the atomic reduction of natural substances to simple, homogeneous parts were, Rivarol maintained, equally "absurd" ideas, and equally pernicious. Both sprang from a misplaced, geometrical manner of thinking (see fig. 7.9). Geometers might reason about the abstract, simple world of forms, but politicians and chemists must keep their eyes fixed upon the irreducibly various worlds of "qualities, affinities, and essences." These were lost to chemical analysis, which would only "destroy endlessly and fruitlessly." One who analyzed, "whether as a chemist or as a reasoner," could never understand but only "suspend" and "kill."[109] In the first year of Napoléon's reign, another critic of the Revolution blamed the events on the spirit of analytical chemistry. Joseph Bernardi wrote that "chemists accustomed to analyzing, that is, ceaselessly destroying," would have been satisfied with nothing less than Armageddon: "The dissolution of the entire globe

107. For a discussion of philosophical conspiracy theories of the Revolution, see Doyle, *Origins* (1989), 24.

108. Rivarol, *Discours* (1797), 189–90.

109. Ibid., 155, 189–90. Lavoisier and his co-authors did not, in fact, put forth an atomic theory of matter. The elements in their system were instrumentally defined as the smallest units attainable in a laboratory. But Rivarol's objections to the analytical spirit of modern chemistry certainly referred to their program as well as to atomic reduction.

would be, for them, nothing but a great chemical experiment, and a means of knowing the *caput mortuum* of all things."[110]

In response, Pierre-Louis Roederer, a follower of the Physiocrats and member of the successor to the Academy of Sciences, the new National Institute, defended analytical natural science. Roederer granted that "the spirit of analysis" had now become widespread but maintained that in natural science, analysis was inseparable from synthesis, like decomposition from composition in nature: "Half the world perishes as the other half grows," he wrote, "and analysis consists not only of decomposing, but also of recomposing." Creation followed close upon the heels of destruction: "The chemist analyzes natural substances only to learn to compose remedies. . . . Politicians, moralists, what do they decompose? . . . Ideas and words, there! and for what? To compose judgments."[111] Roederer's view of things, quieter and less popular, was all but drowned out, while Bernardi's and Rivarol's quickly caught on.

Rivarol found an admirer in the English parliamentarian and political philosopher Edmund Burke. Having read Rivarol's political journal of the first year of the Revolution, Burke pronounced it "brilliant" and reported that there was "a great resemblance in our manners of thinking." The journal may even have inspired Burke to write his own influential reflections on the Revolution.[112] Like Rivarol, Burke used the language of sentimental-empiricist natural philosophers to blame the Revolution on natural science. He identified a dangerous insensibility in the analytical abstractions of "the geometricians, and the chemists." Their dry diagrams and sooty experiments bred "dispositions that make them worse than indifferent about those feelings and habitudes which are the support of the moral world." Contemporary chemistry provided an illustration of the violence inherent in this analytical imperviousness: "These philosophers consider men in their experiments no more than they do mice in an air pump." They were "fanatics" who would "sacrifice the whole human race to the slightest of their experiments."[113]

At the same time, Burke also drew his model for wise administration from contemporary natural science. Again, the model was informed by sensibilist

---

110. Bernardi, *De l'influence* (1800), 230. See also Robison, *Proofs* (1797), 328–39, 367–69, 391–92, 397–99.

111. Roederer, *De la Philosophie* (1799–1800), 4–5. Like Roederer, Rivarol also allowed that if chemists "decompose[d] bodies only to recompose them," they might expand the "repertoire of physics" beyond "putrefaction." He praised French chemistry for having recognized the importance of recompositions paired with decompositions. See Rivarol, *Discours* (1797), 157.

112. For the possibility that Rivarol's journal inspired Burke to write his *Reflections*, see Godechot, *Counter-Revolution* (1981), 33. On Burke's view of the association between Revolutionary chemistry and Revolutionary politics, see Crosland, "Image of Science" (1987).

113. Burke, *Lettre . . . sur les affaires de France* (1791); Burke, *Reflections* (1790), 211; and Burke, "Letter to a Noble Lord" (1795), 520–21.

Figure 7.9. Not only geometry, but mathematics in general was suspect. The series of anonymous anti-Revolutionary engravings entitled *The Patriotic Calculator* drew a graphic connection between a mathematical cast of mind and Revolutionary violence. Bibliothèque nationale de France.

doctrine. Nature was "plastic," Burke maintained: though it could not be reduced, it could be shaped and manipulated. Philosophers who recognized nature's plasticity did not dally in analytical abstractions but instead sought "solid" principles to exploit, qualities and properties of matter. Burke recommended the same approach to government. Equality was an analytical abstraction, while "inheritance" was a solid, political principle. To eliminate inheritance, for example by abolishing corporate bodies, was "tantamount, in the moral world, to the destruction of the apparently active properties of bodies in the material." It would be like trying to exterminate from nature "the expansive force of fixed air in nitre, or the power of steam, or of electricity, or of magnetism."[114]

Thanks largely to Joseph Priestley, Burke could show a close knowledge of contemporary chemistry. He had followed the progress of chemical research on gases, particularly Priestley's work, since the early 1770s. In 1782 Priestley wrote to Burke describing his latest experiments on airs and thanking Burke for his interest in the project. As late as 1789, Burke successfully advocated Priestley's attempts to dedicate the latest edition of *Experiments and Observations on Different Kinds of Air* to the king. So Burke was acquainted with recent exchanges between British and French chemists in which the question of analytical reduction and its evils figured prominently.[115]

Ironically, it was Priestley who became the chief target of Burke's later abomination of chemists—Priestley, who himself, as we have seen, railed against the spirit of system and opposed Lavoisier and his collaborators' new chemistry as system-building. His diatribes sounded much like Burke's own. Priestley warned that systematic natural science was the source of all error, prejudice, usurped authority, and corrupt establishments. Nevertheless, as a Dissenting minister, a liberal political theorist, and a supporter of the French Revolution as well as a chemist, Priestley soon captured Burke's political imagination. In his agitated state, Burke took Priestley to personify the union of radical politics and analytical science. Priestley and his fellow radicals "despise[d] experience." These "metaphysicians descend[ed] from their airy speculations" only to eat and to pay for their destructive habits. Their theories constituted an underground mine that would "blow up at one grand explosion all examples of antiquity, all precedents, charters, and acts of parliament."[116]

Early hints of Burke's disapproval of abstract, analytical approaches to science and government flavored his speeches on the Stamp Act and American taxation during the 1760s and '70s. But such disapproval became central to his political philosophy during the 1789–90 session of Parliament, when Burke

114. Burke, *Reflections* (1790), 282.

115. For Priestley's report to Burke, see Priestley to Burke, 11 December 1782, in Priestley, *A Scientific Autobiography of Joseph Priestley*, 216; For Burke's royal advocacy of Priestley's book, see Crosland, "Image of Science" (1987), 282, 304.

116. Burke, *Reflections* (1790), 268, 370. For Priestley's view of systems, see Priestley, *Experiments and Observations* (1777), 3:xvii; 1:xiv. For a discussion of Priestley's religious and political views, see Schofield, *Enlightenment of Joseph Priestley* (1998); Gibbs, *Joseph Priestley* (1965); and for their connections to his scientific work, Kramnick, "Eighteenth-Century Science" (1992).

drove himself into political isolation by his unremitting insistence that "the vice of the current age is owing to an abstract way of thinking." He was called to order eight times during his speech on the Quebec Government Bill for deflecting the discussion to the subject of his monomania, reminding his weary listeners that systematic abstractions and universals were dangerous because "circumstances are infinite, are infinitely combined, are variable and transient." [117]

Drawing his images from chemistry, Burke compared the enunciation of a single abstract principle in political philosophy, such as the "spirit of liberty," to the explosive release of carbon dioxide from combination: "The wild *gas*, the fixed air is plainly broke loose." A principle was safe only in combination; Burke would suspend his "congratulations on the new liberty of France, until [he] knew how it had been combined with government, with public force," with disciplined armies, effective tax collection, and all the other elements of a compound social body. Society like nature lived by synthesis, so legislators must strive for "excellence of composition" over "excellence of simplicity." In their simplistic abstraction, universal rights flattened the social world while the chemistry of gunpowder leveled buildings to a homogeneous rubble. "As the [geometricians] have set an eye on his Grace's lands," Burke wrote in his "Letter to a Noble Lord" (1795), "the chemists are not the less taken with his buildings." He was referring to Morveau's crash program to extract saltpeter from the excrescences on stone walls. Burke perceived in this program a plan to reduce churches and noble houses into magazines of gunpowder. They would be "equalized" into a "common rubbish" and then transformed into "democratic . . . nitre." [118]

Through such equalizations, chemistry made manifest the unity of the general will and became its chief expression: "There is nothing, on which the leaders of the republic, one and indivisible, value themselves, more than on the chemical operations, by which, through science, they convert the pride of the aristocracy to an instrument of its own destruction." While his buildings fed the chemistry of gunpowder, the Lord himself would provide practitioners of "experimental philosophy" the matter for "an extensive analysis, in all the branches of their science." Burke meant this literally. Analytical chemistry and Revolutionary politics were actual, not metaphorical, equivalents. The coldly abstract manner of thought shared by chemists and constitutionalists rendered them equally and identically dangerous. "While the Morveaux

---

117. Burke, "Speech on the Report of the Army" (8 May 1789), *Parl. Hist. of Engl.*, 28: col. 1031; "Debates in the Commons on the Quebec Government Bill" (21 April 1791), *Parl. Hist. of Engl.*, 29: col. 1394. On Burke's growing suspicion of philosophical systems, of natural philosophy, and especially of chemistry, see Crosland, "Image of Science" (1987); Beretta, "Chemists in the Storm" (1993); and Kramnick, "Eighteenth-Century Science" (1992).

118. Burke, *Reflections* (1790), 90, 148, 282; and Burke, "Letter to a Noble Lord" (1795), 533.

and the Priestleys are proceeding with [their] experiments" on his house, the aristocrat himself would fall victim to "the Sièyes, and the rest of the analytical legislators, and constitution-vendors, [who] are quite as busy in their trade of decomposing organizations."[119]

The arguments of opponents of the analytical new chemistry and its systematic language thus informed influential, contemporary repudiations of the Revolution, notably Rivarol's and Burke's. These in turn spawned a historiographical tradition that has continued to evolve over the course of two and a half centuries. Sentimental empiricism thereby generated one of the oldest, hardiest, and most powerful critical accounts of the French Revolution.

We can follow the sentimental-empiricist view of the Revolution into the nineteenth century. It is plainly visible in Alexis de Tocqueville's assessment that a reductive enthusiasm for the "simple" and "uniform" had engendered in "men of letters" an absolutist insensitivity to the nuances of public administration. When these system-builders came suddenly into political authority, to fill the power vacuum after Louis XIV's despotic reign, according to Tocqueville, their abstract thinking and "scorn for facts" had had catastrophic results that had culminated in the Terror.[120] A generation after Tocqueville published his account of the Revolution, Burke's was picked up and developed by Hippolyte Taine, who wrote that "scientific acquisitions" were the "first element" in the "composition of the Revolutionary Spirit." The other element was the method of analysis, which Taine also called the "mathematical method." Together, science and mathematical analysis formed a "mischievous" and "venomous compound" and resulted in "destructive explosions." The problem with the spirit of analysis, Taine explained, lay in its unwillingness "to embrace the plenitude and complexity of actualities":

> In the eighteenth century, the portrayal of living realities, an actual individual, just as he is in nature and in history . . . an undefined unit, a rich plexus, a complete organism of peculiarities and traits, superposed, commingled and co-ordinated, is improper. . . . Whatever can be discarded is cast aside. . . . [N]othing is left at last but a condensed extract, an evaporated residuum, an almost empty name, in short, a hollow abstraction.[121]

---

119. Burke, "Letter to a Noble Lord" (1795), 520–23, 533.

120. Tocqueville, *Old Regime* (1858), bk. III, chap. I, 229, 239–40.

121. Taine, *Ancien régime* (1856), xiii, 201, 182–83, 170, 191, 197. Daniel Mornet drew attention to the irony in Taine's arguments. Taine believed, Mornet wrote, that the "theory of *milieu*" with which he countered the "abstract systems and theoretical rationalism" of the philosophes had been, before him, but a "dispersed and fleeting idea." Mornet objected that, on the contrary, "it was precisely in the eighteenth century that [this theory of milieu] was formed. . . . [It was] a common idea, long discussed and perfected." Around the middle of the eighteenth century, this theory

Taine's contemporary the sociologist Augustin Cochin similarly charged Enlightenment thinkers with a tendency toward philosophical abstraction, which he said constituted "a destruction. It consist[ed] in sum of eliminating, reducing." Cochin identified a natural philosophical revolution prior and equivalent in violence to the political one: "Before the bloody Terror of 1793, there was, from 1765 to 1780, in the republic of letters, a dry terror, of which the *Encyclopédie* was the Committee of Public Safety, and d'Alembert the Robespierre." Cochin was so sensitive to the dangers of philosophical reduction that he charged even Taine with exhibiting an analytical tendency, sounding much like Taine himself in his criticism of Taine's account of Jacobinism:

> A Jacobin is a man the way a plant is a chemical compound. One can do the psychology of the one as the chemical analysis of the other. But . . . [i]t is not enough to know that a plant dissolves into such proportions of oxygen, hydrogen, carbon and azote—or that the Jacobin is a compound of abstract virtue and practical ambition. The union of these elements, their synthesis . . . is not within the purview of chemistry or psychology.[122]

According to Cochin, Taine's psychological analysis of Jacobinism was insensitive to the living reality in just the same way that chemical analysis failed to capture the essence of a living plant.

There is a hiatus in the early twentieth century, when a Marxist historiography of the Revolution prevailed, according to which social conditions rather than philosophical tendencies were the causes of crisis.[123] But post-Marxist revisionist historians of the 1960s, '70s, and '80s, from Alfred Cobban to François Furet, revived the sentimental-empiricist historiography of the Revolution. Furet modeled his account on Tocqueville's and Cochin's. In his reading of Rousseau's political philosophy, Furet associates analytical thinking with

---

produced plenty of "discussion and dissertations in which it often seems to be Taine himself doing the reasoning." Mornet, *Pensée* (1947), 70–71.

122. Cochin, *L'Esprit de Jacobinisme* (1920), 41, 35, 132–33.

123. Daniel Mornet questioned the relative importance of ideas among the various causes of political transformations in his *Les Origines intellectuelles de la Révolution française* (1933). An emphasis upon economic over philosophical causes of political change was central to the Marxist historiography of the Revolution, of which Georges Lefebvres's *Quatre-Vingt-Neuf* (Paris, 1939) was the seminal work. At the same time, Pierre Trahard argued that historians, shifting their attention from the intellectual origins of the Revolution to its economic and social causes, had skipped over the role of sensibility. Trahard's account was the complement of Taine's and Cochin's, in the sense that Trahard viewed the Terror not as the result of intellectualism carried to an extreme, but as the consequence of sensibility gone awry: "Violence itself became the bitter form of sensibility, its dramatic expression." Trahard, *Sensibilité révolutionnaire* (1936), 251; see also chap. 1. In 1978 François Furet's *Penser la Révolution française* marked the beginning of a renewed focus upon the role of ideas.

absolutism: "It is no coincidence that the philosophe who felt and theorized ... the autonomy of the self, would be the same who conceived this abstract figure of a totally unified social world." According to Furet's interpretation, Rousseau's "atomization" of the social whole into uniformly sovereign individuals carried on its flip side an absolutist conception of their union. In order for each individual to remain sovereign while also obeying the general will, there must be an "absolute" symmetry between each individual will and the general will. This "game of mirrors" precluded any intermediary alienation of sovereignty in the form, for example, of political representation. While Furet was careful to emphasize that "Rousseau [was] not in the least 'responsible' for the French Revolution," he did find that Rousseau "unknowingly constructed the cultural materials of the Revolutionary conscience and practice": the concept of the general will, which allowed for no middle term between anarchy and despotism.[124]

Thus, more than two centuries after the Revolution, the sentimental-empiricist thesis that the Terror originated in a form of insensibility—in analytical, abstract thinking—remained vigorous in new formulations. Revisionists such as Cobban and Furet have called for histories of the Revolution that would subject the actors' own categories to critical scrutiny. These revisionists argued that representing the Revolution as the natural outcome of economic and social oppression amounted to re-rendering uncritically actors' own accounts.[125] Yet the recent revisionist view has had its own origins in actors' accounts. In turning from a primarily social to a primarily intellectual and political understanding of the Revolution, historians returned to an older historiographical tradition initiated by Rivarol, Burke, and their contemporaries, associating political absolutism with reductive science. The injunction to scrutinize actors' categories and their transmission is thus as important to the newer intellectual and cultural historiography as it was for the earlier social historiography.

The problem of actors' categories is in fact crucial to this chapter's—and this book's—argument. I have been suggesting that actors' categories such as "system-builder" contributed crucially to the view, passed down to posterity,

124. Furet, *Penser* (1978), 50–51. Cf. Starobinski, *Jean-Jacques Rousseau* (1971), 256–57. Starobinski chose transparent rather than reflective glass, crystal rather than mirrors, as the organizing metaphor of Rousseau's political philosophy. While Furet attributed the absolutism of social unity in Rousseau's ideal polity to his "atomization" of the social world, Starobinski emphasized instead Rousseau's rejection of atomic reduction. Rousseau, according to Starobinski, invoked the "immediate union" of fluids, whose absolute oneness was measured by their transparency, as a better model than "pulverized crystal," the inferiority of whose inner unity was revealed by its opacity.

125. Cobban, *Social Interpretation* (1964), 8; Furet, *Penser* (1978), 45, 89; and for a discussion of the history of the historiography of the Revolution, see Baker, *Inventing the French Revolution* (1990), 2.

that Enlightenment science and Revolutionary politics were driven by an excessively rationalist tendency. Enlightenment thinkers' and Revolutionary actors' extreme sensitivity to the possibility of rationalist excess led them to write their own epitaph. They incessantly charged themselves and one another with exhibiting a dangerously abstract, analytical manner of thinking, the dreaded "spirit of system." Rather than taking these charges as straightforward descriptions of reality, I believe we should understand them as expressions of a fundamental ambivalence at the heart of the Enlightenment and the Revolution, and therefore as evidence not only of their thoroughgoing rationalism, but also of an equally important countering tendency that I have been calling sentimentalism.

How, then, should intellectual and cultural historians treat actors' categories such as "system-builder"? If we cannot simply adopt them, we cannot dismiss them, either, since they are our raw materials. Actors' categories are both descriptions of a period and elements of it, and we must treat them in both ways simultaneously. Treating them as descriptions, we may accept what they tell us. But treating them as elements of the period they describe, we discover that their own position in that period undermines some of what they say. For example, according to actors' accounts, dogmatic system-building was a central tendency of the Enlightenment and the Revolution. But these same actors' accounts, by their very existence, indicate that an antipathy for rational systems was equally important. My own interpretive categories—sentimental and instrumental empiricism—are derived both from and against actors' categories. I have accepted that "system-builder" corresponded to an approach to science and government, exhibited by the authors of the new nomenclature and by Revolutionary actors such as Condorcet. On the other hand, "system-builder" was also rhetoric, an epithet deployed by people who advocated a different approach. We cannot understand either approach to science and government, or the meaning of the actors' category "system-builder," without taking into account the difference of opinion that created the term. Thus, the terms I propose, sentimental and instrumental empiricism, identify both the so-called system-builders and those who labeled them as such, by means of the disagreement that produced the label.

The new chemical nomenclature was not the result of a tendency toward "system-building" on the part of Enlightenment natural philosophers. Rather, "system-builder" was the sentimental empiricists' response to the nomenclators' instrumental brand of empiricism, in which they treated science as a socially collaborative project, the construction of knowledge from experience by means of social conventions such as technical languages. Sentimental empiricists, advocating a different kind of engagement with nature—a more spontaneous and emotionally intimate connection—characterized the new chemists' approach as overly abstract. Ken Alder has pointed out that the

Revolutionary project to standardize weights and measures, one of the most often-cited examples of Revolutionary rationalism, was fundamentally motivated not by a proclivity for pristine abstractions, but by the most concrete of political and economic interests.[126] This is not to say that Enlightenment natural philosophers did not like to produce rational systems of knowledge, or that Revolutionary reformers did not profess to seek rational foundations for their new civic institutions. It is only to say that our understanding of the context in which they did so, and of the implications of their having done so, has been shaped by a historiography that took its cue from the polemical cries of sentimental empiricists doing battle against instrumental empiricists. We seem to be still engaged in a centuries-old dispute.

126. Alder, "A Revolution to Measure" (1995).

Figure 8.1. The Jacobin calendar is often taken as the epitome of the Revolution's scientific rationalism run amok. Its sleeping Harvester, representing Messidor (June–July), shows the inseparability of rationalism and sentimentalism, empiricism and sensualism, in their Enlightenment pas de deux. © Photothèque des musées de la ville de Paris, negative: Briant.

Chapter Eight

CONCLUSION: THE LEGACY OF THE SENTIMENTAL EMPIRICISTS

Modern science is a self-consciously dispassionate endeavor. Its reputation for sobriety rests largely upon its empirical methods of experiment and observation. Yet these very methods were cooked in the crucible of Enlightenment emotionalism. Sensibility transformed the practice of the sciences as well as the arts and literature. The natural sciences (physics, chemistry, physiology, and medicine) and moral sciences (political economy, civic education, law, and administration) were infused with this distinctively eighteenth-century notion that married physical sensation with emotion and moral sentiment. The sciences, the Enlightenment, and the modern French intellect share a reputation for rationalism. Yet French Enlightenment science was deeply sentimental, and its sentimentalism has had a legacy.

The claims of sentimental empiricists have shaped our understanding of the Enlightenment, the Revolution, French culture, and the history and nature of modern science. In each case, sentimental empiricists have persuaded us that a chilly, arid rationalism dominated and continues to dominate. "Our century," d'Alembert wrote in 1751, "seems to want to introduce cold and didactic discussions into things of sentiment,"[1] and posterity has believed him. Nineteenth- and twentieth-century students of eighteenth-century sensibilists have decried the "emotional dryness and hard individualism" of the Enlightenment.[2] They have seen the Revolution as the triumph of an abstract, analytical, and mathematical cast of mind.[3] Modern heirs to sentimental empiricism have lamented a French lack of "soul": "The Frenchman naturally believes ... in the superiority of intellectualism and of reason. ... He rises up

---

1. D'Alembert, "Discours préliminaire" (1751), in Diderot and d'Alembert, eds., *Encyclopédie* (1751–80), I:xxxi.
2. Wilson, "Sensibility in France in the Eighteenth Century" (1931), 41.
3. Tocqueville, *Ancien régime* (1856); Taine, *Ancient Regime* (1876); Cochin, *L'Esprit de Jacobinisme* (1920); Furet, *Penser* (1978).

against bad works that are born of sentiment, but gladly tolerates the excesses of intellectualism."[4] And the charges of sentimental empiricists against the sciences of their day must surely have contributed to our belief that the history of modern science has been the straightforward unfolding of an impersonal, dispassionate way of knowing the world.

This is not to say that a deliberate rationalism was not a defining characteristic of the Enlightenment and Revolution, or that it has not shaped modern French culture or guided the genesis of the modern sciences. But this aspect of their history developed in tension with another aspect, a persistent appeal to an emotional engagement with nature as crucial to empirical understanding. Enlightenment science included rationalist system-building and instrumental empiricism. But these very tendencies took shape by their dialectical engagement with another element, also acting within the sciences, sentimentalism. (For an image that captures the engagement of rationalist system-building and sentimentalism in the French Enlightenment and Revolution, see fig. 8.1.)

The same questions that caused such productive strife more than two hundred years ago still retain their volatility. During the so-called science wars of the 1990s, scientists and humanists argued about the legitimacy of the modern sciences' claims to epistemological authority. Over the previous three decades, philosophers, historians, anthropologists, and cultural and literary critics had assailed a picture of science that they understood as a product of the Enlightenment. According to this picture, science was universalist, value-neutral, and the product of sober empirical observation and rational theory construction. The attack on this picture had several fronts. Critics denied the claim of universality by pointing to rival, incompatible traditions of scientific inquiry with, they maintained, equal claims to validity. They argued that modern science was the expression of a particular culture, that of the post-Enlightenment West. Other cultures, and marginalized groups within Western culture, had produced other ways of knowing the world, which included emotional and intuitive modes of understanding. The critics also denied value-neutrality by arguing that the sciences had been shaped by social interests. Scientists responded by assuming what they took to be the mantle of the Enlightenment and defending the same picture of science that the critics had attacked. These self-styled defenders of the Enlightenment fought to save what they understood as its legacy, namely, scientific rationalism, against what they saw as a neo-romanticism that threatened not only science, but the Enlightenment political ideals of freedom and equality.[5]

It is a striking feature of the science wars that, despite their rancor, neither

4. Trahard, Les Maîtres de la sensibilité française (1931–33), 4:289–90.

5. For accounts of the science wars from humanists' perspectives, see Ross, ed., Science Wars (1996); Sardar, Thomas Kuhn and the Science Wars (2000); and Segerstråle, ed., Beyond the Science Wars

side questioned the view of Enlightenment science that the other put forward. That is, all the disputants treated Enlightenment science as strictly rationalist, instrumentalist, and universalist. But, as I hope this book has shown, the view that sciences are cultural expressions, that intuition and emotion are crucial to understanding, and that rationalism, in its claims to absolute truth, is dogmatic, arrogant, and oppressive, all played crucial roles in Enlightenment science itself. Thus both sides of the science wars can trace their roots to the Enlightenment and, more specifically, to a profound ambivalence among practitioners of the sciences of the eighteenth century about their own methods. Indeed, I would say that if one could name a unifying feature of the Enlightenment as a whole, it would not be rationalism, but instead a pervasive ambivalence about rationalism, created by the very project of self-conscious inquiry into reason's nature and limits. This ambivalence, not a complacent commitment to either rationalism or sentimentalism, defined Enlightenment science and continues to define its legacy.

Here, for the fun of it, is another example of the persistence of the struggles this book has described, this one drawn from French rather than American pre-occupations of the late second millennium. Nineteen eighty-nine, the year that effectively closed the twentieth century, as well as the bicentenary of the Revolution that ended the eighteenth, included among its lower-profile uprisings an attempted coup led by ten French schoolteachers. Fighting to depose the *accent circonflexe* that had oppressed schoolchildren for four centuries, the teachers were backed by the Socialist leadership of their largest union and most of its membership. Their movement ultimately aimed higher than the abolition of the circumflex, to a broad reform of irregular and nonphonetic spellings.[6]

Orthographic rationalists, invoking the spirit of 1789, charged that spelling was being taught "like a religion" and urged that the long-deferred death of superstition at last be delivered.[7] As in the Revolution-era linguistic controversies, modern debate focused upon the function of language in two related projects: pedagogy and scientific and technical progress. Modern linguistic "*sans-culottes*"[8] argued that illogical rules of spelling and grammar hindered both learning and science. A rationalized French would "provide a true men-

---

(2000). For the other side, see Gross and Levitt, *Higher Superstition* (1994) and Gross, Levitt, and Lewis, eds., *Flight from Science and Reason* (1996).

6. Philippe Bernard, "Les Maux de l'orthographe," *Le Monde* (31 December 1990); A Correspondent in France, "Fonetic French," *Economist* (2 September 1989). For a detailed account of the proposed reforms, see "Les Propositions d''ajustement' de l'orthographe . . . ," *Le Monde* (20 June 1990); "La Nouvelle orthographe le velum ruissele," *Le Monde* (28 June 1990).

7. Nina Catach, *Les Joies de l'orthographe* (Paris, 1989); quoted in Juliet Towhidi, "French Grave Over Acute Language Crisis," Reuters (31 August 1989).

8. Steven Greenhouse, "A Morphological Compromise in France's War of Words," *New York Times* (27 January 1991).

tal discipline" and prepare students for the exigencies of the information age: the supremely rational rules governing communication with computers.[9] Furthermore, rational reforms would democratize French by rendering the universal language of justice more accessible to foreigners, immigrants, and especially francophone Africans.[10]

But the circumflex had powerful defenders who denounced the schoolteachers' proposal as "totalitarian."[11] Deploying a battery of romantic images, a united front of conservative politicians and members of the Académie française defeated the rationalists. Jean d'Ormesson of the Académie compared the reforms to "operating on a living body."[12] Spelling, mused an opposition politician, "is a little like food and love; it's the complications that make it beautiful."[13] In the end the traditionalists triumphed, and French orthography retained its seductive irrationality.

In 1994 it was the government's turn to take the initiative for linguistic reform. The National Assembly passed the Loi Toubon,[14] which imposed fines of up to twenty thousand francs for the use of foreign (read: English) terms in academic conferences, work contracts, advertising, and government documents.[15] To supply indigenous alternatives, terminology commissions in the government ministries collaborated on a *Dictionnaire des Termes Officiels*, from the graphic "*remue-meninges*" (brainstorming),[16] to the unwieldy "*le marge brut d'autofinancement*" (cash flow).[17] Research scientists were primary targets and opponents of the Loi Toubon.[18] While advocates of the law saw international

---

9. Nina Catach, *L'Orthographe en débat* (Paris: Editions Nathan, 1991), 9.

10. A Correspondent, "Fonetic French" (1989).

11. François Bayrou; quoted in Greenhouse, "A Morphological Compromise," §4, 6.

12. Jean d'Ormesson; cited in Towhidi, "French Grave" (1989).

13. Jean-Louis Borloo; quoted in Towhidi, "French Grave" (1989). A lyrical tone was typical of anti-spelling reform writing. Other examples: "L'accent, c'est . . . le detail artistique et le stigmate de l'Histoire. Le supprimer, c'est rendre les mots plus triste, et la tristesse ne convient pas à la pédagogie." [The accent is . . . the artistic detail, the stigmata of History. To remove it is to make words sadder, and sadness is not conducive to pedagogy.] Gilbert Caillat, "Si l'on avait demandé aux profs leur avis," *Le Monde* (5 September 1990); "L'Accent circonflexe peut faire aussi rever." [The circumflex too can inspire dreams.] Jacqueline Bouchet, "L'Imagination décoiffée," *Le Monde* (5 September 1990).

14. The law was named for the Gaullist Minister of Culture, Jacques Toubon, who had proposed it the previous February.

15. Jean Pierre Peroncel Hugoz, "Le Projet de loi sur l'emploi du français en France," *Le Monde* (20 January 1994).

16. David Buchan, "The French Choose Their Words Carefully," *Financial Times* (6 August 1994), xviii; Sharon Waxman, "The French Tell English Users: Just Say 'Non,'" *Chicago Tribune* (14 June 1994), 1; Christopher Burns, "New Dictionary Lays Down the Law, but Will It Stick?" Associated Press, International News (26 April 1994).

17. Buchan, "The French Choose" (1994).

18. Frédéric Bobin, "L'Assemblée nationale adopte le projet de M. Toubon privilègient les publications scientifiques en français," *Le Monde* (6 May 1994).

technical vocabularies as a threat to the cultural *patrimoine*, academic scientists countered that the practice of science is internationally collaborative, its content culturally neutral, and its language, therefore, properly and necessarily universal.[19] Moreover, the commitment to a universal language was entirely in the French tradition, witness the word "*oxygène*," coined two centuries earlier, which, like "*tuiteur*" (tweeter, another entry in the official government dictionary), combined a technical meaning, an ideological origin, an idiosyncratic derivation, and an esoteric sound.

The ubiquitous appeals to the Revolution during the language debates of 1989 and 1994 tended to obscure their lineage. The self-consciousness with which French intellectuals and politicians invoke Revolutionary lessons in the service of modern programs can belie the genuine persistence of earlier debates within modern ones. Noisy charges of superstition and totalitarianism, and celebrations of liberty and fraternity, seem to imply that the continuing relevance of the Revolution is so self-evident as hardly to require analysis. But the powerful catchwords of the appeals to the Revolution drown out the quieter historical echoes in, for example, d'Ormesson's declaration that a language is an organic, evolving entity, or in the conflict between opponents of the Loi Toubon, who treated language as an instrument of scientific and social collaboration, and its proponents, who saw language as the expression of a national culture.

Sentimental empiricism was powerful in 1789, and it has remained powerful. It has provided the medium for an ongoing interaction between natural and moral science, and between scientific ideas and social concerns. The ideals of sensibility engaged the physiology of the senses with the problem of moral solipsism; electrical physics with the economy of the grain trade; chemical nomenclature with civic educational policy; the authority of an observed fact with the authority of a town magistracy; scientific with political revolution. The very existence of the sentimental empiricists gives the lie to the story they tell. If we look for a moment not just at what they said, but at them, and at their striking ability to persuade, it becomes evident that sentimentalism shaped Enlightenment science, guided Revolutionary policy, and has been a persistent force in French cultural politics ever since. Ironically, by being so convincing in their jeremiads, sentimental empiricists erased their own traces and wrote themselves out of history. The project of this book has been to write them back in.

---

19. Paul Germain, chief secretary of the Academy of Sciences, warned that the law would mean the banishment of international conferences from French soil. Waxman, "Just Say 'Non'" (1994); see also Buchan, "The French Choose" (1994).

BIBLIOGRAPHY

I. ARCHIVAL MATERIALS

Académie des Sciences, Paris (AS)
————. Fonds Lavoisier.
————. Fonds Chabrol (materials concerning Lavoisier).
————. Dossiers (of members).
————. Registre des procès-verbaux.
————. Dossiers des séances.
————. Fonds généraux.
Archives Nationales, Paris (AN)
————. Series AD VIII, Instruction publique, lettres, sciences et arts.
————. Series DXXXVIII, Comité d'instruction publique.
————. Series F17, Instruction publique.
————. Series H, Administration locales. 187, Commerce des grains en Bourgogne.
————. Series F11, Subsistances. 265, Commerce des grains.
————. Series AB XIX, Archives privées. 327 (Lavoisier); 333 (Lakanal).
————. Series F7, Police générale. 4770, 4774 (documents concerning the arrest of Lavoisier).
————. Series X1, Parlement civil. B 8966-67, Déliberations du Parlement de Paris, 1775–76.
————. Series F12, Lettres du controleur général. 151–52 (Turgot's correspondence).
Bibliothèque de l'Institut de France, Paris (BIF)
————. MSS 2222, Notes de l'abbé de la Roche.
————. MSS 852, 860–61, Papiers de Condorcet.
Archives Générales de Pas-de-Calais, Arras (AGPC)
————. Collection Barbier (Coll. Bar.) (papers of Antoine-Joseph Buissart concerning the lightning rod affair of Saint-Omer).
Benjamin Franklin Papers Archives, Yale University, New Haven (FPA) (especially French correspondence, 1776–90).
John M. Olin Library, Cornell University, Ithaca
————. Lavoisier Collection.

II. PRIMARY WORKS

Académie des sciences. *Mémoires de l'Académie des Sciences* (MAS).

Alembert, Jean Lerond d'. "Aveugle." 1751. In *Encyclopédie*, ed. Diderot and d'Alembert, 1:870–73.

———. *Essai sur les éléments de philosophie, ou sur les principes des connaissances humaines.* 1759. Paris: Fayard, 1986.

———. "Méthode générale pour déterminer les orbites et les mouvemens de toutes les planètes, en ayant égard à leur action mutuelle." *Mémoires de l'Académie des Sciences* (1745): 365.

———. *Oeuvres complètes de d'Alembert.* Edited by A. Belin. 5 vols. Paris, 1821–22.

———. *Oeuvres et correspondances inédites.* Edited by Charles Henry. Geneva: Slatkine, 1967.

———. *Réflexions sur la théorie de la résistance des fluides; Lues dans l'Assemblée publique de l'Académie des Sciences du 13 novembre 1751.* Paris, 1751.

Anonymous. "Affaire du para-tonnerre de Saint-Omer." *Causes Célèbres* 99 (1782): 1–110.

Anonymous. "Lettre aux auteurs du *Journal de Physique*, sur la nouvelle nomenclature chimique." *Journal de Physique* 31 (1787): 418–32.

Anonymous. "Programme du Lycée Républicain." Paris, n.d.

Anonymous. *Réflexions impartiales sur le magnétisme animal.* Geneva, 1784.

Anonymous. "Suite de l'affaire du para-tonnerre de Saint-Omer." *Causes Célèbres* 117 (1784): 145–88.

Anti-Berkeley. "Errors in the New Theory of Vision Supposed to have been Wrote by Dr. Berkeley, the Present Bishop of Cloyne: First Publish'd in 1709, and Afterwards with the Minute Philosopher in 1732." *Gentleman's Magazine* 22 (1752): 11–13.

Aréjula, Juan Manuel. "Réflexions sur la nouvelle nomenclature chimique pour servir à la Traduction espagnole de cette nomenclature." *Observations et mémoires sur la physique* 33 (1788): 262–86.

Aristotle. *The Basic Works of Aristotle.* Edited by Richard McKeon. New York: Random House, 1941.

———. *De Anima.* In *The Basic Works*, 533–603.

———. *On Generation and Corruption.* In *The Basic Works*, 467–531.

Bachaumont, Louis Petit de. *Mémoires secrets pour servir à l'histoire de la république de lettres en France.* 9 vols. London, 1780–89.

Bailly, Jean-Sylvain. "Rapport secret sur le Mesmérisme, ou Magnétisme animal." 11 August 1784. In D. I. Duveen and H. S. Klickstein, "Documentation." *Annals of Science* 13, no. 1 (March 1957): 42–46.

Bailly, Jean-Sylvain, et al. *Exposé des expériences qui ont été faites pour l'examen du magnétisme animal.* Paris, 1784.

Baker, Henry. *An Attempt Towards a Natural History of the Polype.* London: R. Dodsley, 1743.

Barbier de Tinan, *Mémoires sur les conducteurs pour préserver les édifices de la foudre, par Mr. l'abbé Joseph Toaldo . . . traduits de l'italien avec des Notes et des Additions par M. Barbier de Tinan, de l'Académie des Sciences, Arts et Belles-Lettres de Dijon.* Strasbourg, 1779.

Baumé, Antoine, and Antoine Alexis Cadet de Vaux, Jean Darcet, and Georges Balthazar Sage. "Rapport sur la nouvelle nomenclature." 13 June 1787. In "A propos de *Méthode*," ed. Bensaude-Vincent, 238–52.

Baxter, Andrew. *An Enquiry into the Nature of the Human Soul.* 2nd ed. London: A. Millar, 1737.

Beccaria, Cesare. *On Crimes and Punishments.* 1764. Translated and edited by David Young. Indianapolis: Hackett, 1986.

Beccaria, Giambattista. *Dell'elettricismo artificiale e naturale libri due.* Turin, 1753.

Beer, Georg Josef. *Das Auge, oder Versuch das edelste Geschenk der Schöpfung vor dem höchstverderblichen Einfluß unseres Zeitalters zu sichern.* Wien: In der Camesinaschen Buchhandlung, 1813.

Bergasse, Nicolas. *Considérations sur le magnétisme animal, ou sur la théorie du monde et des êtres organisés.* The Hague, 1784.

———. *Observations de M. Bergasse sur un écrit du docteur Mesmer.* London, 1785.

———. *Receuil des pièces les plus intéressantes sur le magnétisme animal.* Paris, 1784.

Bergman, Torbern. *Opuscules chymiques et physiques.* 2 vols. Translated by Louis-Bernard Guyton de Morveau. Dijon: L. N. Frantin, 1780.

———. *Physical and Chemical Essays . . . to Which Are Added Notes by the Translator.* Translated by Edmund Cullen. London: J. Murray, 1784.

Berkeley, George. *An Essay Towards a New Theory of Vision.* 1709. In *The Works of George Berkeley, Bishop of Cloyne,* edited by A. A. Luce and T. E. Jessop, 1:159–239. London: Thomas Nelson and Sons, 1948.

Bernardi, Joseph. *De l'influence de la philosophie sur les forfaits de la Revolution.* Paris, 1800.

Berthollet, Claude-Louis. *An Essay on Chemical Statics.* Paris, 1803.

Bertholon, Pierre. *Nouvelles preuves de l'efficacité des para-tonnerres.* Montpellier: J. Martel aîné, 1783.

Besenval, Pierre Victor de. *Mémoires du Baron de Besenval sur la cour de France. Introduction et notes de Ghislain de Diesbach.* 1805–7. Paris: Mercure de France, 1987.

Beyer, M. *Aux amateurs de physique sur l'utilité des paratonnerres.* Paris: Institution des sourds-muets, 1809.

Bichat, Xavier. 1801. *Anatomie générale.* 2nd ed. 2 vols. Paris, 1812.

Boerhaave, Hermann. *Praelectiones de morbis nervorum.* Leiden, 1730–35.

Boncerf, Pierre-François (under the pseudonym Francaleu). *Les Inconvénients des droits féodaux.* London: Valade, 1776.

Bonnefoy, J. B. *Analyse raisonnée des rapports des commissaires chargés par le Roi de l'examen du magnétisme animal.* Lyon, 1784.

Bonnet, Charles. *Considérations sur les corps organisés.* 1762. Amsterdam: M. M. Rey, 1768.

———. *Contemplation de la nature.* 1764. In *Oeuvres d'histoire naturelle,* vol. 4.

———. *Essai analytique sur les facultés de l'âme.* Copenhagen, 1760.

———. *Essai de psychologie.* 1754. In *Oeuvres d'histoire naturelle,* vol. 8.

———. *La Palingenésie philosophique.* Geneve: Philibert et Chirol, 1769.

———. *Oeuvres d'histoire naturelle et de philosophie de Charles Bonnet.* 8 vols. Neuchâtel: S. Fauche, 1779–83.

Boullier, D. R. *Essai philosophique sur l'âme des bêtes.* 1728. Amsterdam, 1737.

Brisseau, Michel. *Traité de la cataracte et du glaucoma.* Paris: D'Houry, 1709.

Brisson, Mathurin-Jacques. *Dictionnaire raisonné de physique.* Paris, an viii [1799–1800].

Buffon, Georges. *De l'homme.* 1749. Edited by Michèle Duchet. Paris: François Maspero, 1971.

———. *De la manière d'étudier et de traiter l'histoire naturelle.* 1749. Paris: Société des Amis de la Bibliothèque Nationale, 1954.

———. *Histoire naturelle, générale et particulière.* 3 vols. La Haye, 1750. [First published in Paris in 1749.]

———. *Oeuvres philosophiques de Buffon.* Edited by Jean Piveteau. Paris, 1954.

———. "Réflexions sur la loi de l'attraction." *Mémoires de l'Académie des Sciences* (1745): 493.

Buissart, Antoine-Joseph. *Mémoire signifié pour Me. Charles-Dominique Vyssery de Bois-Valé, Avocat en Parlement, demeurant en Ville de Saint-Omer, Défendeur et Appellant. Contre le Petit-Bailly de la même Ville, Partie publique, Demandeur et intimé.* Arras, 1782.

Burke, Edmund. *Lettre de M. Burke sur les affaires de France et des Pays-Bas, adressée a M. le vicomte de Rivarol.* Paris, 1791.

———. "Letter to a Noble Lord." 1795. In *Selected Writings by Edmund Burke,* ed. W. J. Bate. New York: Modern Library, 1960.

———. *Reflections on the Revolution in France.* 1790. Edited with an introduction by Conor Cruise O'Brien. London: Penguin, 1968.

Cabanis, Pierre-Jean-Georges. *Rapports du physique et du moral de l'homme.* Paris: Caille et Ravier, 1815.

Carra, Jean Louis. *Dissertation élémentaire sur la nature de la lumière, de la chaleur, du feu et de l'électricité.* Paris, 1787.

Cavendish, Henry. *The Scientific Papers of the Honourable Henry Cavendish, F.R.S.* 2 vols. Cambridge: Cambridge University Press, 1921.

Cerfvol, Chevalier de. *L'Aveugle qui refuse de voir.* London, 1771.

Cheselden, William. "An Account of some Observations made by a young Gentleman, who was born blind, or lost his Sight so early, that he had no Remembrance of every having seen, and was couch'd between 13 and 14 Years of Age." *Philosophical Transactions* 35, no. 402 (London Royal Society, April–June 1728): 447–50.

Clairaut, Alexis. "Du système du monde dans les principes de la gravitation universelle." *Mémoires de l'Académie des Sciences* (1745): 329.

———. "Lettre de M. Clairaut aux Messieurs les Auteurs du Journal des Scavans." *Journal des Scavans* 33 (1758): 291–324.

———. "Recherches sur differens points importans du Système du Monde, par M. Dalembert." *Journal des Scavans* 28 (1757): 31–55.

———. "Réponse au nouveau mémoire de M. de Buffon." *Mémoires de l'Académie des Sciences* (1745): 583.

Clément, Pierre. *Les Cinq années littéraires.* Berlin, 1755.

Comité d'Instruction Publique. *Procès-verbaux du Comité d'instruction publique de l'Assemblée législative.* Edited by James Guillaume. Paris, 1889.

———. *Procès-verbaux du Comité d'instruction publique de la Conventional nationale.* 7 vols. Edited by James Guillaume. Paris, 1891–1959.

Condillac, Etienne Bonnot de. *Le Commerce et le gouvernement considérés relativement l'un à l'autre.* 1776. In *Mélanges,* ed. Daire, 1:247–448.

———. *Essai sur l'origine des connaissances humaines.* Amsterdam: P. Mortier, 1746.

———. "Extrait raisonné du traité des sensations." In *Traité des sensations et Traité des animaux,* 284–307. 1798. Paris: Fayard, 1984.

———. *La Logique, ou les premiers développemens de l'art de penser.* 1780. In *Oeuvres complètes,* vol. 22.

———. *Oeuvres complètes.* 23 vols. Edited by G. Arnoux and Gabriel Bonnot, abbé de Mably. Paris: Houel, 1798.

————. *Oeuvres philosophiques.* 3 vols. Edited by Georges Le Roy. Paris: Presses Universitaires de France, 1947.

————. *Traité des sensations.* 1754. London, 1788.

————. *Traité des systèmes.* 1749. Paris: Fayard, 1991.

Condorcet, Marie-Jean-Antoine-Nicolas Caritat, marquis de. *Arithmétique politique, Textes rares ou inédites (1767–1789).* Edited by Bernard Bru and Pierre Crépel. Paris: Presses Universitaires de France, 1994.

————. *Condorcet: Selected Writings.* Edited by Keith Michael Baker. Indianapolis: Bobbs-Merrill, 1976.

————. *Correspondance inédite de Condorcet et Turgot.* Edited by Charles Henry. Paris, 1882.

————. *De l'influence de la Revolution de l'Amerique sur l'Europe.* 1786. In *Mélanges,* ed. Daire, vol. 1, 554–65.

————. *Ecrits sur l'instruction publique.* 2 vols. Edited by Charles Coutel and Catherine Kintzler. Paris: Edilig, 1989.

————. "Eloge de Franklin." 13 November 1790. In *L'Apothéose de Benjamin Franklin,* ed. Chinard, 129–43.

————. *Esquisse d'un tableau historique des progrès de l'esprit humain.* 1793. Paris: Flammarion, 1988.

————. *Oeuvres.* Edited by Arthur Condorcet O'Conner and François Arago. Paris: Didot, 1847.

Cuvier, Georges. *Histoire des progrès des sciences naturelles depuis 1789 jusqu'à ce jour.* Paris: Baudouin, 1826.

————. *Leçons sur l'anatomie comparée.* 5 vols. Paris, 1805.

Daire, Eugène, ed. *Economistes financiers du XVIIIᵉ siècle.* Paris, 1843.

————, ed. *Mélanges d'économie politique. Précédés de notices historiques sur chaque auteur, et accompagnés de commentaires et de notes explicatives, par Eugène Daire et G. de Molinari.* 2 vols. Paris, 1847–48.

————, ed. *Physiocrates. Quesnay, Dupont de Nemours, Mercier de la Rivière, l'abbé Baudeau, le Trosne, avec une introduction sur la doctrine des physiocrates, des commentaires et des notices historiques.* 2 vols. Paris, 1846.

Darcet, Jean. "Mémoire sur la calcination de la pierre calçaire et sur sa vitrification, soit seule, soit combinée avec d'autres terres. Lu à rentrée du Collège Royale de France." Paris, 11 November 1782.

Darwin, Erasmus. *Zoonomia: Or, the Laws of Organic Life.* Vol. 1. London: Johnson, 1794.

Daviel, J. "Lettre de M. Daviel, chirurgien ordinaire et oculiste du Roi en survivance par quartier, à M. de Joyeuse, docteur en médecine de l'Université de Montpellier, aggregé au Collège des Médecins de Marseille, et médecin des Hôpitaux des Galéres." *Mercure de France* (septembre 1748): 198–221.

————. "Réponse de M. Daviel, chirurgien ordinaire, et oculiste du Roi, à la lettre de M. le baron de Haller, du 11 novembre 1761, insérée dans le Mercure de France du mois de février 1762, page 145." *Journal de médecine, chirurgie, et pharmacie* 16 (March 1762): 245–50.

————. "Sur une nouvelle méthode de guérir la cataracte par l'extraction du cristalin." *Mémoires de l'Académie Royale de Chirurgie* 2 (1769): 337–54.

Davy, Humphry. "On some of the combinations of Oxymuriatic Gas and Oxygene." 1810. *The Elementary Nature of Chlorine.* Chicago: University of Chicago Press, 1911.

Decremps, Henri. *La Science Sansculotisée; premier essai sur les moyens de faciliter l'étude de l'astronomie, tant aux amateurs et aux gens de lettres, qu'aux marins de la République française, et d'opérer une RÉVOLUTION dans l'enseignement.* Paris, an 2 [1793–94].

Delaroche, Daniel. *Analyse des fonctions du système nerveux.* 2 vols. Geneva: Villard Fils et Nouffier, 1778.

De Machy, Jean-François. *Examen physique et chimique d'une eau minerale trouvée chez M. de Calsabigi a Passy; comparée aux Eaux du même Coteau, connues sous le nom des nouvelles Eaux Minerales de Madame Belami.* Paris, 1783.

———. [I] "Examen impartial d'un ouvrage intitulé *Traité élémentaire de chimie.*" *Tribut de la Société des Neuf Soeurs* (July 1790): 116.

———. [II] "Suite de l'examen impartial … " *Tribut de la Société des Neuf Soeurs* (January 1791): 69.

———. [III] "Suite de l'examen de la chymie moderne." *Tribut de la Société des Neuf Soeurs* (July 1791): 46.

———. [IV] "Précis élémentaire de chymie." *Tribut de la Société des Neuf Soeurs* (January 1792): 39.

Descartes, René. *La Dioptrique.* 1637. In *Discours de la méthode,* 71–208. Paris: Fayard, 1987.

———. *Discourse on Method.* 1637. In *The Philosophical Writings of Descartes.* Translated by John Cottingham, Robert Stoothoff, and Dugald Murdoch, 1:109–75. Cambridge: Cambridge University Press, 1985.

———. *Regulae ad directionem ingenii / Rules for the Direction of the Natural Intelligence* [bilingual edition]. 1628. Edited and translated by George Heffernan. Atlanta: Rodopi, 1998.

Des Essarts [Nicolas-Toussaint Le Moyne]. *Causes célèbres, curieuses et intéressantes, de toutes les cours souveraines du Royaume, avec les jugements qui les ont décidées.* Paris, 1775–89.

Deslon, Charles. *Observations sur les deux Rapports de MM. les Commissaires nommés par Sa Majesté pour l'Examen du Magnétisme animal.* Paris, 1784.

Destutt de Tracy, Antoine Louis Claude. *Eléments d'idéologie.* 1796. Paris, Courcier, 1804.

Dickens, Charles. *Hard Times.* 1854. New York: Penguin, 1980.

Dictionnaire de l'Académie française. Paris: Coignard, 1694.

Diderot, Denis. *De l'interprétation de la nature.* 1753. In *Oeuvres philosophiques,* 167–244.

———. *De la poésie dramatique.* 1758. In *Oeuvres esthétiques,* 179–287.

———. *Eléments de physiologie.* 1784. Edited by Jean Mayer. Paris: Didier, 1964.

———. "Eloge de Richardson." 1761. In *Oeuvres esthétiques,* 23–48.

———. "Entretien entre d'Alembert et Diderot." 1767. In *Oeuvres philosophiques,* 249–84.

———. *Lettre sur les aveugles.* 1749. In *Oeuvres philosophiques,* 75–146.

———. *Lettre sur les sourds et muets, à l'usage de ceux qui entendent et qui parlent.* In *Oeuvres complètes,* edited by Herbert Dieckmann, Jacques Proust, Jean Varloot et al., 4:111–233. Paris: Hermann, 1975.

———. *Lettres à Sophie Volland.* 2 vols. Edited by André Babelon. Paris: Gallimard, 1950.

———. *Oeuvres complètes.* 20 vols. Edited by Jules Assezat and Maurice Toyrneux. Paris: Garnier, 1875–77.

———. *Oeuvres esthétiques.* Edited by Paul Vernière. Paris: Garnier, 1966.

———. *Oeuvres philosophiques.* Edited by Paul Vernière. Paris: Garnier, 1998.

———. *Réfutation suivie de l'ouvrage d'Helvétius intitulé L'Homme.* 1774. In *Oeuvres complètes,* 2:275–346.

———. *Le Rêve de d'Alembert.* 1767. In *Oeuvres philosophiques,* 285–371.

————, ed. *Nouveau dictionnaire, pour servir de Supplément aux Dictionnaires des sciences, des arts et des métiers.* 4 vols. Paris, 1777.

Diderot, Denis, and Jean d'Alembert, eds. *Encyclopédie, ou, Dictionnaire raisonné des sciences, des arts et des métiers.* Paris, 1751–72.

Dufau, Pierre-Armand. *Des aveugles.* Paris: Renouard, 1850.

Dumas, C. L. *Principes de la physiologie.* 4 vols. Paris, 1800.

Dupont de Nemours, Pierre Samuel. "De l'exportation et de l'importation des grains. Mémoire lu à la Société royale d'agriculture de Soissons." 1764. In *Oeuvres politiques et économiques,* 1:61–139.

————. *De l'origine et des progrès d'une science nouvelle.* 1768. In *Oeuvres politiques et économiques,* 1:525–611.

————. *Oeuvres politiques et économiques.* 2 vols. Nendeln: KTO, 1979.

Durand, Abbé. *Le Franklinisme refuté, ou remarques sur la théorie de l'électricité à l'occasion du Système de plusieurs Physiciens sur ce Sujet.* Paris, 1788.

Durand de Maillane, Pierre-Toussaint. "Opinion sur les écoles primaires." 12 December 1792. In Condorcet, *Ecrits sur l'instruction publique,* 2:222–29.

Epée, Charles-Michel de l'. *La Véritable manière d'instruire les sourds et muets: Confirmée par une longue experience.* 2nd ed. 1784. Paris: Fayard, 1984 (facsimile of the 1784 edition).

Les Ephémérides du citoyen, Ou bibliothèque raisonnée des sciences morales et politiques [journal]. Paris, 1765–72.

Fontenelle, B. le Bovier de. "Sur les cataractes des yeux." *Histoire de l'Académie Royale des Sciences pour l'année 1707,* 22–25.

Fourcroy, Antoine-François. *Mémoires et observations de chimie.* Paris, 1784.

————. "Mémoire pour servir à l'explication du Tableau de Nomenclature." 1787. In "A propos de *Méthode,*" ed. Bensaude-Vincent, 75–106.

————. *Philosophie chimique.* Paris, 1792.

Franklin, Benjamin. *Benjamin Franklin: Writings.* Edited by J. A. Leo Lemay. New York: Library of America, 1987.

————. *Benjamin Franklin's Experiments.* Edited by I. Bernard Cohen. Cambridge: Harvard University Press, 1941.

————. "Opinions and Conjectures concerning the Properties and Effects of the electrical Matter, arising from Experiments and Observations, made at Philadelphia." 1750. In *Benjamin Franklin's Experiments,* 213–36.

————. *Oeuvres de M. Franklin,* ed. Jacques Barbeu-Dubourg. Paris, 1773.

————. *The Papers of Benjamin Franklin* (BFP). Edited by Leonard Labaree, William Wilcox, Claude-Anne Lopez, and Barbara Oberg. 35 vols. New Haven: Yale University Press, 1959–.

————. *The Writings of Benjamin Franklin.* Edited by Albert Henry Smyth. 10 vols. New York: Macmillan, 1907.

Franklin, Benjamin, Antoine Lavoisier et al. *Rapport des commissaires chargés par le Roi de l'examen du magnétisme animal.* Paris, 1784.

Gall, Franz Joseph. *On the Functions of the Brain and Each of Its Parts,* 1825. Vol. 6. Boston: Marsh, Capen and Lyon, 1835.

Galliod, Jean-François, and François-Auguste Groscoeur. *Notice historique sur l'établissement des jeunes aveugles.* Paris, 1829.

Gleize, Jean-François. *Règlement de Vie, ou comment doivent se gouverner ceux qui sont affligés de la faiblesse de la vue, avec les moyens de s'en preserver.* Orléans: Jacob l'aîné, 1787.

Godart, Guillaume-Lambert. *La physique de l'âme*. Berlin, 1755.

Goethe, Johann Wolfgang von. *Theory of Colors*. 1810. Translated by C. L. Eastlake. London: John Murray, 1840.

Grandidier, Philippe André. *Histoire de l'église et des evêques-princes de Strasbourg*. Strasbourg, 1776–78.

Grant, R. [pseud.] *A Full and True Account of a Miraculous Cure of a Young Man in Newington, that was Born Blind, and was in Five Minutes brought to Perfect Sight*. London, 1709.

Graslin, Jean-Joseph-Louis. *Essai analytique sur la richesse et sur l'impôt, où l'on refute la nouvelle doctrine économique qui a fourni à la Société Royale d'Agriculture de Limoges les principes d'un programme qu'elle a publié sur l'effet des Impôts indirects*. London, 1767.

Grégoire, Henri Baptiste. "Rapport sur l'établissement d'un Conservatoire des arts et métiers. 8 vendémiaire, an 3 [29 September 1794]." In *Convention Nationale, vendémiaire, an III*. Paris, an III [1794].

Guillié, Sébastien. *Essai sur l'instruction des aveugles*. Paris, 1817.

———. *Notice historique sur l'institution des jeunes aveugles*. Paris, 1819.

Guyton de Morveau, Louis-Bernard. *Digressions académiques, ou Essais sur quelques sujets de physique, de chymie et d'histoire naturelle*. Dijon, 1762.

———. *Discours publics, et Eloges*. 3 vols. Paris, 1775–82.

———. *Elémens de chymie théorique et pratique*. 3 vols. Dijon: L. N. Frantin, 1777.

———. "Mémoire sur les dénominations chimiques, la nécessité d'en perfectionner le système, et les règles pour y parvenir." *Observations et mémoires sur la physique* 19 (1782): 370–82.

———. "Mémoire sur le développement des principes de la nomenclature chimique, lu à l'Académie, le 2 Mai 1787." In "A propos de *Méthode*," ed. Bensaude-Vincent, 26–74.

Guyton de Morveau, Louis-Bernard, Hugues Maret, Jean Pierre François Guillot Duhamel, and Antoine-François Fourcroy, eds. *Encyclopédie méthodique: Chymie, pharmacie, et métallurgie*. 5 vols. Paris: Panckoucke, 1786–1815.

Guyton de Morveau, Louis-Bernard, Antoine-François Lavoisier, Claude Louis Berthollet, and Antoine-François Fourcroy, *Méthode de nomenclature chimique*. Paris, 1787.

Haller, Albrecht von. *Elementa physiologiae corporis humani*. 8 vols. Lausanne, 1757–66.

———. *First Lines of Physiology*. 2 vols. Edited by William Cullen. Edinburgh, 1786. Reprinted in one volume with an introduction by Lester S. King. New York: Johnson Reprint Corporation, 1966.

———. "On the Sensible and Irritable Parts of Animals." 1755. In *The Natural Philosophy of Albrecht von Haller*, ed. Shirley Roe, 651–99, i–xxxii. New York: Arno, 1981.

Hartley, David. *Observations on Man, His Frame, His Duty, and His Expectations*. 2 vols. London: S. Richardson, 1749.

Haüy, Valentin. "Adresse du citoyen Haüy . . . aux 48 sections de Paris," 13 December 1792.

———. *Essay on the Education of the Blind*. Paris: Les Enfans-Aveugles, 1786.

———. Lettre. *Journal de Paris*, no. 111 (1784): 488.

———. Lettre. *Journal de Paris*, no. 274 (1784): 1158.

———. "Mémoire de M. Haüy sur l'éducation des aveugles." In *Mémoires lus dans la Séance publique du Bureau académique d'écriture* (Paris, 1784).

————. *Trois notes du citoyen Haüy*. Paris, 1801.

Helvétius, Claude-Adrien. *De l'esprit*. Paris: Durand, 1758.

————. *De l'homme*. London, 1773.

Holbach, Paul Henri Thiry, d'. *La Morale universelle, ou Les Devoirs de l'homme fondés sur sa nature*. 3 vols. Amsterdam: M. M. Rey, 1776.

————. *Système de la nature, ou, Des loix du monde physique et du monde moral*. 2 vols. London: [no pub.], 1771.

Home, E. "An Account of Two Children born with Cataracts in their Eyes, to shew that their Sight was Obscured in Very Different Degrees: With Experiments to Determine the Proportional Knowledge of Objects Acquired by them Immediately after the Cataracts were Removed." *Philosophical Transactions of the Royal Society of London* 97 (1807): 83–92.

Hume, David. *Enquiries concerning Human Understanding and concerning the Principles of Morals*. 1777. Edited by L. A. Selby-Bigge and P. H. Nidditch. Oxford: Oxford University Press, 1992. [From 3rd edition. The 1st edition of the *Enquiry into the Human Understanding* was 1748; the 1st edition of the *Enquiry concerning Morals* was 1751.]

————. *The History of England from the Invasion of Julius Caesar to the Abdication of James the Second, 1688*. 1754–62. 6 vols. Boston: Phillips, Sampson & Co., 1856.

————. "Of Commerce." 1752. In *Essays Moral, Political and Literary*, ed. T. H. Green and T. H. Grose, 1:287–99. London: Longmans, 1898.

————. *A Treatise of Human Nature, being an attempt to introduce the experimental method of reasoning into moral subjects; and dialogues concerning natural religion*. 1739–40. Edited by T. H. Green and T. H. Grose. London: Longmans, 1886.

Hutcheson, Francis. *On the Nature and Conduct of the Passions*. 1728. Introduced and annotated by Andrew Ward. Manchester: Clinamen, 1999.

————. "Original Letter from Dr. Francis Hutcheson to Mr. William Mace, Professor at Gresham College, 6 September 1727." *European Magazine and London Review* (September 1788): 158–60.

Hutton, James. "A Chymical Dissertation concerning Phlogiston, or the Principle of Fire." In *Dissertations on Different Subjects in Natural Philosophy*, 171–270. Edinburgh, 1792.

Itard, Jean. "The Wild Boy of Aveyron." 1799–1806. In *Wolf Children*, ed. Lucien Malson, 89–179. New York: Monthly Review Press, 1972.

Janin, J. *Mémoires et observations anatomiques, physiologiques et physiques sur l'oeil, et sur les maladies qui affectent cet organe: avec un précis des opérations et des remèdes qu'on doit pratiquer pour les guérir*. Paris: Didot, 1772.

Jaucourt, Louis de. "Tact." 1765. In *Encyclopédie*, ed. Diderot and d'Alembert, 15:819b–20b.

Jefferson, Thomas. *Notes on the State of Virginia*. 1784–85. New York: Penguin, 1999.

————. *Writings*. New York: Library of America, 1984.

Jurin, James. "Dr. Jurin's Solution of Mr. Molyneux's Problem." In R. Smith, *A Compleat System of Opticks in Four Books*, 2:27–29 (remarks upon bk. 1, chap. V, art. 132, §§160–70). Cambridge, 1738.

Jussieu, Bernard de. *Rapport de l'un des commissaires chargés par le Roi de l'examen du magnétisme animal*. 12 September 1784. In Alexandre Bertrand, *Du magnétisme animal en France, et des jugements qu'en portés les sociétés savants*, 151–206. Paris, 1826.

Kirwan, Richard. *An Essay on Phlogiston and the Constitution of Acids.* 2nd ed. London: J. Johnson, 1789. Reprint, London: Frank Cass and Company, 1968. [1st ed., 1784.]

Krüger, Johann Gottlob. *Naturlehre.* 3 vols. Halle, 1740–49.

———. *Versuch einer Experimental-Seelenlehre.* Halle, 1756.

Lacépède, Bernard Germain Etienne de La Ville, comte de. *Vues sur l'enseignement public.* Paris, 1790.

Laclos, Pierre Choderlos de. *Les Liaisons dangereuses.* 1782. Edited by René Pomeau. Paris: Flammarion, 1996.

Lamarck, Jean Baptiste. *Philosophie zoologique.* Paris: Dentu, 1809.

La Métherie, Jean-Claude de. "Essai sur la nomenclature chimique." *Observations et mémoires sur la physique* 31 (1787): 270–85.

La Mettrie, Julien Offray de. *L'Anti-Sénèque ou le souverain bien.* 1750. In *De la volupté,* 27–112.

———. *De la volupté: L'Anti-Sénèque ou le souverain bien; L'Ecole de la volupté; Le Système d'Epicure.* Edited by Ann Thomson. Paris: Desjonquières, 1996.

———. *L'Ecole de la volupté.* 1746. In *De la volupté,* 115–53.

———. *Histoire naturelle de l'âme, traduite de l'anglois de M. Charpe, par feu M.H** de l'Académie des Sciences &c.* La Haye: Jean Néaulme, 1745.

———. *L'Homme-machine.* 1748. Edited by Paul-Laurent Assoun. Paris: Denoël/ Gonthier, 1981.

———. *Le Système d'Epicure.* 1750. In *De la volupté,* 157–91.

Lanteires, J. *Essai sur le Tonnerre considéré dans ses effets moraux.* Paris, 1789.

Lavoisier, Antoine Laurent. *Elements of Chemistry.* 1789. Translated by Robert Kerr (1790). Introduction by Douglas McKie. New York: Dover, 1965.

———. "Mémoire sur la nécessité de réformer et de perfectionner la nomenclature de la chimie." In Guyton de Morveau et al., *Méthode* (1787).

———. *Oeuvres de Lavoisier.* 6 vols. Edited by J. B. Dumas (vols. 1–4) and Edouard Grimaux (vols. 5–6). Paris: Imprimerie Impériale, 1862–93.

———. *Oeuvres de Lavoisier—Correspondance.* Vol. 7 of *Oeuvres.* Fascs. 1–3 edited by René Fric. Paris: Albin Michel, 1955–64. Fascs. 4–5 edited by Michelle Goupil. Paris: Académie des Sciences, 1986–93.

———. *Traité élémentaire de chimie.* 1789. In *Oeuvres,* 1:1–407.

Le Camus, Antoine. *La Médecine de l'esprit.* 1753. Paris: Ganeau, 1769.

Le Cat, Claude Nicolas. *Traité des sens.* Rouen, 1740.

———. *Traité des sensations et des passions en général.* 2 vols. Paris, 1767.

Lee, Henry. *Anti-Scepticism: or, Notes upon each Chapter of Locke's Essay concerning Humane Understanding.* London, R. Clavel and C. Harper, 1702.

Leibniz, Gottfried Wilhelm. *Nouveaux essais sur l'entendement humain.* 1705. Paris: Garnier-Flammarion, 1966.

Le Roy, Jean-Baptiste. "Mémoire sur l'électricité. [1] Où l'on montre . . . qu'il y a deux espèces d'électricité . . . et qu'elles ont chacune des phénomènes particuliers qui les caractérisent parfaitement." *Mémoires de l'Académie des Sciences* (1753): 447–59.

———. "Mémoire sur l'électricité. [2] Où l'on rapporte les expériences qui confirment l'existence des deux électricités par condensation et par rarefaction." *Mémoires de l'Académie des Sciences* (1753): 459–68.

———. "Mémoire sur l'électricité. [3] Supplément au mémoire précédent." *Mémoires de l'Académie des Sciences* (1753): 468–74.

Lignac, Joseph Adrien Lelarge de. *Lettres à un Américain sur l'histoire générale et particulière de M. de Buffon*. Hamburg, 1751.

Linguet, Simon-Nicolas-Henri. *Annales politiques, civiles et littéraires du dix-huitième siècle* [periodical]. London, 1777–91.

Locke, John. *The Correspondence of John Locke*. 8 vols. Edited by E. S. De Beer. Oxford: Clarendon Press, 1976–89.

———. *An Essay concerning Human Understanding*. 1690. Edited by Peter H. Nidditch. Oxford: Clarendon, 1975.

Louis, A. "Cataracte." In *Encyclopédie*, ed. Diderot and d'Alembert, vol. 2 (1752): 770–71.

Macquer, Pierre-Joseph. *Dictionnaire de chymie*. 2 vols. Paris: Lacombe, 1766. [2nd ed., 1779.]

Mairan, Dortous de. *Dissertation sur la Glace, ou Explication physique de la formation de la Glace, et de ses divers phénomènes*. Paris, 1749.

Maître-Jan, A. *Traité des maladies de l'oeil et des remèdes propres pour leur guérison*. Troyes: Le Febvre, 1707.

Manuel, E. B. *Etude de la nature en général et de l'homme en particulier, considerée dans ses rapports avec l'instruction publique*. Paris, 1793.

Marat, Jean-Paul. *Les Charlatans modernes; ou Lettres sur le charlatanisme académique, publiée par M. Marat, l'Ami du peuple*. Paris, 1791.

———. "Dénonciation de Lavoisier." *L'Ami du Peuple, ou le Publiciste*, no. 353 (27 January 1791): 5.

———. *Recherches physiques sur le feu*. Paris: C.-A. Jombert, 1780.

Marivetz, Baron de. "Lettre à M. de la Métherie sur la nomenclature chimique." *Observations et mémoires sur la physique* 32 (1788): 61–63.

Mercier de la Rivière, P. F. J. H. *L'Ordre naturel et essentiel des sociétés politiques*. 1767. In *Physiocrates*, ed. Daire, 2:445–638.

Mérian, Jean-Bernard. *Sur le problème de Molyneux*. 1772–82. Edited by Francine Markovits. Paris: Flammarion, 1984.

Mesmer, Franz Anton. "Discourse by Mesmer on Magnetism." 1784. In *Mesmerism*, 23–38.

———. "Dissertation on the Discovery of Animal Magnetism." 1779. In *Mesmerism*, 43–76.

———. "Dissertation by F. A. Mesmer, Doctor of Medicine, on His Discoveries." 1799. In *Mesmerism*, 89–130.

———. *Mesmerism, A Translation of the Original Scientific Writings of F. A. Mesmer*. Translated by George Bloch. Los Altos: William Kaufman, 1980.

———. "Physical-Medical Treatise on the Influence of the Planets." 1766. In *Mesmerism*, 3–20.

———. "Précis historique des faits relatifs au magnétisme animal jusqu'en avril 1781." 1781. In *Le Magnétisme Animal. Oeuvres publiés par Robert Amadou*. Notes and commentary by A. Pattie and J. Vinchon, 93–194. Paris: Payot, 1971.

Métra, François. *Correspondance secrete, politique et littéraire, ou, Mémoires pour servir à l'histoire des Cours, des Sociétés et de la Littérature en France, depuis la mort de Louis XV*. London, 1787.

Mirabeau, Victor de Riquetti, marquis de, with François Quesnay. *Philosophie rurale, ou économie générale et politique de l'agriculture, reduite à l'ordre immuable des loix physiques & morales, qui affurent la prospérité des empires*. Amsterdam, 1764.

Molyneux, William. *Dioptrica nova: A Treatise of Dioptricks.* London, 1692.

———. "A Problem Proposed to the Author of the Essai Philosophique concernant l'Entendement." 7 July 1688. Oxford, Bodleian Library, ms. Locke c. 16, fol. 92 recto.

Monnet, Antoine Grimoald. *Demonstration de la fausseté des principes des nouveaux chimistes; pour servir de supplément de la dissolution des metaux.* Paris, an vi [1797–98].

Montesquieu, Charles de Secondat de. *De l'esprit des lois.* 4 vols. Edited by J. Brethe de la Gressaye, Livres 1–9. Paris: Les Belles-Lettres, 1950–61.

———. *Mes pensées.* 1720–55. In *Oeuvres complètes,* 853–1082.

———. *Oeuvres complètes.* Edited by Georges Vedel and Daniel Oster. Paris: Editions du Seuil, 1964.

Montjoie, Galart de. *Lettre sur le magnétisme animal.* Paris, 1784.

Morand, S. F. "Eloge de M. Cheselden." 1757. *Mémoires de l'Académie Royale de Chirurgie,* 7:168–90.

Morellet, André. *Mémoires inédits de l'abbé Morellet de l'Académie Française sur le dix-huitième siècle et sur la Révolution.* Edited by Jean-Pierre Guicciardi. Paris, 1988.

Necker, Jacques. *Sur la législation et le commerce des grains.* 1775. In *Mélanges,* ed. Daire, 2:211–361.

Needham, John Turberville. *Nouvelles observations microscopiques.* Paris, 1747.

Newton, Isaac. *Opticks: or, A treatise of the reflections, refractions, inflections & colours of light.* 1730. 4th ed. New York: Dover, 1952.

———. *Philosophiae Naturalis Principia Mathematica.* 1686. 2 vols. Translated by Andrew Motte in 1729. Edited by Florian Cajori. Berkeley: University of California Press, 1934.

Nicholson, William. *A Dictionary of Chemistry.* London: Robinson, 1795.

———. *First Principles of Chemistry.* London: Robinson, 1792.

Nollet, Jean-Antoine. "Conjectures sur les causes de l'électricité des corps." *Mémoires de l'Académie des Sciences* (1745): 107–51.

———. *Correspondance entre l'abbé Nollet et le physicien genevois Jean Jallabert.* Edited by Isaac Benguigui. Geneva, n.d.

———. *Essai sur l'électricité des corps.* Paris, 1746. 2nd ed. Paris, 1750.

———. "Examen de deux questions concernans l'électricité." *Mémoires de l'Académie des Sciences* (1753): 475–502.

———. *Leçons de physique expérimentale.* 6 vols. Paris, 1743–48 (1st ed.); 1754–65 (4th ed.); 1759–66 (5th ed.).

———. *Lettres sur l'électricité.* Paris, 1753–67. Vol. 1, *Dans lesquelles on examine les dernières découvertes qui ont été faites sur cette matière, et les conséquences que l'ont peut tirer* (1753); vol. 2, *Dans lesquelles on soutient le principe des Effluences et Affluences simultanées contre la doctrine de M. Franklin, et contre les nouvelles prétensions de ses partisans* (1760); vol. 3, *Dans lesquelles on trouvera les principaux phénomènes qui ont été découverts depuis 1760, avec les discussions sur les conséquences qu'on en peut tirer* (1767).

———. "Réponse au supplément d'un Mémoire lu à l'Académie par M. le Roy." *Mémoires de l'Académie des Sciences* (1753): 503–14.

Parlement de Paris. *Remontrances du Parlement de Paris au XVIIIe siècle.* Edited by Jules Flammermont. 3 vols. Paris, 1888–98.

Parliament of England. *Cobbett's Parliamentary History of England from the Norman Conquest to the Year 1803.* 36 vols. London: Hansard, 1806–20.

Poissonier, Pierre-Isaac, Nicolas Louis de La Caille et al. *Rapport des Commissaires de la Société Royale de Médecine, nommés par LE ROI pour faire l'examen du Magnétisme animal.* Paris, 1784.

Pons, Alain, ed. *Encyclopédie, ou, Dictionnaire raisonné des sciences, des arts et des métiers (articles choisis).* 2 vols. Paris: Flammarion, 1986.

Priestley, Joseph. *Considerations on the Doctrine of Phlogiston, and the Decomposition of Water.* 1796. In *Lectures on Combustion,* ed. William Foster, 17–41. Princeton: Princeton University Press, 1929.

———. *Experiments and Observations on Different Kinds of Air.* 3 vols. London: Joseph Johnson, 1777.

———. *Experiments and Observations Relating to Various Branches of Natural Philosophy.* 3 vols. London, 1779–86. Reprint, New York: Kraus, 1977.

———. *The History and Present State of Electricity.* 3rd ed. 2 vols. London: J. Johnson, 1775. Reprinted with an introduction by Robert E. Schofield, New York: Johnson Reprint Corporation, 1966.

———. *A Scientific Autobiography of Joseph Priestley, 1733–1804.* Edited by Robert E. Schofield. Cambridge: M.I.T. Press, 1966.

Puységur, Armand Marie-Jacques de Chastenet, marquis de. *Mémoires pour servir à l'histoire et à l'établissement du magnétisme animal.* 1786. Edited by Georges Lapassade and Philippe Pédelahore. Bordeaux: Editions Privat, 1986.

Quesnay, François. *François Quesnay et la Physiocratie.* Vol. 2, *Textes Annotés.* Paris: Institut National d'Etudes Démographiques, 1958.

———. *Oeuvres économiques et philosophiques.* Edited by August Oncken. Frankfurt, 1888.

Rabaut Saint-Etienne, Jean-Paul. "Projet d'Education Nationale." 21 December 1792. In Condorcet, *Ecrits sur l'instruction publique,* 2:231–35.

Reid, Thomas. *An Inquiry into the Human Mind: On the Principles of Common Sense.* 1764. Edited by Derek R. Brookes. University Park: Pennsylvania State University Press, 1997.

Régnier, Edmond. "Aux auteurs du Journal." *Journal de Paris,* no. 51 (20 February 1784): 234.

Richerand, Anthelme. *Nouveaux élémens de physiologie.* 1801. 3rd ed. 2 vols. Paris, 1804.

Rivarol, Antoine. *Discours préliminaire du nouveau dictionnaire de la langue française.* Paris, 1797.

Robespierre, Maximilien. *Oeuvres complètes.* Edited by Charles Vellay. Paris, 1910.

———. *Plaidoyers pour le Sieur de Vissery de Bois-Valé, appellant d'un jugement des Echevins de Saint-Omer, qui avoit ordonné la déstruction d'un Par-à-Tonnerre élevé sur sa maison.* Arras, 1783.

———. *Robespierre: Ecrits.* Edited by Claude Mazauric. Paris, 1989.

Robison, John. *Proofs of a Conspiracy Against all the Religions and Governments of Europe, carried on in the secret meetings of Free Masons, Illuminati and Reading Societies.* Edinburgh, 1797.

Rodenbach, Alexandre. *Lettre sur les aveugles faisant suite à celle de Diderot.* Brussels, 1828.

Roederer, Pierre-Louis. *De la Philosophie Moderne et de la part qu'elle a eue à la Révolution française.* Paris, an viii [1799–1800].

Romas, Jacques de. *Oeuvres inédites de J. de Romas . . . avec une notice biographique et bibliographique par Paul Courteault.* Edited by J. Bergonie. Bordeaux, 1911.

Rouelle, Guillaume-François. *Prospectus du cours d'expériences chymiques.* Paris, 1759.

Rousseau, Jean-Jacques. *Les Confessions*. 1770. Edited by Michel Launay. Paris: Flammarion, 1968.

———. *Du Contrat Social*. 1762. Edited by B. de Jouvenel. Geneva: C. Bourquin, 1947.

———. "Discours sur les sciences et les arts." 1750. In *Oeuvres complètes*, vol. 3.

———. *Emile, ou de l'Education*. 1762. In *Oeuvres complètes*, vol. 4.

———. *Essai sur l'origine des langues*. 1781. Paris: Hatier, 1983.

———. *Julie, ou la Nouvelle Héloïse*. Edited by Daniel Mornet. Paris: Hachette, 1925.

———. *Lettres Morales*. 1757–58. In *Oeuvres complètes*, 4:1092–93.

———. *Oeuvres complètes*. Edited by Bernard Gagnebin and Marcel Raymond. Paris: Pléiade, 1959–95.

Sade, D. A. F., marquis de. "Pensée inédite." 1782. In *Le Surréalisme au service de la Révolution* 4 (December 1931): 1–3.

Sage, Georges Balthazar. "Lettre à M. de La Métherie sur la nouvelle nomenclature." *Observations et mémoires sur la physique* 33 (1788): 478–79.

———. *Opuscules historiques et physiques*. Paris, 1816.

———. *Opuscules de physique*. Paris, 1813; 2nd ed., Paris, 1815.

Saunderson, Nicholas. *The Elements of Algebra*. Cambridge, 1740.

Schwenger, A. G., ed. *Mémoires sur les aveugles, sur la vue et la vision, suivis de la description d'un télégraphe très simple*. Paris: Didot, 1800.

Servan, Antoine. *Doutes d'un Provincial, Proposés à M. M. les Médecins-Commissaires, chargés par le Roi, de l'examen du Magnétisme animal*. Lyon, 1784.

Sigaud de la Fond, Joseph Aignan. *Précis historique et expérimental des phénomènes électriques, depuis l'origine jusqu'à ce jour*. Paris, 1788.

Smith, Adam. *An Inquiry into the Nature and Causes of the Wealth of Nations*. 1776. 2 vols. Edited by R. H. Campbell and A. S. Skinner. Oxford: Clarendon Press, 1976.

Steele, R. "An Account of Mr. Grant's Operation of 29 June 1709." *Tatler* 1, no. 55 (Tuesday, 16 August 1709). Reprinted in *The Tatler*, edited by G. A. Aitken, 2:41–46. London: Duckworth, 1898.

Stewart, Dugald. "Account of the Life and Writings of Adam Smith, LL.D." 1793. In *Collected Works of Dugald Stewart*, ed. Sir William Hamilton. Edinburgh: Thomas Constable, 1854–60, 10:1–98.

———. "Some Account of a Boy Born Blind and Deaf." *Transactions of the Royal Society of Edinburgh* 7 (1815). Reprinted in *The Collected Works of Dugald Stewart*, edited by Sir William Hamilton, 4:300–70, 388–89. Edinburgh: T. and T. Clark, 1854.

Swedenborg, Emanuel. *The Brain, considered anatomically, physiologically and philosophically*. ca. 1740. Edited, translated, and annotated by R. L. Tafel. 2 vols. London: Speirs, 1882–87.

Thibaudeau, A. C. *Mémoires sur la Convention et le Directoire*. 2 vols. Paris, 1824.

Trublet, Nicolas Charles Joseph. *Mémoire sur la vie et les oeuvres de Fontenelle*. Amsterdam, 1759.

Turgot, Anne-Robert-Jacques. *Oeuvres de Turgot et documents le concernant*. 5 vols. Edited by Gustave Schelle. Paris: Felix Alcan, 1913. Reprint, Darmstadt: Bläsche & Ducke, 1972.

———. *Réflexions sur la formation des richesses*. 1766. In *Formation et distribution des richesses*. Edited by Joël-Thomas Ravix and Paul-Marie Romani. Paris: Flammarion, 1997.

Vandeul, Marie-Angélique de. *Mémoires pour servir à l'histoire de la vie et des ouvrages de Diderot*. Paris, 1830–31.

Venel, Gabriel-François. "Chymie." 1753. In *Encyclopédie*, ed. Diderot and d'Alembert, 1:408–37.

Véri, abbé Joseph Alphonse de. *Journal de l'abbé de Véri*. 1774–81. Edited by Baron Jehan De Witte. 2 vols. Paris, 1928.

Voltaire, François-Marie Arouet de. *Candide, or Optimism.* 1759. In *Candide, Zadig, and Selected Stories.* Translated and edited by Donald M. Frame. New York: Signet, 1961.

———. *Dictionnaire philosophique.* 1764. Edited by René Pomeau. Paris: Flammarion, 1964.

———. *Eléments de la philosophie de Newton.* 1738. In *The Complete Works of Voltaire.* Edited by Robert L. Walters and W. H. Barber, vol. 15. Oxford: Voltaire Foundation, 1992.

———. *Letters concerning the English Nation.* 1733. Edited by Nicholas Cronk. Oxford: Oxford University Press, 1994.

———. *Voltaire's Correspondence.* Edited by Theodore Besterman. Geneva, 1964.

Wardrop, J. "Case of a Lady Born Blind, Who Received Sight at an Advanced Age by the Formation of an Artificial Pupil." *Philosophical Transactions of the Royal Society of London* 116 (1826): 529–40.

Ware, J. "Case of a Young Gentleman, who Recovered his Sight when Seven Years of Age, after having been Deprived of it by Cataracts, before he was a Year Old: With Remarks." *Philosophical Transactions of the Royal Society of London* 91 (1801), 382–96.

Wolff, Christian. *Philosophia rationalis sive logica.* 3rd ed. Frankfurt am Main, 1740.

III. SECONDARY WORKS

Acton, H. B. "The Philosophy of Language in Revolutionary France." In *Studies in Philosophy,* edited by J. N. Findlay, 143–68. London: Oxford University Press, 1966.

Airiau, Jean. *L'Opposition aux Physiocrates a la fin de l'ancien régime.* Paris: Libraire générale de droit et de jurisprudence, 1965.

Albury, W. R. "The Logic of Condillac and the Structure of the French Chemical and Biological Theory, 1780–1801." Ph.D. diss., Johns Hopkins University, 1972.

———. "The Order of Ideas: Condillac's Method of Analysis as a Political Instrument in the French Revolution." In *The Politics and Rhetoric of Scientific Method,* ed. John A. Schuster and Richard R. Yeo, 203–25. Dordrecht: D. Reidel, 1986.

Alder, Ken. "A Revolution to Measure: The Political Economy of the Metric System in France." In *The Values of Precision,* ed. M. Norton Wise, 39–71. Princeton: Princeton University Press, 1995.

———. "To Tell the Truth: The Polygraph Exam and the Marketing of American Expertise." In *Evidence and the Law,* ed. Robert A. Nye. *Historical Reflections* 24, no. 3 (1998): 487–525.

Aldridge, Alfred Owen. *Franklin and His French Contemporaries.* New York: New York University Press, 1957.

Anderson, Douglas. *The Radical Enlightenments of Benjamin Franklin.* Baltimore: Johns Hopkins University Press, 1997.

Anderson, Lorin. *Charles Bonnet and the Order of the Known.* Dordrecht: Reidel, 1982.

Anderson, Wilda. *Between the Library and the Laboratory: The Language of Chemistry in Eighteenth-Century France.* Baltimore: Johns Hopkins University Press, 1984.

———. "Morale de la Terreur." *Le Genre humain* 29 (June 1995): 91–110.

————. "Scientific Nomenclature and Revolutionary Rhetoric." *Rhetorica* 7, no. 1 (winter 1989): 45–53.

Andrews, Richard Mowery. *Law, Magistracy and Crime in Old Regime Paris, 1735–1789.* Vol. 1, *The System of Criminal Justice.* Cambridge: Cambridge University Press, 1994.

Auroux, Sylvain. *La Sémiotique des encyclopédistes.* Paris: Payot, 1979.

Ayers, Michael. *Locke.* Vol. 1, *Epistemology.* London: Routledge, 1991.

Aykroyd, W. R. *Three Philosophers (Lavoisier, Priestley and Cavendish).* London: William Heineman, 1935.

Badinter, Elisabeth, and Robert Badinter. *Condorcet: Un Intellectuel en politique.* Paris: Fayard, 1988.

Baker, Keith Michael. *Condorcet: From Natural Philosophy to Social Mathematics.* Chicago: University of Chicago Press, 1975.

————. "Enlightenment and the Institution of Society: Notes for a Conceptual History." In *Main Trends in Cultural History,* ed. W. F. B. Melching and W. R. E. Velema, 95–120. Amsterdam: Rodopi, 1994.

————. *Inventing the French Revolution: Essays on French Political Culture.* Cambridge: Cambridge University Press, 1990.

Baker, Keith Michael, Colin Lucas, François Furet, and Mona Ozouf, eds. *The French Revolution and the Creation of Modern Political Culture.* 3 vols. Oxford: Pergamon, 1987.

Ballanche, Pierre-Simon. *Du sentiment considéré dans ses rapports avec la littérature et les arts.* Lyon: Ballanche et Barret, 1801.

Barker-Benfield, G. J. *The Culture of Sensibility: Sex and Society in Eighteenth-Century Britain.* Chicago: University of Chicago Press, 1992.

Barre, Raymond. *Economie politique.* Paris: Presses Universitaires de France, 1966.

Barthélemy, Guy. *Les Savants sous la Révolution.* Le Man: Editions Cénomane, 1988.

Bate, Walter Jackson. *From Classic to Romantic: Premises of Taste in Eighteenth-Century England.* New York: Harper and Row, 1961.

Becker, Carl L. *The Heavenly City of the Eighteenth-Century Philosophers.* New Haven: Yale University Press, 1932.

Beer, M. *An Inquiry into Physiocracy.* London, 1939.

Bell, David. *Lawyers and Citizens: The Making of a Political Elite in Old Regime France.* New York: Oxford University Press, 1994.

————. "Lawyers into Demagogues: Chancellor Maupeou and the Transformation of Legal Practice in France 1771–1789." *Past and Present* 130 (1991): 107–41.

Bensaude-Vincent, Bernadette. "The Balance: Between Chemistry and Politics." *The Eighteenth Century: Theory and Interpretation* 33, no. 3 (fall 1992): 217–37.

————. "A propos de *Méthode de nomenclature chimique,* Esquisse suivie du texte de 1787." *Cahiers d'Histoire et de Philosophie de Sciences* 6 (1983).

Bensaude-Vincent, Bernadette, and Ferdinando Abbri, eds. *Lavoisier in European Context: Negotiating a New Language for Chemistry.* Canton: Science History Publications, 1995.

Beretta, Marco. "Chemists in the Storm: Lavoisier, Priestley and the French Revolution." *Nuncio* 8, no. 1 (1993): 75–104.

————. *The Enlightenment of Matter: The Definition of Chemistry from Agricola to Lavoisier.* Canton: Watson, 1993.

Berlanstein, Lenard R., *The Barristers of Toulouse in the Eighteenth Century (1740–1793).* Baltimore: Johns Hopkins University Press, 1975.

Berlin, Isaiah. "Giambattista Vico and Cultural History." In *The Crooked Timber of Humanity*, 49–69. New York: Knopf, 1991.

Berthelot, M. *La Révolution chimique: Lavoisier*. Paris: Félix Alcan, 1890.

Biagioli, Mario. *Galileo, Courtier: The Practice of Science in the Culture of Absolutism*. Chicago: University of Chicago Press, 1993.

Birkhead, Edith. "Sentiment and Sensibility in the Eighteenth-Century Novel." 1925. In *Essays and Studies by Members of the English Association*, 32 vols., ed. Oliver Elton, 11:92–116. London: Wm. Dawson, 1966.

Blaug, Mark. *Economic Theory in Retrospect*. Homewood: Irwin, 1968.

Blunt, Wilfred. *The Compleat Naturalist: A Life of Linnaeus*. New York: Viking, 1971.

Bobé, Louis, ed. *Mémoires de Charles Claude Flahaut, comte de la Billarderie d'Angivillier: Notes sur les mémoires de Marmontel publiés d'apres le manuscrit*. Copenhagen: Levin and Munksgaard, 1933.

Bogdanor, Vernon, ed. *Science and Politics. The Herbert Spencer Lectures, 1982*. Oxford: Clarendon, 1984.

Bongie, L. L. *Diderot's femme savante*. Oxford: Voltaire Foundation, 1977.

———. "A New Condillac Letter and the Genesis of the Traité des sensations." *Journal of the History of Philosophy* 16 (1978).

Bolton Brandt, M. "The Real Molyneux Question and the Basis of Locke's Answer." In *Locke's Philosophy*, ed. G. A. J. Rogers, 75–99. New York: Oxford University Press, 1994.

Bourde, André. *The Influence of England on the French Agronomes, 1750–1789*. Cambridge: Cambridge University Press, 1953.

Brasart, Patrick. *Paroles de la Révolution: Les Assemblées parlementaires 1789–1794*. Paris: Minerve, 1988.

Bredvold, Louis I. *The Natural History of Sensibility*. Detroit: Wayne State University Press, 1962.

Breitweiser, Mitchell. *Cotton Mather and Benjamin Franklin: The Price of Representative Personality*. Cambridge: Cambridge University Press, 1984.

Brewer, John. *The Pleasures of the Imagination: English Culture in the Eighteenth Century*. Chicago: University of Chicago Press, 1997.

Brian, Eric. *La Mesure de l'Etat. Administrateurs et géomètres au XVIIIᵉ siècle*. Paris: Albin Michel, 1994.

Brissenden, R. F. *Virtue in Distress: Studies in the Novel of Sentiment from Richardson to Sade*. London: Macmillan, 1974.

Brock, William H. *The Norton History of Chemistry*. New York: W. W. Norton, 1993.

Brown, Marshall. *Preromanticism*. Stanford: Stanford University Press, 1991.

Brunet, Pierre. "La Vie et l'oeuvre de Clairaut." *Revue d'histoire des sciences* 5 (1952): 334–49.

Buchdahl, Gerd. *The Image of Newton and Locke in the Age of Enlightenment*. London: Sheed and Ward, 1961.

Burkhardt, Richard W. *The Spirit of System: Lamarck and Evolutionary Biology*. Cambridge: Harvard University Press, 1977.

Burwick, Frederick. "Romantic Drama: From Optics to Illusion." In *Literature and Science, Theory and Practice*, ed. Stuart Peterfreund, 167–208. Boston: Northeastern University Press, 1990.

Bynum, William F., and Roy Porter, eds. *Medical Fringe and Medical Orthodoxy, 1750–1850*. London: Croom Helm, 1987.

Cajori, Florian. *Mathematics in Liberal Education*. Boston: Christopher Publishing, 1928.

Canguilhem, Georges. *La Formation du concept de réflexe aux XVIIe et XVIIIe siècles*. Paris: Vrin, 1977.

Cassirer, Ernst. *The Philosophy of the Enlightenment*. Translated by Fritz Koelln and James Pettegrove. Princeton: Princeton University Press, 1951.

Certaux, M. de, D. Julia, and J. Revel. *Une Politique de la Langue—La Révolution française et les patois: L'enquête de Grégoire*. Paris, 1975.

Chandler, James, Arnold I. Davidson, and Harry Harootunian, eds. *Questions of Evidence: Proof, Practice, and Persuasion across the Disciplines*. Chicago: University of Chicago Press, 1993.

Chandler, Philip. "Clairaut's Critique of Newtonian Attraction: Some Insights into His Philosophy of Science." *Annals of Science* 32 (1975): 369–78.

Charlton, D. G. "The French Romantic Movement." In *The French Romantics*, 1:1–32. Cambridge: Cambridge University Press, 1984.

———. *New Images of the Natural in France: A Study in European Cultural History, 1750–1800*. Cambridge: Cambridge University Press, 1984.

Chartier, Roger. *The Cultural History of the French Revolution*. Durham: Duke University Press, 1991.

Chartier, Roger, Marie-Madeleine Compère, and Dominique Julia. *L'Education en France du XVIe siècle au XVIIIe siècle*. Paris: CDU et SEDES, 1976.

Chertok, Leon, and Isabelle Stengers. *A Critique of Psychoanalytic Reason: Hypnosis as a Scientific Problem from Lavoisier to Lacan*. Translated by Marie Noel Evans. Stanford: Stanford University Press, 1989.

Chinard, Gilbert, ed. *L'Apothéose de Franklin*. Paris, 1955.

———. *The Correspondence of Jefferson and Du Pont de Nemours, with an Introduction on Jefferson and the Physiocrats*. Baltimore: Johns Hopkins University Press, 1931.

Chisick, Harvey. *The Limits of Reform in the Enlightenment: Attitudes toward the Education of the Lower Classes in Eighteenth-Century France*. Princeton: Princeton University Press, 1981.

Christensen, Paul P. "Fire, Motion and Productivity: The Proto-Energetics of Nature and Economy in François Quesnay." In *Natural Images in Economic Thought*, ed. Philip Mirowski. Cambridge: Cambridge University Press, 1994.

Church, William F., ed. *The Influence of the Enlightenment on the French Revolution*. Lexington: D.C. Heath, 1974.

Clagett, Marshall, ed. *Critical Problems in the History of Science*. Madison: University of Wisconsin Press, 1959.

Clark, William, Jan Golinski, and Simon Schaffer. *The Sciences in Enlightened Europe*. Chicago: University of Chicago Press, 1999.

Cléro, Jean-Pierre. *La Philosophie des passions chez David Hume*. Paris: Klincksieck, 1985.

Cobban, Alfred. *The Social Interpretation of the French Revolution*. Cambridge: Cambridge University Press, 1964.

Cochin, Augustin. *L'Esprit de Jacobinisme*. 1920. Paris, 1979.

Cohen, I. Bernard. *Benjamin Franklin's Science*. Cambridge: Harvard University Press, 1990.

———. *Franklin and Newton*. Philadelphia: American Philosophical Society, 1956.

———. *The Newtonian Revolution, with Illustrations of the Transformations of Scientific Ideas*. Cambridge: Cambridge University Press, 1980.

———. *Science and the Founding Fathers: Science in the Political Thought of Thomas Jefferson, Benjamin Franklin and James Madison*. New York: W. W. Norton, 1995.

Connor, Paul W. *Poor Richard's Politics*. Oxford: Oxford University Press, 1965.

Cope, Zachary. *William Cheselden, 1688–1752*. Edinburgh, E. & S. Livingstone, 1953.

Cornette, Joël. *Histoire de la France: Absolutisme et Lumières, 1652–1783*. Paris: Hachette, 1993.

———. "Turgot, ou la dernière chance de la monarchie." *L'Histoire*, no. 191 (September 1995): 64–69.

Counson, Albert. *Franklin et Robespierre*. Paris: H. Champion, 1930.

Crabtree, Adam. *From Mesmer to Freud: Magnetic Sleep and the Roots of Psychological Healing*. New Haven: Yale University Press, 1993.

Cranston, Maurice. *Philosophers and Pamphleteers: Political Theorists of the Enlightenment*. Oxford: Oxford University Press, 1986.

Crosland, Maurice. *Historical Studies in the Language of Chemistry*. Cambridge: Harvard University Press, 1962.

———. "The Image of Science as a Threat: Burke versus Priestley and the 'Philosophic Revolution.'" *British Journal for the History of Science* 20 (1987): 277–307.

———. *In the Shadow of Lavoisier: The Annales de Chimie and the Establishment of a New Science*. Oxford: Alden, 1994.

———. "'Nature' and Measurement in Eighteenth-Century France." In *Studies on Voltaire and the Eighteenth Century*, ed. Theodore Besterman, 87: 277–309. Banbury: Voltaire Foundation, 1972.

Dagognet, François. *Tableaux et langages de chimie*. Paris: Editions du Seuil, 1969.

Dakin, Douglas. *Turgot and the Ancien Régime in France*. 1939. New York: Octagon, 1965.

Darnton, Robert. *The Forbidden Bestsellers of Pre-Revolutionary France*. New York: W. W. Norton, 1995.

———. *Mesmerism and the End of the Enlightenment in France*. Cambridge: Harvard University Press, 1968.

Daston, Lorraine. "Baconian Facts, Academic Civility, and the Prehistory of Objectivity." *Annals of Scholarship* 8, nos. 3–4 (1991): 337–64.

———. *Classical Probability in the Enlightenment*. Princeton: Princeton University Press, 1988.

———. "Fear and Loathing of the Imagination in Science." *Daedalus* 127, no. 1 (1998): 73–95.

———. "Marvelous Facts and Miraculous Evidence in Early Modern Europe." *Critical Inquiry* 18, no. 1 (fall 1991): 93–124.

———. "The Moral Economy of Science." *Osiris* 10 (1995): 3–24.

———. "Objectivity and the Escape from Perspective." *Social Studies of Science* 22 (November 1992): 597–618.

Daston, Lorraine, and Peter Galison. "The Image of Objectivity." *Representations* 40 (fall 1992): 81–128.

Daumas, Maurice. *Lavoisier: Théoricien et expérimentateur*. Paris: Presses Universitaires de France, 1955.

Davis, Kenneth S. *The Cautionary Scientists: Priestley, Lavoisier and the Founding of Modern Chemistry*. New York: G. Putnam's Sons, 1966.

Davis, Natalie Zemon. *Society and Culture in Early Modern France*. 1965. Stanford: Stanford University Press, 1975.

De Pas, Justin. *A Travers le vieux Saint Omer*. Paris, 1914.

De Vries, Jan. *The Economy of Europe in an Age of Crisis, 1600–1750*. New York: Cambridge University Press, 1976.

Dear, Peter. *Discipline and Experience: The Mathematical Way in the Scientific Revolution*. Chicago: University of Chicago Press, 1995.

————. "From Truth to Disinterestedness in the Seventeenth Century." *Social Studies of Science* 22 (1992): 619–31.

Debus, Allen. *Man and Nature in the Renaissance*. Cambridge: Cambridge University Press, 1978.

Degenaar, Marjolein. *Molyneux's Problem: Three Centuries of Discussion on the Perception of Forms*. Dordrecht: Kluwer, 1996.

Dérathé, Robert. *Jean-Jacques Rousseau et la science politique de son temps*. Paris, 1970.

Dhombres, Nicole, and Jean Dhombres. *Naissance d'un nouveau pouvoir: Sciences et savants en France, 1793–1824*. Paris: Payot, 1989.

Dibner, Bern. *Early Electrical Machines*. Norwalk, Conn.: Burndy Library, 1957.

Didier-Weygand, Zina. "La Cécité et les aveugles dans la société française: Representations et institutions du moyen âge aux premieres années du XIXe siècle." Thèse de doctorat, Université de Paris I–Panthéon Sorbonne U.F.R. d'Histoire, 1997–98.

Dijksterhuis, E. J. *The Mechanization of the World Picture*. Translated by C. Dikshoorn. Oxford: Oxford University Press, 1961.

Dinechin, Bruno Dupont de. *Duhamel du Monceau: Un Savant exemplaire au siècle des lumières*. Luxembourg: Connaissance et mémoires européennes, 1999.

Dobbs, Betty-Jo Teeter. *The Janus Faces of Genius: The Role of Alchemy in Newton's Thought*. Cambridge: Cambridge University Press, 1991.

Donovan, Arthur. *Antoine Lavoisier: Science, Administration and Revolution*. Oxford: Blackwell, 1993.

Doyle, William. *Origins of the French Revolution*. New York: Oxford University Press, 1989.

————. *The Parlement of Bordeaux at the End of the Old Regime, 1771–1790*. New York: St. Martin's, 1974.

Duchesneau, François. *La Physiologie des lumières: Empirisme, modèles et theories*. Boston: M. Nijhoff, 1982.

Duhem, Pierre. *Le Mixte*. Paris: Fayard, 1902.

Duris, Pascal. *Linné et la France (1780–1850)*. Geneva: Droz, 1993.

Duveen, Denis I., and Herbert S. Klickstein. "Benjamin Franklin (1706–1790) and Antoine Laurent Lavoisier (1743–1794)." [I] "Franklin and the New Chemistry." *Annals of Sciences* 2, no. 2 (June 1955): 103–29.

————. [II] "Joint Investigations." *Annals of Science* 2, no. 4 (December 1955): 271–302.

————. [III] "Documentation." *Annals of Science* 13, no. 1 (March 1957): 30–46.

Echeverria, Durand. *Mirage in the West: A History of the French Image of American Society to 1815*. Princeton: Princeton University Press, 1957.

Elias, Norbert. *The Civilizing Process*. Vol. 1, *The History of Manners*. New York: Pantheon, 1978.

Eltis, Walter. "L'Abbé de Condillac and the Physiocrats." *History of Political Economy* 27, no. 2 (1995): 217–36.

Erametsa, Erik. *A Study of the Word "Sentimental" and of Other Linguistic Characteristics of Eighteenth-Century Sentimentalism in England*. Helsinki, 1951.

Evans, G. "Molyneux's Question." In *Collected Papers*, edited by A. Phillips, 364–99. Oxford: Clarendon, 1985.

Faure, Edgar. *La Disgrâce de Turgot*. Paris: Gallimard, 1961.

Figlio, Karl. "Theories of Perception and the Physiology of Mind in the Late Eighteenth Century." *History of Science* 13 (1975): 177–212.

Figuier, Louis. *Les Merveilles de la science, ou Déscription populaire des inventions modernes*. 4 vols. Paris: Furne, Jouvet et cie, 1870.

Fontaine, Philippe. "Turgot's 'Institutional Individualism.'" *History of Political Economy* 19, no. 1 (1997): 1–20.

Ford, Franklin. *Robe and Sword: The Regrouping of the French Aristocracy After Louis XIV*. New York: Harper, 1965.

Foucault, Michel. *Les Mots et les choses: Une Archaeologie des sciences humaines*. Paris, 1966.

Fox-Genovese, Elizabeth. *The Origins of Physiocracy*. Ithaca: Cornell University Press, 1976.

Frängsmyr, Tore, ed. *Solomon's House Revisited: The Organization and Institutionalization of Science*. Nobel Symposium 75. Canton, Mass.: Science History Publications, 1990.

Frängsmyr, Tore, J. L. Heilbron, and Robin E. Rider. *The Quantifying Spirit in the Eighteenth Century*. Berkeley: University of California Press, 1990.

Frasca-Spada, Marina. "The Science and Conversation of Human Nature." In *The Sciences in Enlightened Europe*, ed. Clark, Golinski, and Schaffer, 218–45.

French, Sidney J. *Torch and Crucible: The Life and Death of Antoine Lavoisier*. Princeton: Princeton University Press, 1941.

Frye, Northrop. "Towards Defining an Age of Sensibility." *English Literary History* 23, no. 2 (June 1956): 144–52.

———. "Varieties of Eighteenth-Century Sensibility." *Eighteenth-Century Studies* 24, no. 2 (1990–91): 157–72.

Furbank, P. N. *Diderot: A Critical Biography*. New York: Knopf, 1992.

Furet, François. *Penser la Révolution française*. Paris: Gallimard, 1978.

Furet, François, and Mona Ozouf, eds. *Dictionnaire critique de la Révolution Française: Idées*. Paris: Flammarion, 1992.

Gardiner, H. Norman. *Feeling and Emotion: A History of Theories*. New York: America Book Company, 1937.

Gay, Peter. *The Enlightenment: An Interpretation*. 2 vols. New York: Norton, 1966, 1969.

Gibbs, F. W. *Joseph Priestley: Revolutions of the Eighteenth Century*. New York: Doubleday, 1965.

Gillispie, Charles. "The *Encyclopédie* and the Jacobin Philosophy of Science: A Study in Ideas and Consequences." In *Critical Problems in the History of Science*, ed. Clagett, 255–90.

———. "The Liberating Influence of Science in History." *Aspects of American Liberty, Philosophical, Historical and Political. Memoirs of the American Philosophical Society* 118 (1977): 37–56.

———. "Probability and Politics: Laplace, Condorcet, and Turgot." *Proceedings of the American Philosophical Society* 166, no. 1 (February 1972): 1–19.

————. *Science and Polity at the End of the Old Regime*. Princeton: Princeton University Press, 1980.

Godechot, Jacques. *The Counter-Revolution, Doctrine and Action*. Princeton: Princeton University Press, 1981.

Golan, Tal. "The Authority of Shadows; The Legal Embrace of the X-Ray." In *Evidence and the Law*, ed. Robert A. Nye. *Historical Reflections* 24, no. 3 (1998): 437–58.

————. "The History of Scientific Expert Testimony in the English Courtroom." *Science in Context* 12 (1999): 7–32.

Golan, Tal, and Snait Gissis, eds. *Science and Law. Science in Context* 12, no. 1 (spring 1999).

Goldstein, Jan. "Enthusiasm or Imagination? Eighteenth-Century Smear Words in Comparative National Context." *Huntington Library Quarterly* 60, nos. 1 & 2 (1998): 29–49.

Golinski, Jan. "The Chemical Revolution and the Politics of Language." *The Eighteenth Century Theory and Interpretation* 33, no. 3 (fall 1992): 238–51.

————. *Science as Public Culture: Chemistry and Enlightenment in Britain, 1760–1820*. Cambridge: Cambridge University Press, 1992.

Gordon, Daniel. *Citizens without Sovereignty: Equality and Sociability in French Thought, 1670–1789*. Princeton: Princeton University Press, 1994.

Grimsley, Ronald. "The Idea of Nature in Montesquieu's Lettres Persanes." In *From Montesquieu to Laclos: Studies in the French Enlightenment*. Geneva: Droz, 1974.

Gross, Charles G. *Brain, Vision, Memory: Tales in the History of Neuroscience*. Cambridge: MIT Press, 1999.

Gross, Paul R., and Norman Levitt. *Higher Superstition: The Academic Left and Its Quarrels with Science*. Baltimore: Johns Hopkins University Press, 1994.

Gross, Paul R., Norman Levitt, and Martin W. Lewis, eds. *The Flight from Science and Reason*. New York: American Academy for the Advancement of the Sciences, 1996.

Grunwald Center for the Graphic Arts. *French Caricature and the French Revolution, 1789–1799*. Los Angeles, 1988.

Guerlac, Henry. *Essays and Papers in the History of Modern Science*. Baltimore: Johns Hopkins University Press, 1977.

————. *Lavoisier—The Crucial Year: The Background and Origins of His First Experiments on Combustion in 1772*. Ithaca: Cornell University Press, 1961.

————. "On the Papers of Charles Coulston Gillispie and L. Pearce Williams." In *Critical Problems in the History of Science*, ed. Clagett, 317–20.

Guilhaumou, Jacques. *La Langue politique et la Révolution française*. Paris: Méridiens Klincksieck, 1989.

Gusdorf, Georges. *La Conscience Révolutionnaire, Les Idéologues*. Paris: Payot, 1978.

Hacking, Ian. *The Emergence of Probability: A Philosophical Study of Early Ideas about Probability, Induction and Statistical Inference*. Cambridge: Cambridge University Press, 1975.

Hackman, W. D. *Electricity from Glass: The History of the Frictional Electrical Machine 1600–1850*. Alphen aan den Rijn: Sijthoff & Norrdhoff, 1978.

Hagner, Michael. "Enlightened Monsters." In *The Sciences in Enlightened Europe*, ed. Clark, Golinski, and Schaffer, 175–217.

Hahn, Roger. *The Anatomy of a Scientific Institution: The Paris Academy of Sciences, 1666–1803*. Berkeley: University of California Press, 1971.

————. "Lavoisier et ses collaborateurs: Une équipe au travail." In *Il y a 200 ans Lavoi-*

sier: Actes du Colloque organisé à l'occasion du bicentenaire de la mort d'Antoine Laurent Lavoisier, ed. Christiane Demeulenaere-Douyère. Paris: Académie des Sciences, 1995.

Haigh, Elizabeth L. "Vitalism, the Soul, and Sensibility: The Physiology of Théophile Bordeu." *Journal of the History of Medicine and Allied Sciences* 31 (1976): 30–41.

Hale, Edward E., and Edward E. Hale Jr. *Franklin in France*. 2 vols. Boston: Roberts 1887–88.

Halevi, Ran. *Les Loges maconniques dans la France d'Ancien Régime*. Paris: A. Colin, 1984.

Hall, A. R. *Philosophers at War: The Quarrel between Newton and Leibniz*. Cambridge: Cambridge University Press, 1980.

Hankins, Thomas L. *Jean D'Alembert. Science and the Enlightenment*. Oxford: Clarendon, 1970.

———. *Science and the Enlightenment*. Cambridge, Cambridge University Press, 1985.

Hankins, Thomas L., and Robert J. Silverman. *Instruments and the Imagination*. Princeton: Princeton University Press, 1995.

Hannaway, Owen. *The Chemists and the Word*. Baltimore: Johns Hopkins University Press, 1975.

Hardman, John. *French Politics, 1774–1789: From the Accession of Louis XVI to the Fall of the Bastille*. London: Longman, 1995.

Harris, Robert D. *Necker, Reform Statesman of the Ancien Régime*. Berkeley: University of California Press, 1979.

———. *Necker and the Revolution of 1789*. Lanham, Md.: University Press of America, 1986.

Hatfield, Gary. "Remaking the Science of Mind." In *Inventing Human Science: Eighteenth-Century Domains*, ed. Christopher Fox, Roy Porter, and Robert Wokler. Berkeley: University of California Press, 1995.

Hazard, Paul. *The European Mind 1689–1715*. New Haven: Yale University Press, 1953.

———. *European Thought in the Eighteenth Century, from Montesquieu to Lessing*. London: Hollis and Carter, 1954.

Heilbron, J. L. "Fin de Siècle Physics." In *Science, Technology and Society in the Time of Alfred Nobel*, ed. Carl Gustaf Bernhard, Elisabeth Crawford, and Per Sörbom, 51–73. Oxford: Pergamon, 1982.

———. "Franklin as an Enlightened Natural Philosopher." In *Reappraising Benjamin Franklin: A Bicentennial Perspective*, ed. J. A. Leo Lemay, 8–14. Newark: University of Delaware Press, 1993.

———. "Franklin, Haller and Franklinist History." *Isis* 68, no. 4 (1977): 539–49.

———. *The History of Electricity in the 17th and 18th Centuries*. Berkeley: University of California Press, 1979.

———. "The Measure of Enlightenment." In *The Quantifying Spirit in the Eighteenth Century*, ed. Frängsmyr, Heilbron, and Rider, chap. 7.

Heilbroner, Robert L. *The Worldly Philosophers: The Lives, Times and Ideas of the Great Economic Thinkers*. New York: Simon and Schuster, 1980.

Hesse, Carla. "Enlightenment Epistemology and the Laws of Authorship in Revolutionary France 1777–1793." *Representations* 30 (spring 1990): 109–37.

———. *Publishing and Cultural Politics in Revolutionary Paris: 1789–1810*. Berkeley: University of California Press, 1989.

Higonnet, Patrice. "The Politics of Linguistic Terrorism." *Social History* 5 (1980): 41–69.

Hill, Henry Bertram. "On the Papers of Charles Couston Gillispie and L. Pearce Williams." In *Critical Problems in the History of Science*, ed. Clagett, 309–16.

Hilles, F., and H. Bloom, eds. *From Sensibility to Romanticism*. New York: Oxford University Press, 1965.

Hirschman, Albert O. *The Passions and the Interests: Political Arguments for Capitalism Before Its Triumph*. 1977. Princeton: Princeton University Press, 1997.

Holmes, Frederic Lawrence. *Eighteenth-Century Chemistry as an Investigative Enterprise*. Berkeley Papers in History of Science, vol. 12. Berkeley: University of California at Berkeley, 1989.

———. *Lavoisier and the Chemistry of Life, An Exploration of Scientific Creativity*. Madison: University of Wisconsin Press, 1985.

Home, Roderick. *Electricity and Experimental Physics*. Hampshire: Variorum, 1992.

Hordé, Tristan. "Les Idéologues: Théorie due signe, sciences et enseignement." *Langages* 45 (March 1977): 43–66.

Huet, Marie Hélène. *Monstrous Imagination*. Cambridge: Harvard University Press, 1993.

———. "Thunder and Revolution: Franklin, Robespierre and Sade." *The Eighteenth Century: Theory and Interpretation* (1989, special issue): 13–32.

Hunt, Lynn, ed. *Eroticism and the Body Politic*. Baltimore: Johns Hopkins University Press, 1991.

Ilie, Paul. *The Age of Minerva*, Vol. 2, *Cognitive Discontinuities in Eighteenth Century Thought, From Body to Mind in Physiology and the Arts*. Philadelphia: University of Pennsylvania Press, 1995.

Ingrao, Bruna, and Giorgio Israel. *The Invisible Hand: Economic Equilibrium in the History of Science*. 1990. Reprint, Cambridge: MIT Press, 2000.

Isambert, François André, Decrusy, and Athanase Jean Leger Jourdan, eds. *Recueil Général des Anciennes Lois Françaises depuis l'an 420 jusqu'à la Révolution de 1789*. Paris: Plon Frères, 1822–33.

Jacob, L. "Un Ami de Robespierre, Buissart (d'Arras)." *Revue du Nord* 20, no. 80 (November 1934), 277–94.

Jacobs, Margaret. *The Cultural Meaning of the Scientific Revolution*. Philadelphia: Temple University Press, 1980.

Jaume, Lucine. *Le Discours Jacobin et la démocratie*. Paris: Fayard, 1989.

Jehlen, Myra. "Imitate Jesus and Socrates: The Making of a Good American." *South Atlantic Quarterly* 89, no. 3 (summer 1990): 501–24.

Jessenne, Jean-Pierre, Gilles Deregnaucourt, Jean-Pierre Hirsch, and Hervé Leuwers, eds. *Robespierre: De la Nation artésienne à la Républiques et aux Nations*. Villeneuve d'Asq: Université Charles de Gaulle-Lille III, 1994.

Jones, Chris. *Radical Sensibility: Literature and Ideas in the 1790s*. London: Routledge, 1993.

Josipovici, Jean. *Franz Anton Mesmer, Magnétiseur, Médecin, et Franc-Maçon*. Monaco: Editions du Rocher, 1982.

Kaplan, Lawrence S. *Jefferson and France: An Essay on Politics and Political Ideas*. New Haven: Yale University Press, 1967.

Kaplan, Steven. *Bread, Politics and Political Economy in the Reign of Louis XV*. 2 vols. The Hague: Martinus Nijhoff, 1976.

———. *Provisioning Paris: Merchants and Millers in the Grain and Flour Trade During the Eighteenth Century*. Ithaca: Cornell University Press, 1984.

Kaptchuk, Ted J. "Intentional Ignorance: A History of Blind Assessment and Placebo Controls in Medicine." *Bulletin of the History of Medicine* 72, no. 3 (1998): 389–433.

Keohane, Nannerl. *Philosophy and the State in France, the Renaissance to the Enlightenment.* Princeton: Princeton University Press, 1980.

Kiernan, Colm. *Science and the Enlightenment in Eighteenth-Century France.* Geneva: Institut et musée Voltaire, 1968.

Kintzler, Catherine. *Condorcet: L'Instruction publique et la naissance du citoyen.* Paris: S.F.I.E.D., 1984.

Knight, David. *The Age of Science.* Oxford: Basil Blackwell, 1986.

Knight, Isabelle. *The Geometric Spirit: The Abbé de Condillac and the French Enlightenment.* New Haven: Yale University Press, 1968.

Koyré, Aléxandre. *Etudes newtoniennes.* Paris: Gallimard, 1968.

Koyré, Aléxandre, and I. B. Cohen. "Newton and the Leibniz-Clarke Correspondence." *Archives Internationales d'Histoire des Sciences* 15 (1962): 63–126.

Kramnick, Isaac. "Eighteenth-Century Science and Radical Social Theory: The Case of Joseph Priestley's Scientific Liberalism." In *The Scientific Enterprise,* ed. Edna Ullmann-Margalit, vol. 4. Dordrecht: Kluwer, 1992.

Krieger, Lawrence. *Kings and Philosophers 1689–1789.* New York: Norton, 1970.

Kubrin, D. "Newton and the Cyclical Cosmos: Providence and the Mechanical Philosophy." *Journal of the History of Ideas* 28 (1967): 325–26.

Lambert, Jacques. "Analyse Lavoisienne et Chimie Condillacienne?" In *Condillac et les problèmes du language,* ed. Jean Sgard, 369–77. Geneva: Editions Slatkine, 1982.

Lane, Harlan, ed. *The Wild Boy of Aveyron.* Cambridge: Harvard University Press, 1976.

Langins, Janis. *La République avait besoin des savants.* Paris: Belin, 1987.

———. "Words and Institutions During the French Revolution: The Case of 'Revolutionary' Scientific and Technical Education." In *The Social History of Language,* ed. Peter Burke and Roy Porter, 136–60. Cambridge: Cambridge University Press, 1987.

Larrère, Catherine. *L'Invention de l'économie au XVIIIe siècle.* Paris: Presses Universitaires de France, 1992.

Lartichaux, J.-Y. "Linguistic Politics During the French Revolution." *Diogenes* 97 (1977): 65–84.

Lesch, John E. "Systematics and the Geometrical Spirit." In *The Quantifying Spirit in the Eighteenth Century,* ed. Frängsmyr, Heilbron, and Rider, 73–111.

Levere, Trevor. *Chemists and Chemistry in Nature and Society, 1770–1878.* Aldershot: Variorum, 1994.

———. "Lavoisier: Language, Instruments and the Chemical Revolution." In *Nature, Experiment and the Sciences,* ed. Trevor Levere and W. R. Shea, 207–23. Dordrecht: Kluwer, 1990.

Lingo, Alison Klairmont. "Empirics and Charlatans in Early Modern France: The Genesis of the Classification of the 'Other' in Medical Practice." *Journal of Social History* 19 (1986): 583–604.

Lipatti, Valentin. *Mémoires de Beaumarchais dans l'affaire Goezman.* Paris: Nagel, 1974.

Lopez, Claude-Anne. *Mon Cher Papa: Franklin and the Ladies of Paris.* 1966. New Haven: Yale University Press, 1990.

———. "Saltpetre, Tin and Gunpowder: Addenda to the Correspondence of Lavoisier and Franklin." *Annals of Science* 16, no. 2 (1960): 83–94.

Lough, John. *The Encyclopédie.* New York: David McKay Company, 1971.

Maclauren, Lois. *Franklin's Vocabulary*. New York: Doubleday, 1928.

Mah, Harold. "The Epistemology of the Sentence: Language, Civility and Identity in France and Germany, Diderot to Nietzsche." *Representations* 47 (summer 1994): 64–84.

Manent, Pierre. *An Intellectual History of Liberalism*. Translated by Rebecca Balinski. Princeton: Princeton University Press, 1994.

Manicas, Peter. *A History and Philosophy of the Social Sciences*. New York: Basil Blackwell, 1987.

Marion, Marcel. *Dictionnaire des Institutions aux XVII et XVIIIe siècles*. Paris: Antoine Picard, 1923.

Mathiez, Albert. *La Théophilanthropie et le culte décadaire, 1796–1801; essai sur l'histoire religieuse de la revolution*. Paris: Felix Alcan, 1904.

Maurepas, Arnaud de, and Antoine Boulant. *Les Ministres et les ministères du siècle des lumières (1715–1789)*. Paris: Christian/JAS, 1996.

Maza, Sarah. *Private Lives and Public Affairs: The Causes Célèbres of Prerevolutionary France*. Berkeley: University of California Press, 1993.

———. "Le Tribunal de la nation: Les Mémoires judiciaries et l'opinion publique à la fin de l'ancien régime." *Annales: Economies, Sociétés, Civilisations* 42, no. 1 (1987): 73–90.

Mazzolini, Renato G. *The Iris in Eighteenth-Century Physiology*. Bern: Hans Huber, 1980.

McEvoy, John. "The Chemical Revolution in Context." *The Eighteenth Century: Theory and Interpretation* 33, no. 3 (fall 1992): 198–216.

———. "Continuity and Discontinuity in the Chemical Revolution." *Osiris* 4, 2nd ser., (1988): 195–213.

———. "The Enlightenment and the Chemical Revolution." In *Metaphysics and Philosophy of Science in the Seventeenth and Eighteenth Centuries*, ed. R. S. Woolhouse, 307–25. Dordrecht: Kluwer, 1988.

———. "Priestley Responds to Lavoisier Nomenclature: Language, Liberty and Chemistry in the English Enlightenment." In *Lavoisier in European Context*, ed. Bensaude-Vincent and Abbri, 123–42.

McIntyre, Jane L. "Personal Identity and the Passions." *Journal of the History of Philosophy* 27, no. 4 (1989): 545–56.

McKie, Douglas. *Antoine Lavoisier: Scientist, Economist, Social Reformer*. New York: Harper & Row, 1952.

———. "Introduction." In Lavoisier, *Elements of Chemistry*.

Meek, Ronald. *The Economics of Physiocracy*. Cambridge: Harvard University Press, 1962.

Metzger, Hélène. *Les Doctrines chimiques en France du début du XVIIe siècle à la fin du XVIIIe siècle*. 1923. Paris: Albert Blanchard, 1969.

———. *La Genèse de la sciences des cristaux*. Paris: Felix Alcan, 1918.

———. *Newton, Stahl, Boerhaave et la doctrine chimique*. 1930. Paris: Albert Blanchard, 1974.

Miller, Jonathan. "Going Unconscious." *New York Review of Books* (20 April 1995): 59–65.

Mitchell, Trent A. "The Politics of Experiment in the Eighteenth Century: The Pursuit of Audience and the Manipulation of Consensus in the Debate over Lightning Rods." *Eighteenth-Century Studies* 31, no. 3 (1998): 307–31.

Mnookin, Jennifer. "The Image of Truth: Photographic Evidence and the Power of Analogy." *Yale Journal of Law and the Humanities* 10, no. 1 (1998): 1–74.

Moravia, Sergio. "The Enlightenment and the Sciences of Man." *History of Science* 18 (1980): 247–68.

———. "From 'Homme machine' to 'Homme sensible': Changing Eighteenth-Century Models of Man's Image." *Journal of the History of Ideas* 39, no. 1 (1978): 45–60.

Morgan, Michael. *Molyneux's Question: Vision, Touch, and the Philosophy of Perception.* Cambridge: Cambridge University Press, 1977.

Morilhat, Claude. *La Prise de conscience du capitalisme.* Paris: Méridiens Klincksieck, 1988.

Mornet, Daniel. *Les Origines intellectuelles de la Révolution française.* Paris: Armand Colin, 1933.

———. *Le Sentiment de la nature: De J.-J. Rousseau à Bernardin de Saint-Pierre.* Paris: Hachette, 1907.

———. *La Pensée française au dix-huitième siècle.* Paris: Armand Colin, 1947.

Mousnier, Roland. *The Institutions of France under the Absolute Monarchy 1598–1789.* 2 vols. Translated by Brian Pierce. Chicago: University of Chicago Press, 1979–84.

Mullan, John. *Sentiment and Sociability: The Language of Feeling in the Eighteenth Century.* New York: Oxford University Press, 1988.

Norton, David Fate, ed. *The Cambridge Companion to Hume.* Cambridge: Cambridge University Press, 1993.

Nye, Robert A., ed. *Evidence and the Law. Historical Reflections/Réflexions historiques* 24, no. 3 (fall 1998).

O'Brien, Conor Cruise. *The Long Affair: Thomas Jefferson and the French Revolution, 1785–1800.* Chicago: University of Chicago Press, 1996.

Olson, Richard. *The Emergence of the Social Sciences, 1642–1792.* New York: Maxwell Macmillan International, 1993.

Oncken, August. *Die Maxime Laissez faire et laissez passer.* Bern, 1886.

Outram, Dorinda. "Science and the Terror: The Paris Academy of Sciences and the Terror." *History of Science* 21, part 3, no. 53 (1983), 251–73.

Pack, Spencer J. "Theological (and hence Economic) Implications of Adam Smith's 'Principles which Lead and Direct Philosophical Inquiries.'" *History of Political Economy* 27, no. 2 (1995): 289–307.

Palmer, R. R. *The Improvement of Humanity: Education and the French Revolution.* Princeton: Princeton University Press, 1985.

Pappas, John. "L'Esprit de finesse contre l'esprit de géometrie: Un Débat entre Diderot et d'Alembert." *Studies on Voltaire and the Eighteenth Century* 89 (1972): 1229–53.

Paris, J. A. *La Jeunesse de Robespierre.* Arras, 1870.

Partington, J. R. *Historical Studies on the Phlogiston Theory.* [I] "The Levity of Phlogiston." *Annals of Science* 2, no. 4 (October 1937): 361–404; [II] "The Negative Weight of Phlogiston." *Annals of Science* 3, no. 1 (January 1938): 1–58; [III] "Light and Heat in Combustion." *Annals of Science* 3, no. 4 (October 1938): 337–71; [IV] "Last Phases of the Theory." *Annals of Science* 4, no. 2 (April 1939): 113–49.

———. *A Short History of Chemistry.* 3rd ed. New York: Harper & Row, 1957.

Pastore, Nicholas. *Selective History of Theories of Visual Perception, 1650–1950.* Oxford: Oxford University Press, 1971.

Paulson, William R. *Enlightenment, Romanticism and the Blind in France.* Princeton: Princeton University Press, 1987.

Pinch, Adela. *Strange Fits of Passion: Epistemologies of Emotion, Hume to Austen.* Stanford: Stanford University Press, 1996.

Pitcher, George. "Molyneux's Problem." *Journal of Philosophy* 71, no. 18 (24 October 1974): 637–53.

Poirier, Jean-Pierre. "Antoine-Laurent Lavoisier, théoricien et praticien de l'économie." Thèse de doctorat, Université de Paris II, 1992.

———. *Lavoisier*. Paris: Editions Pygmalion, 1993.

———. *Lavoisier: Chemist, Biologist, Economist*. Translated by Rebecca Balinski. Philadelphia: University of Pennsylvania Press, 1993.

———. *Turgot: Laissez-faire et progrès social*. Paris: Perrin, 1999.

Poovey, Mary. *A History of the Modern Fact: Problems of Knowledge in the Sciences of Wealth and Society*. Chicago: University of Chicago Press, 1998.

Porter, Roy. *Health for Sale: Quackery in England, 1660 – 1850*. Manchester: Manchester University Press, 1989.

Porter, Theodore M. *The Rise of Statistical Thinking, 1820 – 1900*. Princeton: Princeton University Press, 1986.

———. *Trust in Numbers: The Pursuit of Objectivity in Science and Public Life*. Princeton: Princeton University Press, 1995.

————, ed. "The Social History of Objectivity." *Social Studies of Science* 22, no. 4 (1992): 595 – 652.

Proust, Jacques. *L'Encyclopédie*. Paris: A. Colin, 1965.

Py, Gilbert. *Rousseau et les éducateurs: Etude sur la fortune des idées pédagogiques de Jean-Jacques Rousseau en France et en Europe au XVIIIᵉ siècle*. Oxford: Voltaire Foundation, 1997.

Rey, Roselyne. "La partie, le tout et l'individu: Science et philosophie dans l'oeuvre de Charles Bonnet." In *Charles Bonnet: Savant et philosophe (1720 – 1793), Actes du Colloque international de Genève (25 – 27 novembre 1993). Mémoires de la Société Physique et d'Histoire Naturelle de Genève*, ed. Marino Buscaglia, René Sigrist, Jacques Trembley, and Jean Wüest, 47:61–75. Geneva: Editions Passé Présent, 1994.

Richetti, John, ed. *The Cambridge Companion to the Eighteenth-Century Novel*. Cambridge: Cambridge University Press, 1996.

Rider, Robin E. "Measure of Ideas, Rule of Language: Mathematics and Language in the 18th Century." In *The Quantifying Spirit in the Eighteenth Century*, ed. Frängsmyr, Heilbron, and Rider, 113–40.

Ridgeway, R. S. *Voltaire and Sensibility*. Montreal: McGill-Queen University Press, 1973.

Rieu, Alain-Marc. "Le Complexe Nature-Science-Langage chez Condillac." In *Condillac et les Problèmes du Language*, ed. Jean Sgard, 27–46. Geneva: Editions Slatkine, 1982.

Riskin, Jessica. "The Lawyer and the Lightning Rod." *Science in Context* 12, no. 1 (1999): 61–99.

———. "Poor Richard's Leyden Jar: Electricity and Economy in Franklinist France." *Historical Studies in the Physical and Biological Sciences* 28, no. 2 (1998): 301–36.

———. "Rival Idioms for a Revolutionized Science and a Republican Citizenry." *Isis* 89 (1998): 203–32.

Roberts, Lissa. "Condillac, Lavoisier and the Instrumentalization of Science." *The Eighteenth Century Theory and Interpretation* 33, no. 3 (fall 1992): 252–71.

———. "The Death of the Sensuous Chemist: The 'New' Chemistry and the Transformation of Sensuous Technology." *Studies in the History and Philosophy of Science* 26, no. 4 (1995): 503–29.

Roche, Daniel. *Le Siècle des lumières en province: Académies et académiciens provinciaux, 1689–1789.* 2 vols. Paris: Mouton, 1978.

Rodgers, James. "Sensibility, Sympathy, Benevolence: Physiology and Moral Philosophy in 'Tristram Shandy.'" In *Languages of Nature: Critical Essays on Science and Literature,* ed. L. J. Jordanova, 117–58. New Brunswick: Rutgers University Press, 1986.

Roger, Jacques. *Buffon: Un Philosophe au Jardin du Roi.* Paris: Fayard, 1989.

———. *Les Sciences de la vie dans la pensée française au XVIIIᵉ siècle.* Paris: Armand Colin, 1963.

Roll, Eric. *The History of Economic Thought.* London: Faber and Faber, 1954.

Rorty, Amelie Oksenberg. "From Passions to Emotions and Sentiments." *Philosophy* 57 (1982): 159–72.

Rosenfeld, Sophia A. "Language and Deviancy: Deaf Men on Trial in Late Eighteenth-Century France." *Eighteenth-Century Life* 21, no. 2 (1997): 157–75.

———. *A Revolution in Language: The Problem of Signs in Late Eighteenth-Century France.* Stanford: Stanford University Press, 2001.

Ross, Andrew, ed. *Science Wars.* Durham: Duke University Press, 1996.

Rothschild, Emma. *Economic Sentiments: Adam Smith, Condorcet and the Enlightenment.* Cambridge: Harvard University Press, 2001.

Rousseau, G. S. "Discourses of the Nerve." In *Literature and Science as Modes of Expression,* ed. Frederick Amrine, 29–60. Dordrecht: Kluwer, 1989.

———. "Nerves, Spirits, and Fibres: Towards Defining the Origins of Sensibility." *Studies in the Eighteenth Century III: Papers Presented at the Third David Nichol Smith Memorial Seminar.* Toronto: University of Toronto Press, 1976.

———. "Pineapples, Pregnancy, Pica, and the *Peregrine Pickle.*" In *Tobias Smollett: Bicentennial Essays,* ed. G. S. Rousseau and P. G. Boucé. New York: Oxford University Press, 1971.

———. "Science and the Discovery of the Imagination in Enlightened England." *Eighteenth-Century Studies* 3 (1969): 108–35.

———. "Sensibility Reconsidered." *Medical History* 39 (1995): 375–77.

———, ed. *The Languages of Psyche: Mind and Body in Enlightenment Thought.* Berkeley: University of California Press, 1990.

Rousseau, Nicolas. *Connaissance et langage chez Condillac, Histoire des idées et critique littéraire.* Vol. 242. Geneva: Droz, 1986.

Rudé, Georges. *The Crowd in the French Revolution.* Oxford: Oxford University Press, 1959.

———. *The Crowd in History, 1730–1848.* New York: Wiley, 1964.

Salomon, Jean-Jacques. *Science and Politics.* Translated by Noël Lindsay. London: Macmillan, 1973.

Sambrook, James. *The Eighteenth Century: The Intellectual and Cultural Context of English Literature 1700–1789.* London: Longman, 1986.

Sardar, Ziauddin. *Thomas Kuhn and the Science Wars.* New York: Totem Books, 2000.

Schaffer, Simon. "Natural Philosophy and Public Spectacle in the Eighteenth Century." *History of Science* 21, part 1, no. 51 (1983): 1–43.

———. "Self Evidence." 1992. In *Questions of Evidence,* ed. Chandler et al., 56–91.

Schelle, Gustave. *Dupont de Nemours et l'école physiocratique.* Paris: Librairie Guillaumin, 1888.

Schofield, Robert, *The Enlightenment of Joseph Priestley: A Study of His Life and Work from 1733 to 1773*. University Park: Penn State University Press, 1998.

————, ed. *The Scientific Autobiography of Joseph Priestley*. Cambridge: MIT Press, 1966.

Segerstråle, Ullica, ed. *Beyond the Science Wars: The Missing Discourse about Science and Society*. Albany: State University of New York Press, 2000.

Senden, Marius von. *Space and Sight; the Perception of Space and Shape in the Congenitally Blind Before and After Operation*. London: Methuen, 1960.

Sgard, Jean, ed. *Condillac et les problèmes du langage*. Geneva: Slatkine, 1982.

Shackleton, Robert. "The Encyclopaedic Spirit." In *Greene Centennial Studies: Essays Presented to Donald Greene in the Centennial Year of the University of Southern California*, ed. Paul J. Korshin and Robert R. Allen, 377–90. Charlottesville: University Press of Virginia, 1984.

————. *Essays on Montesquieu and on the Enlightenment*. Edited by David Gilson and Martin Smith. Oxford: Alden, 1988.

Shapin, Steven. "Of Gods and Kings: Natural Philosophy and Politics in the Leibniz-Clarke Disputes." *Isis* 72 (1981): 187–215.

————. *A Social History of Truth*. Chicago: University of Chicago Press, 1994.

Shapin, Steven, and Simon Schaffer. *Leviathan and the Air-Pump: Hobbes, Boyle and the Experimental Life*. Princeton: Princeton University Press, 1985.

Shapiro, Barbara. *Probability and Certainty: A Study of the Relationships between Natural Science, Religion, History, Law and Literature*. Princeton: Princeton University Press, 1983.

Shklar, Judith. *Montesquieu*. New York: Oxford University Press, 1987.

Siegfried, Robert. "The Chemical Revolution in the History of Chemistry." In *The Chemical Revolution: Essays in Reinterpretation*, ed. Arthur Donovan. *Osiris* 4, 2nd ser. (1988): 34–50.

Simms, J. G. *William Molyneux of Dublin, 1656–1698*. Blackrock: Irish Academic Press, 1982.

Smeaton, W. A. *Fourcroy: Chemist and Revolutionary*. Cambridge: W. Heffer & Sons, 1962.

Smith, Olivia. *The Politics of Language, 1791–1819*. Oxford: Clarendon, 1984.

Smith, Roger. "The Background of Physiological Psychology in Natural History." *History of Science* 11 (1973): 75–123.

Spary, E. C. "The 'Nature' of Enlightenment." In *The Sciences in Enlightened Europe*, ed. Clark, Golinski, and Schaffer, 272–306.

————. *Utopia's Garden: French Natural History from Old Regime to Revolution*. Chicago: University of Chicago Press, 2000.

Spink, John S. "Marivaux: The 'Mechanism of the Passions' and the 'Metaphysic of Sentiment.'" *Modern Language Review* 73 (1978): 278–90.

————. "'Sentiment,' 'sensible,' 'sensibilité': Les Mots, les idées, d'après les 'moralistes' français et britanniques du début du dix-huitième siècle." *Zagadnienia Rodzagów Literackich* 20 (1977): 33–47.

Stafford, Barbara. *Artful Science: Enlightenment Entertainment and the Eclipse of Visual Education*. Cambridge: MIT Press, 1994.

Stafleu, Frans A. *Linnaeus and the Linnaeans: The Spreading of Their Ideas in Systematic Botany*. Utrecht: A. Oosthoek's Uitgeversmaatschappij N.V., 1971.

Starobinski, Jean. *Jean-Jacques Rousseau: Transparency and Obstruction*. 1971. Translated by Arthur Goldhammer. Chicago: University of Chicago Press, 1988.

Staum, Martin S. *Cabanis: Enlightenment and Medical Philosophy in the French Revolution*. Princeton: Princeton University Press, 1980.

Stephanson, Raymond. "Richardson's 'Nerves': The Physiology of Sensibility in 'Clarissa.'" *Journal of the History of Ideas* 49 (1988): 267–85.

Stewart, Larry R. *The Rise of Public Science: Rhetoric, Technology, and Natural Philosophy in Newtonian Britain, 1660–1750*. New York: Cambridge University Press, 1992.

Stock-Morton, Phyllis. *Moral Education for a Secular Society: The Development of Moral Laïque in 19th Century France*. Albany: SUNY Press, 1988.

Stone, Bailey. *The French Parlements and the Crisis of the Old Regime*. Chapel Hill: University of North Carolina Press, 1986.

———. *The Parlement of Paris, 1774–1789*. Chapel Hill: University of North Carolina Press, 1981.

Stroud, Barry. *Hume*. London: Routledge, 1977.

Sutton, Geoffrey. *Science for a Polite Society: Gender, Culture, and the Demonstration of Enlightenment*. Boulder: Westview, 1995.

Swain, Virginia E. "Lumières et Vision: Reflections on Sight and Seeing in Seventeenth- and Eighteenth-Century France." *L'Esprit Créateur* 28 (1988): 7–16.

Swann, Julian. *Politics and the Parlement of Paris During the Reign of Louis XV*. Cambridge: Cambridge University Press, 1995.

Taine, Hippolyte. *The Ancient Regime*. 1876. Translated by John Durand. New York: Peter Smith, 1931.

Taton, René. *Enseignement et diffusion des sciences en France au XVIIIᵉ siècle*. Paris: Hermann, 1964.

Terrall, Mary. "Metaphysics, Mathematics and the Gendering of Science in Eighteenth-Century France." In *The Sciences in Enlightened Europe*, ed. Clark, Golinski, and Schaffer, 246–71.

Thackray, Arnold. *Atoms and Powers: An Essay on Newtonian Matter-Theory and the Development of Chemistry*. Cambridge: Harvard University Press, 1970.

Thackray, Arnold, and Everett Mendelsohn. *Science and Values*. New York: Humanities Press, 1974.

Tocqueville, Alexis de. *L'Ancien régime et la révolution*. 1856. Edited by J. P. Mayer and G. Rudler. Paris: Gallimard, 1954.

Todd, Janet. *Sensibility: An Introduction*. London: Methuen, 1986.

Toraude, L. G. *De Machy's histoire et contes*. Paris, 1907.

Torlais, Jean. *L'Abbé Nollet (1700–1770) et la Physique Expérimentale au XVIIIᵉ siècle*. Les Conférences du Palais de la Découverte, no. 60. Paris, 1959.

———. *Un Esprit Encyclopédique en dehors de l'Encyclopédie. Réaumur d'apres des documents inédites*. Paris, 1936.

———. *Un Physicien au siècle des lumières. L'Abbé Nollet. 1700–1770*. 1954. Paris: Jonas Editeur, 1987.

Trahard, Pierre. *Les Maîtres de la sensibilité française au XVIIIᵉ siècle*. 4 vols. Paris: Boivin et Cie, 1931–33.

———. *La Sensibilité révolutionnaire (1789–1794)*. Paris: Boivin, 1936.

Trésor de la langue Française. Paris: Gallimard, 1990.

Trottein, Serge. "Diderot et la philosophie du clair-obscur." *L'Esprit Créateur* 28 (1988): 1007–129.

Tuveson, Ernest. *The Imagination as a Means of Grace: Locke and the Aesthetics of Romanticism*. Berkeley: University of California Press, 1960.

Vaggi, Gianni. *The Economics of François Quesnay*. London: Macmillan, 1987.

Vailati, Ezio. *Leibniz and Clarke: A Study of Their Correspondence*. Oxford: Oxford University Press, 1997.

———. "Leibniz and Clarke on Miracles." *Journal of the History of Philosophy* 33, no. 4 (1995): 563–91.

Van Doren, Carl. *Benjamin Franklin*. New York: Viking, 1938.

Van Sant, Ann Jessie. *Eighteenth-Century Sensibility and the Novel: The Senses in Social Context*. Cambridge: Cambridge University Press, 1993.

Vartanian, Aram. *Diderot and Descartes: A Study in Scientific Naturalism in the Enlightenment*. Princeton: Princeton University Press, 1953.

———. *Science and Humanism in the French Enlightenment*. Charlottesville: Rockwood, 1999.

Vellay, Charles. "Benjamin Franklin et le procès du paratonnerre de Saint-Omer (1782–3)." *Revue historique de la Révolution française* 5 (1914): 135–43.

———. "Lettres inédites de Marat à Benjamin Franklin." *Revue historique de la Révolution française* 3 (1912): 353–61.

———. "Mélanges et documents." *Revue historique de la Révolution Française et de l'Empire* 5 (1914): 135–37.

Viel, Claude. "Duhamel du Monceau: Naturaliste, physicien et chemiste." *Revue d'histoire des sciences et de leurs applications* 38 (1985): 55–71.

Vila, Anne C. *Enlightenment and Pathology: Sensibility in the Literature and Medicine of Eighteenth-Century France*. Baltimore: Johns Hopkins University Press, 1998.

Vincent-Buffault, Anne. *The History of Tears: Sensibility and Sentimentality in France*. Translated by Teresa Bridgeman. New York: St. Martin's Press, 1991.

Vinchon, J. *Mesmer et son secret: Textes choisis et présentés par R. de Saussure*. Toulouse: Privat, 1971.

Vovelle, Michel. *La Revolution française 1789–1799: Images et récit*. 5 vols. Paris: Editions Messidor, 1986.

Wade, Nicholas J. *A Natural History of Vision*. Cambridge: MIT Press, 1999.

Walter, Gérard. *Robespierre*. 2 vols. Paris: Gallimard, 1961.

Weiner, Dora B. "The Blind Man and the French Revolution." *Bulletin of the History of Medicine* 48 (1974): 60–89.

———. *The Citizen-Patient in Revolutionary and Imperial Paris*. Baltimore: Johns Hopkins University Press, 1993.

Westfall, Richard S. *The Construction of Modern Science: Mechanisms and Mechanics*. New York: Wiley, 1971.

Weulersse, Georges. *Le Mouvement physiocratique en France*. 2 vols. Paris: Félix Alcan, 1910.

———. *La Physiocratie à l'aube de la Révolution, 1781–1792*. Paris: Ecole des Hautes Etudes en Sciences Sociales, 1985.

———. *La Physiocratie sous les ministères de Turgot et Necker (1774–1781)*. Paris: Presses Universitaires de France, 1950.

Williams, Bernard. *Descartes: The Project of Pure Enquiry*. New York, Penguin, 1978.

Williams, Elizabeth. *The Physical and the Moral: Anthropology, Physiology, and Philosophical Medicine in France, 1750–1850*. Cambridge: Cambridge University Press, 1994.

Williams, L. Pearce. "The Politics of Science in the French Revolution." In *Critical Problems in the History of Science*, ed. Clagett, 291–308.

Williams, Raymond. *Keywords: A Vocabulary of Culture and Society*. New York: Oxford University Press, 1976.

Wilson, Arthur M. "Sensibility in France in the Eighteenth Century: A Study in Word History." *French Quarterly* 13, nos. 1 & 2 (1931): 35–46.

Wilson, Lindsay. *Women and Medicine in the French Enlightenment: The Debate over Maladies des Femmes*. Baltimore: Johns Hopkins University Press, 1993.

Winter, Alison. *Mesmerized: Powers of Mind in Victorian Britain*. Chicago: University of Chicago Press, 1998.

Woloch, Isser. *Eighteenth-Century Europe: Tradition and Progress, 1715–1789*. New York: Norton, 1982.

Yolton, John W. *Locke and French Materialism*. Oxford: Clarendon, 1991.

Zweig, Stefan. *Mental Healers: Franz Anton Mesmer, Mary Baker Eddy, Sigmund Freud*. 1932. New York: Ungar, 1962.

INDEX

Page ranges in **bold** indicate extensive discussion of a topic.